Studies in Systems, Decision and Control

Volume 266

Series Editor

Janusz Kacprzyk, Systems Research Institute, Polish Academy of Sciences, Warsaw, Poland

The series "Studies in Systems, Decision and Control" (SSDC) covers both new developments and advances, as well as the state of the art, in the various areas of broadly perceived systems, decision making and control–quickly, up to date and with a high quality. The intent is to cover the theory, applications, and perspectives on the state of the art and future developments relevant to systems, decision making, control, complex processes and related areas, as embedded in the fields of engineering, computer science, physics, economics, social and life sciences, as well as the paradigms and methodologies behind them. The series contains monographs, textbooks, lecture notes and edited volumes in systems, decision making and control spanning the areas of Cyber-Physical Systems, Autonomous Systems, Sensor Networks, Control Systems, Energy Systems, Automotive Systems, Biological Systems, Vehicular Networking and Connected Vehicles, Aerospace Systems, Automation, Manufacturing, Smart Grids, Nonlinear Systems, Power Systems, Robotics, Social Systems, Economic Systems and other. Of particular value to both the contributors and the readership are the short publication timeframe and the world-wide distribution and exposure which enable both a wide and rapid dissemination of research output.

** Indexing: The books of this series are submitted to ISI, SCOPUS, DBLP, Ulrichs, MathSciNet, Current Mathematical Publications, Mathematical Reviews, Zentralblatt Math: MetaPress and Springerlink.

More information about this series at http://www.springer.com/series/13304

Nilanjan Dey · Parikshit N. Mahalle ·
Pathan Mohd Shafi · Vinod V. Kimabahune ·
Aboul Ella Hassanien
Editors

Internet of Things, Smart Computing and Technology: A Roadmap Ahead

Springer

Editors
Nilanjan Dey
Techno India College of Technology
Kolkata, West Bengal, India

Parikshit N. Mahalle
Smt Kashibai Navale College of Engineering
Pune, Maharashtra, India

Pathan Mohd Shafi
Smt Kashibai Navale College of Engineering
Pune, Maharashtra, India

Vinod V. Kimabahune
Smt Kashibai Navale College of Engineering
Pune, Maharashtra, India

Aboul Ella Hassanien
Department of Information Technology (IT)
Cairo University
Giza, Egypt

ISSN 2198-4182 ISSN 2198-4190 (electronic)
Studies in Systems, Decision and Control
ISBN 978-3-030-39046-4 ISBN 978-3-030-39047-1 (eBook)
https://doi.org/10.1007/978-3-030-39047-1

This Springer imprint is published by the registered company Springer Nature Switzerland AG
The registered company address is: Gewerbestrasse 11, 6330 Cham, Switzerland

Preface

Artificial Intelligence has been used nowadays increasingly used in numerous fields like industries and finance including health care, education, transportation and more. Machine Learning is a subset of AI techniques which includes various algorithms that give machines the ability to learn from data or while interacting with the world without being explicitly programmed. With the advent in Big Data Analytics, IoT and Machine Learning newer opportunities are created for Business organizations to analyze, monitor and mine user-generated contents in real time for business intelligence. Gartner's predictions rise to include 26 billion devices connected to the Internet by 2020 with a global economic added value of $1.9 trillion. Pervasive IoT applications like healthcare applications generate a huge amount of sensor data and imaging data that needs to be handled rightly for further processing. In traditional IoT ecosystem, Cloud computing ensures solution for efficient management of huge data with its ability to access shared resources and provide common infrastructure in a ubiquitous manner. Though these new technologies are overwhelming, they also expose severe IoT security challenges. IoT applications suffer from different types of attacks such as eavesdropping attack, spoofing and false data injection attack, man-in-the-middle attack, replay attack, denial-of-service attack, jamming attack, flooding attacks, etc.

This book enclosed 17 chapters as follows. In chapter "Efficacy of a Classical and a Few Modified Machine Learning Algorithms in Forecasting Financial Time Series", Shilpa et al. proposed a machine learning algorithms in forecasting financial time series. Significant enhancement in forecasting efficacy is obtained with the application of the Modified GDM methods in all the data sets: training, testing and out-of-sample. Chapter "Analysis of Facial Expression Recognition of Visible, Thermal and Fused Imaginary in Indoor and Outdoor Environment", by Ravindra Patil et al. analysis of facial expression is carried out for indoor and outdoor environment. The experimental result analysis shows that fused images give better results as compared to visible images. The accuracy of smile expression is better than anger and disgust facial expression. Chapter "AI—Assisted Chatbot for E-Commerce to Address Selection of Products from Multiple Products", by Pathan Mohd Shafi et al. explains selection of products from multiple products using AI-Assisted Chatbot. E-commerce websites are trending nowadays due to online shopping that makes customer's life easier. Similar to this, Chatter Robots, i.e. ChatBots are providing better customer service through the Internet. Chapter "A Novel Approach to Detect Microcalcification for Accurate Detection for Diagnosis of Breast Cancer", P. P. Patil and S. Kotrappa proposes a novel approach to detect microcalcification for accurate detection for diagnosis of Breast Cancer. Projected work was evaluated with the assistance of the two most accepted digitized dataset referred to as MIAS and DDSM and capable digital dataset. Work shows the different topology modelling is extremely necessary for accurate estimation of the microcalcification. Chapter "Ensemble Classifier for Praise or Complaint Classification and Visualization from Big Data", by Sujata Khedkar and Subhash Shinde proposes a classifier for praise or complaint classification and visualization from big data. This chapter proposes the Ensemble classifier using linguistic features for praise or complaint classification from big customer review datasets and visualization of it. The Praise and Complaint sentences are further classified based on aspect and analysis at the aspect level is presented from business intelligence point of view. The performance of the four different supervised machine learning classifiers, namely Random forest, SVC, KNeighbours, MLP with linguistic hybrid features and Ensemble of above algorithms is evaluated on Hotel and Amazon product reviews dataset using parameters such as Accuracy, Precision, Recall, and F1-score. Chapter "Big Data & Disruptive Computing Platforms Braced Internet of Things: Facets & Trends" by Naveenkumar Jayakumar and Dhanashri P. Joshi explains about big data and disruptive computing platforms braced Internet of Things is presented. This chapter discusses the fundamental facets of various computing paradigms and approaches that may help to address issues of big data through creating IoT ecosystems. Chapter "Healthcare Information Technology for Rural Healthcare Development: Insight into Bioinformatics Techniques" by Satya Narayan Sahu et al. presented insight into healthcare information technology for rural healthcare development. With the help of this, doctor can easily predict the diseases and refer to the patient for early treatment. The health hazards in rural areas can be minimized with constructive application of this method in healthcare system. Chapter "Green Internet of Things Schemes and Techniques for Adaptive Energy Saving in Emergency Services" by Rachana Borawake-Satao and Rajesh Prasad proposed green internet of things schemes and techniques for adaptive energy saving in emergency services. In this chapter, various techniques and approaches used for energy saving in Green–Internet of Things (G-IoT) are discussed in detail. This chapter discusses the various vital applications of IoT (Internet of Things) such as border security and battlefield systems. This chapter also elaborates energy-saving approaches and techniques proposed by various researchers for such applications for ubiquitous environments. Chapter "Risk-Driven Analytics for Banking IoT Strategy", F. Khanboubi and A. Boulmakoul presented risk-driven analytics for banking IoT strategy. As the digital transformation driven by IoT begins to take root, new business models and products emerge. This opens new frontiers for innovation that can change customer behavior in the banking industry. The objective of this chapter is to highlight and illustrate the different uses of IoT in banking. Chapter "Unmanned Aerial Vehicle and IoT as Enabling Technologies for 5G: Frameworks, Applications and Challenges" by R. Sokullu and M. A. Akkaş demonstrated Unmanned Aerial Vehicle (UAV) and IoT as Enabling Technologies for 5G. A short introduction into the topic covering the main functionality, advantages and challenges of UAVs is followed by discussing the complementary roles that drones and IoT devices can have in the communication canopy. The second part concentrates on specific architectural issues, discussing different frameworks and network models that can accommodate UAV and IoT integration. Chapter "Multi-criteria Decision Making for Routing Process in MANET", by M. Deshmukh et al. presented multi-criteria decision-making for routing process in MANET. The proposed schemes perform best in various network densities as compared to Ad hoc On-demand Distance Vector (AODV) routing by increasing the lifetime of network and decreasing the delay, packet dropping and energy consumption, while achieving a good packet delivery ratio. Chapter "Development of an IOT-Based Atmospheric Fine Dust Monitoring System" by N. Kavitha and P. Madhumathy demonstrated the development of an IOT-Based atmospheric Fine Dust Monitoring System. After measuring the dust particle it analyses the dust levels in real-world scenario in addition it tests and collects the pattern change of dust at different locations. The analysed data is given to the users in the form of instant alerts to the subscribers. Preventive measures can be taken prior based on the instant alert of the analysed data given to the subscribers. Chapter "Toward Smart and Secure IoT Based Healthcare System" by Smita Sanjay Ambarkar and Narendra Shekokar proposes a secure healthcare system based on IoT. Here the authors attempt to analyze the smart healthcare architecture, its threats, vulnerabilities and the security measures to provide a secure smart healthcare system. Chapter "Lightweight Secure Technology Future of Internet of Things" by Aruna Gawade and Narendra Shekokar presented lightweight secure technology future of Internet of Things. Currently protocols like PRESENT, Clefia, TEA, etc., are designed for lightweight encryption but still suffers from Differential Attack, Related Key Attack, etc. Lightweight integrity is also achieved with SHA2, SHA 3, Light-Weight One-way Cryptographic Hash Algorithm (LOCHA), Timestamp-Defined Hash Algorithm (TDHA). So there is a need for a whole concrete lightweight system which will remove the shortcomings in the current lightweight encryption and integrity protocols and provides better security. In Chapter "Secure Data Transmission in Underwater Sensor Network: Survey and Discussion" by Pooja A. Shelar et al. survey on secure data transmission in underwater Sensor Network have been done . Underwater data security is a necessity as UWSN is prone to various security threats and malicious attacks in addition to problems of dynamic underwater environment and communication mediums. Earth is a water planet and around 71% of the earth's surface is covered by water. In spite of this water to land ratio of the world the "secure underwater data transmission" is a technologically lesser explored area. Therefore, this survey work will intrigue many researchers to work on underwater sensor networks and to find innovative solutions for transferring digital data securely in the underwater environment. Chapter "Rethinking Decentralised Identifiers and Verifiable Credentials for the Internet of Things" by Parikshit N. Mahalle et al. revisits the decentralized identifiers and verifiable credentials for the Internet of Things. In this chapter, various identification methods are analysed. Further, Decentralized Identifiers and Verifiable Credentials are discussed, the discussion focus on whether DID and VC are applicable to the IoT as IoT includes resource-constrained devices at the base level. It also presents smart home use case utilizing DID and VCs in order to evaluate its applicability. Chapter "The Invisible Eye—A Security Architecture to Protect Motorways" by Neel Patel et al. proposed a security architecture to protect motorways. This chapter introduces an intelligent cop assisting system TIE (The Invisible Eye). The system senses parameters that aren't directly knowable by anyone at their first glance on the vehicle like weight, etc. and make them available to the security personnel deployed to check those vehicles. Also, the system would provide a vulnerability index and it can be deployed in an unnoticeable way.

The objective of this book is to bring several innovative studies in Machine learning, Big data and Internet of Things. It supports the researchers, engineers and designers in several interdisciplinary domains to support applied applications. The book presents an overview of the different algorithms with focusing on the advantages and disadvantages of each algorithm in the field of Machine learning and Big data. It supplies the researchers with the outstanding state-of-the-art studies. The book emphasizes a brief outline of the some of the well-known and established techniques and their integration with different computational intelligence paradigms. It includes next-generation computing paradigms which are expected to support wireless networking with high data transfer rates and autonomous decision-making capabilities. The book provides a global outstanding research and recent progress in the said field. The book also reports the challenges and future directions in the IoT. This book aims to sustenance engineers, professionals, researchers and designers with an application oriented resource in innumerable interdisciplinary areas. This book aims to sustenance engineers, professionals, researchers and designers with an application oriented resource in innumerable interdisciplinary areas.

The editors are appreciating the outstanding authors for their valuable contributions. In addition, we are thankful to the book series editor for the endless support. Lastly, but not the last, no words can express our sincere gratitude to the team members of Springer, who are always supportive as usual.

Kolkata, India — Nilanjan Dey
Pune, India — Parikshit N. Mahalle
Pune, India — Pathan Mohd Shafi
Pune, India — Vinod V. Kimabahune
Giza, Egypt — Aboul Ella Hassanien
September 2019

About This Book

This book includes all related topics to Machine learning, Big data, Internet of things and security in internet of things. The objective of this book is to bring several innovative studies in Machine learning, Big data and Internet of Things. It supports the researchers, engineers and designers in several interdisciplinary domains to support applied applications. The book presents an overview of the different algorithms with focusing on the advantages and disadvantages of each algorithm in the field of Machine learning and Big data. It supplies the researchers with the outstanding state of the art studies. The book emphasizes a brief outlines of the some of the well-known and established techniques and their integration with the different computational intelligence paradigms. It includes next-generation computing paradigms which are expected to support wireless networking with high data transfer rates and autonomous decision-making capabilities. The book provides global outstanding research and recent progress in the said field. The book also reports the challenges and the future directions in the IoT. This book aims to sustenance engineers, professionals, researchers and designers with an application oriented resource in innumerable interdisciplinary areas.

Book includes IoT applications like healthcare applications which generates a huge amount of sensor data and imaging data that needs to be handled rightly for further processing. In traditional IoT ecosystem, Cloud computing ensures solution for efficient management of huge data with its ability to access shared resources and provide common infrastructure in a ubiquitous manner. Though these new technologies are overwhelming, they also expose severe IoT security challenges. IoT applications suffer from different types attacks such as eavesdropping attack, spoofing and false data injection attack, man-in-the-middle attack, replay attack, denial-of-service attack, jamming attack, flooding attacks, etc. In this way, it addresses the security issues in Internet of Things.

Contents

Internet of Things

Internet of Things Security

About the Editors

Nilanjan Dey is an Assistant Professor in Department of Information Technology at Techno India College of Technology, Kolkata, India. He is a visiting fellow of the University of Reading, UK. He is a Visiting Professor at Wenzhou Medical University, China and Duy Tan University, Vietnam. He was an honorary Visiting Scientist at Global Biomedical Technologies Inc., CA, USA (2012–2015). He was awarded his Ph.D. from Jadavpur University in 2015.

He has authored/edited more than 60 books with Elsevier, Wiley, CRC Press and Springer, and published more than 300 papers. He is the Editor-in-Chief of International Journal of Ambient Computing and Intelligence, IGI Global, Associated Editor of IEEE Access and International Journal of Information Technology, Springer. He is the Series Co-Editor of Springer Tracts in Nature-Inspired Computing, Springer, Series Co-Editor of Advances in Ubiquitous Sensing Applications for Healthcare, Elsevier, Series Editor of Computational Intelligence in Engineering Problem Solving and Intelligent Signal processing and data analysis, CRC.

His main research interests include Medical Imaging, Machine learning, Computer-Aided Diagnosis, Data Mining etc. He is the Indian Ambassador of International Federation for Information Processing—Young ICT Group. Recently, he has been awarded as one among the top 10 most published academics in the field of Computer Science in India (2015–17).

Parikshit N. Mahalle obtained his B.E. in Computer Science and Engineering from Sant Gadge Baba Amravati University, Amravati, India, and an M.E. in Computer Engineering from Savitribai Phule Pune University, Pune, India (SPPU). He completed his Ph.D. in Computer Science and Engineering (specializing in Wireless Communication) from Aalborg University, Aalborg, Denmark. He has more than 16 years of teaching and research experience. He has been a member of the Board of Studies in Computer Engineering, SPPU. He is a member of the Board of Studies Coordination Committee in Computer Engineering, SPPU. He is also a member of the Technical Committee, SPPU. He is an IEEE member, ACM member and a life member of CSI and ISTE. He is a reviewer for the Journal of Wireless Personal Communications (Springer), a reviewer for the Journal of Applied Computing and Informatics (Elsevier), a member of the Editorial Review Board for IGI Global—International Journal of Ambient Computing and Intelligence (IJACI), a member of the Editorial Review Board for Journal of Global Research in Computer Science, a reviewer for IGI Global—International Journal of Rough Sets and Data Analysis (IJRSDA), an Associate Editor for IGI Global—International Journal of Synthetic Emotions (IJSE) and Inderscience International Journal of Grid and Utility Computing (IJGUC). He has also a member of the Technical Program Committee for international conferences and symposia such as IEEE ICC—2014, IEEE ICACCI 2013, IEEE ICC 2015—SAC-Communication for Smart Grid, IEEE ICC 2015—SAC-Social Networking, IEEE ICC 2014—Selected Areas in Communication Symposium, IEEE INDICON 2014, CSI ACC 2014, IEEE GCWSN 2014, GWS 2015, GLOBECOMM 2015, ICCUBEA 2015, ICCUBEA 2016. He has published 56 research publications at national and international journals and conferences with 177 citations. He has authored 8 books on Identity Management for Internet of Things, River Publishers; Identity Management Framework for Internet of Things, Aalborg University Press; Data Structures and Algorithms, Cengage Publications; Theory of Computations, Gigatech Publications; Fundamentals x

Editors 9781138339798PRE.3D 11 [1–16] 14.3.2019 12:19 PM of Programming Languages—I, Gigatech Publications; Fundamentals of Programming Languages —II, Gigatech Publications; Design and Analysis of Algorithms: A Problem Solving Approach, (in press)—Cambridge University Press. He is also the recipient of Best Faculty Award by STES and Cognizant Technologies Solutions. He has also delivered an invited talk on "Identity Management in IoT" to Symantec Research Lab, Mountain View, California. Currently, he is working as Professor and Head of Department of Computer Engineering at STES's Smt. Kashibai Navale College of Engineering, Pune, India. He has guided more than 100 plus undergraduate students and 20 plus post-graduate students for projects. His recent research interests include Algorithms, Internet of Things, Identity Management and Security. He has travelled to Denmark, Sweden, Germany, Austria, Norway, China and Switzerland.

Pathan Mohd Shafi is a Professor at Smt. Kashibai Navale College of Engineering, Pune. He completed his Ph.D. (CSE) from JNTU Anantapur, India. He has completed a university- funded research project on "Public key cryptography for cross-realm authentication in Kerberos" and he has worked as the resource person for workshops and seminars. He is a reviewer of many national and international journals and conferences. He has worked as Head of the Publicity Committee for International Conference at the Global ICT Standardization Forum for India. He has worked as organizing secretary for an international conference on Internet of Things, Next- Generation Networks and Cloud Computing, held at SKNCOE, Pune in 2016 and 2017. He was guest editor for a special issue of ICINC 2016 by IGI Global International Journal of Rough Data Sets and Analytics. He has authored four books and one chapter for Springer, and he has published more than 40 research articles in national and international journals. He is a life member of ISTE and CSI.

Vinod V. Kimabahune is Associate Professor at Smt. Kashibai Navale College of Engineering, Pune. He has completed his Ph.D. (Computer Engineering) from Savtribai Phule Pune University Pune, India. He has worked as the resource person for workshops and seminars. He is a reviewer for many national and international journals and conferences. He has published more than 50 research articles in various national and international conferences and Journals. He is having two books on his credit. He has worked as Head of the Publicity Committee for International Conference at the Global ICT Standardization Forum for India. He has worked as organizing secretary for an international conference on Internet of Things, Next-Generation Networks and Cloud Computing, held at SKNCOE, Pune in 2016, 2017 and 2018. He is a life member of ISTE and CSI.

Aboul Ella Hassanien is the Founder and Head of the Egyptian Scientific Research Group (SRGE) and a Professor of Information Technology at the Faculty of Computer and Information, Cairo University. Professor Hassanien is ex-dean of the faculty of computers and information, Beni Suef University. Professor Hassanien is a collaborative researcher member of the Computational Intelligence Laboratory at the Department of Electrical and Computer Engineering, University of Manitoba. He also holds the Chair of Computer Science and Information Technology at the Egyptian Syndicate of Scientific Professions (ESSP). Dr. Hassanien is the Founder and Head of Africa Scholars Association in Information and Communication Technology. Professor Hassanien has more than 650 scientific research papers published in prestigious international journals and conferences and over 40 books covering such diverse topics as data mining, medical images, Big Data analysis, virtual reality, intelligent systems, social networks and smart environment. His other research areas include computational intelligence, medical image analysis, security, animal identification and multimedia data mining.

Machine Learning

Efficacy of a Classical and a Few Modified Machine Learning Algorithms in Forecasting Financial Time Series

Shilpa Amit Verma, G. T. Thampi and Madhuri Rao

Abstract Financial markets and economy forecast are closely related to each other. Forecast of prices of financial assets is therefore of importance for any economy-planning be it global, national or individual. There are various global, local and psychological factors that affect financial markets making its forecasting a non-trivial, complex problem. Numerous machine learning techniques have been applied by various researchers for a last few decades for making forecasts in various fields including the financial one, with varying degree of success. In the present article, time-series data of NIFTY50 of the National Stock Exchange (NSE) of India is considered as a reference data. Forecasting of its prices is done by applying the classical Gradient Descent Method (GDM) and by a few herein proposed modifications of it. The modifications are essentially using variants of the mean square error function of the classical GDM. All the proposed variants, Mean median (MMD) error function, Minkowski (MKW) error function, Logcosh (LCH) error function and Cauchy (CCY) error function, result in significant improvement in all the efficacy parameters of forecasting. Two widely varying time horizons, monthly and daily, have been considered. Significant enhancement in forecasting efficacy is obtained with the application of the Modified GDM methods in all the data sets: training, testing and out-of-sample.

Keywords Gradient descent method · Forecasting · Machine learning · Stock market · NIFTY50 · Time series

S. A. Verma (✉)
Computer Department, Mumbai University Thadomal Shahani Engineering College, Bandra, Mumbai, India
e-mail: shilpaverma65@gmail.com

G. T. Thampi · M. Rao
IT Department, Mumbai University Thadomal Shahani Engineering College, Bandra, Mumbai, India
e-mail: gtthampi@yahoo.com

M. Rao
e-mail: my_rao@yahoo.com

N. Dey et al. (eds.), *Internet of Things, Smart Computing and Technology: A Roadmap Ahead*, Studies in Systems, Decision and Control 266,
https://doi.org/10.1007/978-3-030-39047-1_1

1 Introduction

Machine learning is a procedure or an algorithm where machines directly perform the task at hand with minimal human intervention. There are several aspects of the real world problems which require algorithms to 'learn' and improve on its performance based on experience which is referred to as 'training'. The main aim of training in any machine learning algorithm would therefore be to 'adapt', given a certain degree of complexity. There are many complex tasks like speech recognition, image processing, transliteration, healthcare, time series forecasting etc. which require state of the art techniques (machine learning algorithms) that can learn from experience and perform the task at hand. In an interesting study, Jain and Bhatnagar have effectively and efficiently employed AI for monitoring the health and security of patient in a hospital environment [1]. In this method a warning system is activated during unwanted scenarios. In another significant study, Kamal et al. have undertaken the study and prediction of coding regions from diseases infected biological data employing the classifiers namely the support vector machine (SVM), principal component analysis (PCA) technique, Fisher's discriminant analysis (FDA) and compared the results with the neural mapping skyline filtering (NMSF) [2]. The results obtained from NMSF surpassed others emphasizing the importance of this methodology. In a quite detailed study, Dey N. et al. have employed machine learning methods for smart applications like transliteration for the multilingual support in printing of utility bills, data mining, IOT and security and achieved a high degree of accuracy [3]. Dey N. et al. have also employed IOT and big data driven technologies for next generation healthcare and amply demonstrated the efficacy and importance of these methods for future healthcare industry [4].

Financial markets facilitate, the world around, exchange of financial assets like equities, bonds, commodities and currencies etc. The term equity may be used to describe one's ownership in a company. The three terms equity price, stock price and share price are often used interchangeably and they would carry the same meaning in the present work. Financial markets forecasts are generally required for economics planning. Stock market forecasting therefore attracts worldwide attention of researchers, investors and traders. Although financial markets of developed countries have been quite comprehensively studied, those of developing countries like India have not been considered to that extent. Moreover these markets are generally more volatile and hence more difficult to forecast.

Share price forecasting may be done in two widely differing ways: fundamental analysis and technical analysis. Fundamental analysis of a company is done by evaluating its past economic performance, expected future demand of its products, credibility of its management, government policies and local and global economics scenario. In technical analysis, on the other hand, future prices of stocks are forecasted solely on the basis of the trends/patterns of their past prices. Historically, according to the proponents of Efficient Market Hypothesis (EMH) and Random Walk Hypothesis (RWH) day-to-day share prices of a company are nothing but random fluctuations around a central value and therefore, in the short run, it is not possible to forecast the markets with more than 50% accuracy. Subsequently however there have been numerous studies providing evidence contrary to EMH and RWH. A large accumulated and growing experience also suggests that these hypotheses may not be strictly applicable as some analysts regularly make forecasts with more than 50% accuracy. Since last many decades, technical analysts have been observing graphs of share prices determining their support and resistance points based on some rather rough empirical rules. Academicians, corporate houses and traders have been making a lot of efforts to forecast stock prices on various time horizons. Stock price movements are nonlinear, complex and at times discontinuous and, in the short run, they are largely driven by sentiments, fear and greed, of numerous investors and traders and by crowd psychology rendering the forecast a difficult task. Any shift in demand and supply of a company's share results in changes of its market price which often deviates from its intrinsic value. There are no formal mathematical models/equations to describe movements of stock prices. Artificial intelligence (AI) based methods are inherently suited to make forecasts in such scenarios. Advantages of AI methods lie in their ability to model complex and nonlinear stock prices without any prior knowledge of the processes generating them [5]. This advantage has recently been accentuated due to availability of large and fast digital computers and substantial progress in research in numerical computational methods. AI methods can be classified roughly into three categories: Artificial Neural Network (ANN), Fuzzy logic (FL) and Genetic algorithm (GA). Owing to their characteristics of being extremely powerful in extracting trends and patterns in unknown environments, ANNs have become preferred tools for prediction of the financial markets. ANNs are frequently referred to as universal approximators as they are capable of approximating any function. Trained ANNs can be considered experts in the domain of their use [6]. However within the ANN domain, choosing the best model for the problem in hand is an important task [7]. The ultimate test for the best choice is naturally the highest forecasting accuracy. A lot of research work has been done and published in open scientific literature on application of AI methods in forecasting financial markets of developed countries. In the present work focus has been on an Indian stock market index NIFTY50 and efforts have been made to enhance forecasting efficacy by deploying a few error functions different from the one used in the classical GDM. Forecasting efficacy parameters are used to measure degree of success in forecasting. Most popular efficacy parameters are briefly described below [8].

2 Efficacy Parameters of Market Forecasting

Let y_{d_t} and y_{f_t} denote the actual (desired) and forecasted prices of the stock/index at time 't' respectively. The forecast error is then defined by $e_t = (y_{d_t} - y_{f_t})$. Let the total number of input-output sets, constructed from the stock prices, be denoted by 'n'. Let y_{mean} denote the mean of all 'n' actual prices of the stock. Efficacy parameters of price forecasting generally used to investigate efficiency of AI methods are listed below in Eqs. 1–8.

(1) Mean Error (ME)

$$ME = \frac{1}{n}\sum_{t=1}^{n} e_t \tag{1}$$

ME is defined as the average deviation of the calculated values from the actual values. The mean error can be regarded as an informal value that generally refers to the mean or average of all the errors in a set. An error in this context is an uncertainty in a measurement, or the difference between the measured value and true/correct value. It may also be called forecast bias as it shows the direction of error. Positive and negative values of individual e_t may nullify each other to give deceptively small value of ME.

(2) Mean Absolute Deviation (MAD)

$$MAD = \frac{1}{n}\sum_{t=1}^{n} |e_t| \tag{2}$$

In its simplest form the MAD of a dataset is just the average distance between each data point contained in the data and the data mean. It denotes a simple measure about the variability or the spread of the dataset. This parameter adds up absolute errors in forecasting. Positive and negative errors do not cancel each other. It does not provide any idea about the direction of errors. MAD should be as small as possible for a good forecast.

(3) Mean Square Error (MSE)

$$MSE = \frac{1}{n}\sum_{t=1}^{n} (e_t)^2 \tag{3}$$

MSE denotes the average squared difference between the calculated and the actual value. In strict sense, this error parameter can be regarded as a risk function, which corresponds to the expected value of the square error loss. This parameter measures average squared deviation of forecasted values. It is significantly affected by large individual errors. It does not provide any indication of the direction of the overall error.

(4) Root Mean Square Error (RMSE)

$$\mathrm{RMSE} = \sqrt{\mathrm{MSE}} \tag{4}$$

As denoted above in Eq. 4, the RMSE is the standard deviation of the residuals in the data. Residuals are denoted as a measure of spread of the regression line form the actual data points. This parameter is frequently computed and reported. All the properties of MSE hold for RMSE also.

(5) Mean Percent Forecast Error (MPE)

$$\mathrm{MPE} = \frac{1}{n}\sum_{t=1}^{n}\left(\frac{e_t}{y_{d_t}}\right) * 100 \tag{5}$$

In its most general form MPE is regarded as the computed average of percentage errors by which the calculated values of a model differ from actual values for which the quantity is being forecast. This parameter gives the percentage of average forecasting error and indicates its direction also. Opposite signed individual errors cancel out each other. It is independent of the scale of measurement. Its value should be small for a good forecast.

(6) Mean Absolute Percent Forecast Error (MAPE)

$$\mathrm{MAPE} = \frac{1}{n}\sum_{t=1}^{n}\left|\left(\frac{e_t}{y_{d_t}}\right) * 100\right| \tag{6}$$

MAPE is defined as a quantity to predict the accuracy of a forecasting methods. However, it is pertinent to mention that MAPE has also been employed as a measure of the loss function for regression problems in machine learning algorithms. MAPE gives the percentage of average absolute forecast error. It does not depend on the scale of the data. Opposite signed individual errors do not cancel each other. Sufficiently small value of this parameter ensure a good forecast.

(7) Mean Percent Accuracy (MPA)

$$\mathrm{MPA} = 100 - \mathrm{MAPE} \tag{7}$$

MPA is a measure of the accuracy as a percentage value. It can also be denoted as the absolute average percent error for each time period minus the actual values and divided by actual value itself. It depends directly on MAPE. As it is the mean percent accuracy as opposed to mean percent error it may be more appealing to some users.

(8) Coefficient of Determination (R^2)

$$R^2 = 1 - \frac{\sum_{t=1}^{n}\left(y_{d_t} - y_{f_t}\right)^2}{\sum_{t=1}^{n}\left(y_{d_t} - y_{mean}\right)^2} \tag{8}$$

R^2 can be understood as the variance proportion of the dependent data variable which has been predicted from the independent variable. This parameter is a key output of the regression analysis methods and represents the goodness of fit of a model. Large values of R^2 suggest a good fit to the historical data. It is however not applicable to forecast of out of sample data [8].

3 Review of Forecasting Efficiencies Achieved by Researchers

AI techniques have frequently been applied in forecasting financial assets like stocks, commodities etc. giving a wide range of forecasting accuracies. Some of them are briefly mentioned here. Genetic Algorithm (GA) approach for feature selection in ANN was used by Kyoung-jae Kim et al. for forecasting the direction of movement of the Tokyo stock exchange prices index (TOPIX) achieving accuracies ranging between 52 and 62% [9]. A support vector machine (SVM) algorithm was used by Kyoung-jae Kim to forecast direction of daily price change in the Korean Stock market. The prediction accuracy of SVM used therein was better (58%) than the other two back propagation methods (55%) and case-based reasoning (CBR) method (52%) [10]. ANN-GA method was used by Kyoung-jae Kim to predict direction of movement of South Korean market index (KOSPI) and achieved accuracies ranging between 59 and 65% [11]. An ANFIS model was used by Ebrahim Abbasi et al. to investigate current trend of stock prices of "Iran Khodro Corporation" at Tehran Stock Exchange achieving MAPE as low as 0.9 [12]. An ANFIS expert system was employed by George S. Atsalakis et al. to predict changes in trends of the stock prices of National Bank of Greece and General Electric. Accuracy of about 68% were achieved [13]. An ANN combined with a genetic fuzzy system (GFS) was used by Melek Acar Boyacioglu et al. forecasting monthly returns of National 100 Index of Istanbul Stock Exchange (ISE) with an accuracy rate of 98.3% [14]. An ANN- GFS was used by Esmaeil Hadavandi et al. predicting direction of movement of share prices of IBM, Dell, British airline and Ryan airline achieving MAPE in the range 0.63–1.9 [15]. An ANN model was used by Yakup Kara et al. to predict direction of Istanbul Stock market index (ISE) with an accuracy of about 76%. An SVM model was also used giving an accuracy of 72% [16]. A Bayesian regularized ANN model was used by Jonathan L. Ticknor to predict one day future closing prices of Microsoft and Goldman Sachs and achieved MAPE ranging between 1.06 and 1.3 [17]. Amin Hedayati Moghaddam et al. used a back propagation network (BPNN) to forecasting daily index of NASDAQ stock exchange obtaining 0.9622 for determination coefficient (R^2) [18]. Aparna Nayak et al. used a supervised machine learning algorithms using historical prices combining them with sentiments from social media obtaining up to 70% of accuracy stock market trends [19]. Fuzzy inference rules were used by Siham Abdulmalik et al. to forecast Russian Trading System Index (RTSI) achieving a confidence level of about 90% with incomplete initial data [20]. Standard Backpropagation algorithm of ANN was employed by Devadoss to forecast closing share prices of Tata Consultancy Services Ltd, Wipro Ltd, Dr. Reddy's Laboratories Ltd and Sun Pharmaceutical Ltd. listed in the Bombay Stock Exchange (BSE) achieving MAPE having a range between 1.1 and 11.7% [21]. An ANN model was used by Yunus Yetis et al. to predict NASDAQ's stock values with an error of 2% [22]. BPN-GA was used by Mingyue Qiu et al., to predict returns of Neikki 225 index. Small numbers of MSE 0.0043 were attained [23]. Sachin Kamley et al. reviewed various machine learning techniques applied to forecasting of share markets by various researchers during the period 2000–2015 and observed that forecasting accuracies ranging from 88 to 97% were obtained [24].

4 Machine Learning Algorithms

Machine learning is an interdisciplinary field having many things in common with mathematics and statistics, information theory and most importantly optimization. The ability to scan and process huge databases allows machine learning programs to detect patterns that are outside the scope of human perception. Machine learning, by its very definition, implies machines that can learn, which points us to techniques like ANN, FS, GA etc. For the present we shall restrict our focus to ANN methods and the error back propagation methodology contained therein.

The backpropagation (BP) algorithm that was introduced by Rumelhart is a well-known method for training multilayer feed-forward artificial neural networks [25]. It makes use of the classical GDM technique. The Multi-layer perceptron (MLP) networks trained by BP algorithm have been a popular choice in ANN approach in financial applications [26]. Figure 1 illustrates the basic multilevel perceptron (MLP) network without using biases.

The MLP consists of three types of layers. The first layer is the input layer and corresponds to the input variables of the problem with one node for each input variable. For 'p' input variables, 'p' nodes are required in the input layer. The second layer is the hidden layer, containing 'q' nodes, and it helps to capture non-linear relationships among variables. The third layer is the output layer. The number of nodes required in this layer is equal to the number of outputs for each set of 'p' inputs. In the present problem there is only one output and therefore only one node is required in the output layer. The relationship between the output y and the input vector x is given by Eq. 9 below, where $w_{i,j}$ ($i = 0, 1, 2, \ldots, p$; $j = 1, 2, \ldots, q$) and v_j ($j = 0,1,2, \ldots, q$) are the connection weights.

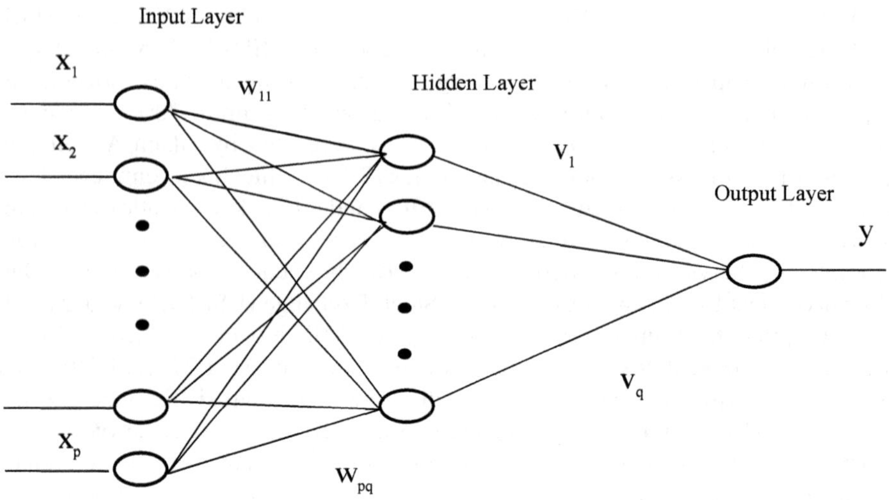

Fig. 1 ANN/MLP network

$$y = \sum_{j=1}^{q} v_j \cdot f\left(\sum_{i=1}^{p} w_{i,j} \cdot x_i\right) \tag{9}$$

The nonlinear activation function ‘f’ enables the network to learn nonlinear features of the input-output interdependence. The most widely used activation functions for the output layer are the sigmoid and hyperbolic functions. In this paper, the hyperbolic tangent transfer function is employed and is defined as in Eq. 10 below.

$$\tanh(r) = \frac{e^r - e^{-r}}{e^r + e^{-r}} \tag{10}$$

The MLP is trained using the BP algorithm and the weights are optimized. The objective function to be minimized is the sum of the squares of the differences between the desirable output (y_d) and the forecasted output (y_f) and is given by Eq. 11 below.

$$E = 0.5 \sum_{t=1}^{n} \left(y_{d_t} - y_{f_t}\right)^2 = 0.5 \sum_{t=1}^{n} e_t^2 \tag{11}$$

The training of the network is performed by BP algorithm trained with the steepest descent algorithm given in Eq. 12, m denoting the iteration index.

$$\Delta w_m = -\alpha_m \cdot g_m \tag{12}$$

The BP algorithm suffers from slow convergence and traps in local minima of error function. Several modifications have been introduced by eminent researchers for improving the local minima problem. These are: gradient descent method with momentum, gradient descent method with adaptive learning rate, gradient descent with both momentum and adaptive learning rate and resilient backpropagation [27]. Some other methods are based on variations of numerical optimization technique itself. These variations formed bases of Conjugate-Gradient, Quasi-Newton, and Levenberg–Marquardt algorithms. Conjugate-Gradient algorithms perform searches along conjugate directions requiring calculation of second derivatives. Nevertheless its convergence is generally faster than that of the steepest descent direction method [28]. Quasi-Newton method may converge still faster than conjugate gradient methods since it avoids calculation of second derivatives. It approximates the Hessian matrix and updates it at each iteration. The Levenberg–Marquardt method interpolates between the Gauss–Newton and the gradient descent methods. It also avoids calculation of the Hessian Matrix by replacing it by the Jacobian needing lesser computation [29]. Some details of these methods can be found in the work done by Dhar et al. [7] and Lahmiri [30].

In all the methods mentioned above, the optimisation function E given in Eq. 11, is the quadratic sum of the errors, the difference between the actual and predicted values of the variable of interest. It is envisaged in the present work that variants of the objective function *E* may improve the forecasting accuracy MPA and other efficacy parameters. The classical GDM method was chosen for this investigation.

4.1 *Modified GDM Method for Forecast*

The most general and the default error (optimization) function employed in training of gradient descent method is the generalized quadratic error function as given in Eq. 3. In the present work, four variants of the optimization function listed below in Eqs. 13–16 are tested in the GDM algorithm [31]. Motivation to use them comes from their characteristic advantages mentioned below briefly.

1. Mean median (MMD) error function

The Mean median error function as given in Eq. 13 has the advantage of both the Mean error function and Median error function. Hence reduces the influence of large errors but at the same time retains its convexity.

$$E = \sum_{t=1}^{n}\left(2 * \left(\sqrt{1 + (e_t^2/2)}\right) - 1\right) \tag{13}$$

2. Minkowski (MKW) error function

Minkowski error function as given in Eq. 14 can be treated as a special case of MSE error and reduces to the same form if we put $r = 2$. Variations of r around 2 have been observed to have significant influence on accuracy of forecast. Here r has been chosen to be 0.4.

$$E = \sum_{n} |e_i|^r \tag{14}$$

3. Log cosh (LCH) error function

Log cosh error function as given in Eq. 15 below is the logarithm of the hyperbolic cosine of the prediction error obtained. This error function is an approximation of $x^2/2$ for small values of x and approximately equal to absolute value of x for large x.

$$E = \sum_{n} \ln(\cosh(e_i^2)) \tag{15}$$

4. Cauchy (CCY) error function:

Cauchy error function as given in Eq. 16 below is known to show robustness against outliers.

$$E = \sum_{n} c^2/2 \ln\left(1 + \left(\frac{e_i}{c}\right)^2\right) \text{where } c = 2.38 \tag{16}$$

The modified algorithms incorporating the above functions are referred to here as GDM_MMD, GDM_MKW, GDM_LCH and GDM_CCY respectively. Forecasting accuracies (MPAs) and errors (MSEs) of NIFTY50 obtained using the above listed error functions in the two time horizons (monthly and daily) are compared with those of the classical ones. Other efficacy parameters are also evaluated for the algorithm giving the best MPA and are compared with those of the classical GDM algorithm. Influence of the network structure on these parameters is also examined. In general a significant improvement in all the efficacy parameters is obtained when using modified algorithms. The results are briefly discussed in the sections below.

5 Results and Discussion

Two kinds of forecasting time horizons have been considered in the present work to make the research useful for varied users. These are monthly and daily time horizons. Actual closing price data of NIFTY50 is downloaded from a site of NSE

[32]. Prices are normalized between 0.1 and 0.9. Time series formation is done by making sets of 5 days/months prices as input vector and the 6th day/month price as the corresponding output. Time series so formed are divided in three categories: training, testing and validation. Training and testing series constitute about 80% of the total. About 70% of these series are randomly selected for training the network and the remaining 30% are used for testing. Validation series correspond to chronologically the latest data and constitute about 20% of the total. The validation data is thus not contained in the training and testing data and is referred to, here, as out-of-sample (OS) data. The work done can be categorized as below:

(a) Forecast of monthly closing share prices of the index NIFTY50 of NSE of India using the classical GDM technique
(b) Forecast of daily closing prices of NIFTY50 using the classical GDM technique
(c) Implementation of modified GDM methods to increase forecasting efficiency and comparison of the results obtained by the classical and the modified GDM algorithms for both monthly and daily time horizons.

5.1 NIFTY50 Monthly Forecast

NIFTY50 price data of 18 years, from Jan 2000 to Dec 2017, has been considered for this study. The data consists of closing prices of NIFTY50 for the first trading day of every month for the entire time period mentioned. The prices variation as a function of the time period is plotted in the Fig. 2.

In this study an attempt has been made to forecast one month ahead price based on the previous 5 month prices. Efficacy parameters of monthly forecasts of NIFTY50 using the classical GDM method and modified GDM methods for various ANN architectures and for testing and validation phases are summarized below in Tables 1, 2, 3, 4, 5, 6, 7 and 8. The first row of the table gives the ANN architecture corresponding to three different architectures used i.e. 3, 5 and 7 neurons corresponding to 1 hidden layer. Important observations are mentioned below each table.

It can be seen from Table 1 that when the classical GDM is employed, MPA of forecast in testing phase decreases as the number of hidden layer neurons is increased from 3 to 7 in the network. The best MPA obtained is 89.42% with an architecture of 05-03-01. This parameter needs to be compared with the values obtained from other methods which is demonstrated below in Tables 2, 3, 4 and 5.

It can be seen from Table 2 that when GDM_MKW method is employed the best MPA (91.18%), in the testing phase, is obtained with the architecture 05-05-01 and this MPA is significantly larger than the best given by the Classical GDM. This is an advantage of the GDM_MKW method in comparison to the classical GDM.

Next, the same data as above is considered and modified backpropagation method employing the log cosh error function (GDM_LCH), instead of conventionally used

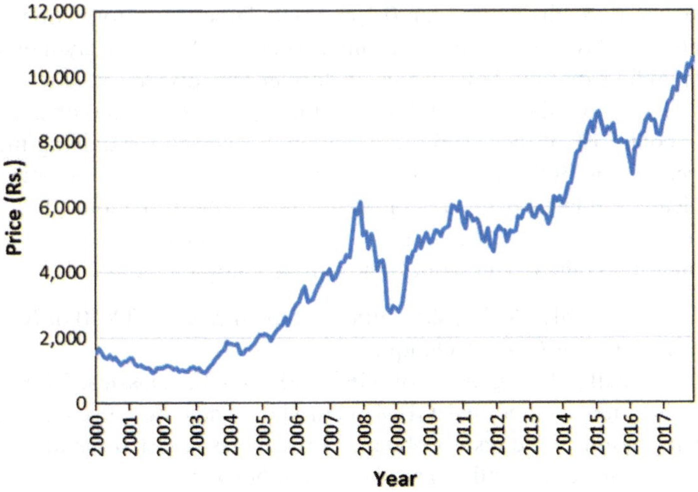

Fig. 2 Monthly price variation of NIFTY50

Table 1 Efficacy parameters of NIFTY50 monthly forecasting using the classical GDM in testing phase

ANN architecture	05-03-01	05-05-01	05-07-01
ME	−6.95E–03	1.72E–03	−4.84E–03
MAD	3.63E–02	4.11E–02	4.47E–02
MSE	2.65E–03	2.48E–03	3.54E–03
RMSE	5.15E–02	4.98E–02	5.95E–02
MPE%	−2.92	−4.31	−2.57
MAPE%	10.58	15.40	15.70
MPA%	89.42	84.60	84.30
R^2	0.9268	0.9315	0.9022

Table 2 Efficacy parameters of NIFTY50 monthly forecasting using GDM_MKW in testing phase

ANN architecture	05-03-01	05-05-01	05-07-01
ME	6.63E–03	3.60E–03	1.45E–02
MAD	3.73E–02	2.64E–02	3.98E–02
MSE	2.84E–03	1.42E–03	2.54E–03
RMSE	5.33E–02	3.76E–02	5.04E–02
MPE%	−0.3	−0.02	2.54
MAPE%	13.02	8.82	14.64
MPA%	86.98	91.18	85.36
R^2	0.9214	0.9609	0.9298

Table 3 Efficacy parameters of NIFTY50 monthly forecasting using GDM_LCH in testing phase

ANN architecture	05-03-01	05-05-01	05-07-01
ME	−3.64E−03	−3.51E−03	6.37E−05
MAD	3.57E−02	3.21E−02	2.77E−02
MSE	2.76E−03	1.94E−03	1.32E−03
RMSE	5.26E−02	4.41E−02	3.63E−02
MPE%	−1.17	−0.25	−0.44
MAPE%	11.4	11.81	10.35
MPA%	88.6	88.19	89.65
R^2	0.9236	0.9463	0.9635

Table 4 Efficacy parameters of NIFTY50 monthly forecasting using GDM_MMD in testing phase

ANN architecture	05-03-01	05-05-01	05-07-01
ME	−1.73E−03	−4.87E−03	−6.06E−03
MAD	2.34E−02	2.37E−02	2.36E−02
MSE	1.17E−03	1.19E−03	1.07E−03
RMSE	3.42E−02	3.45E−02	3.27E−02
MPE%	−0.32	−1.35	−1.32
MAPE%	7.13	6.99	7.89
MPA%	92.87	93.01	92.11
R^2	0.9677	0.9671	0.9705

Table 5 Efficacy parameters of NIFTY50 monthly forecasting using GDM_CCY in testing phase

ANN architecture	05-03-01	05-05-01	05-07-01
ME	2.03E−03	−3.01E−03	−2.01E−03
MAD	4.48E−02	1.78E−02	2.23E−02
MSE	3.72E−03	6.70E−04	1.08E−03
RMSE	6.10E−02	2.59E−02	3.29E−02
MPE%	−1.41	−0.53	−1.39
MAPE%	15.46	5.6	7.77
MPA%	84.54	94.40	92.23
R^2	0.8973	0.9815	0.9701

quadratic error function, is used for forecasting. These results are tabulated in Table 3.

It can be seen from Table 3 that while using the algorithm GDM_LCH the best MPA (89.65%) is obtained with the structure 05-07-01. In general, GDM_LCH gives larger MPA than that of the Classical GDM. This is an advantage of the GDM_LCH method in comparison to the classical GDM.

Next, the same data as above is considered and modified backpropagation method employing the mean median error function (GDM_MMD), instead of

Table 6 MPA and MSE summary for NIFTY50 monthly forecast in testing phase

ANN architecture	Method	MSE	MPA%
05-03-01	Classical GDM	2.65E–03	89.42
	GDM_MKW	2.84E–03	86.98
	GDM_LCH	2.77E–03	88.60
	GDM_CCY	3.72E–03	84.54
	GDM_MMD	1.17E–03	92.87
05-05-01	Classical GDM	2.48E–03	84.60
	GDM_MKW	1.42E–03	91.18
	GDM_LCH	1.94E–03	88.19
	GDM_CCY	6.70E–04	94.40
	GDM_MMD	1.19E–03	93.01
05-07-01	Classical GDM	3.53E–03	84.30
	GDM_MKW	2.54E–03	85.36
	GDM_LCH	1.32E–03	89.65
	GDM_CCY	1.10E–03	92.23
	GDM_MMD	1.07E–03	92.11

Table 7 Efficacy parameters of classical GDM and GDM_CCY in testing phase of NIFTY50 monthly data with a 05-05-01 ANN

Method	Classical GDM	GDM_CCY
ME	1.72E–03	−3.01E–03
MAD	4.11E–02	1.78E–02
MSE	2.48E–03	6.70E–04
MPE%	−4.31	−0.53
MAPE%	15.4	5.6
MPA%	84.6	94.4
R^2	0.9315	0.9815

Table 8 Efficacy parameters of NIFTY50 monthly forecasting using the classical GDM and GDM_CCY in validation phase

Method	Classical GDM	GDM_CCY
ME	6.59E–02	7.45E–03
MAD	7.15E–02	3.05E–02
MSE	6.56E–03	1.47E–03
RMSE	8.10E–02	3.83E–02
MPE%	8.84	0.73
MAPE%	9.77	4.34
MPA%	90.23	95.66

conventionally used quadratic error function, is used for forecasting. These results are tabulated in Table 4.

It can be seen from Table 4 that while using the algorithm GDM_MMD the best MPA (93.01%) is obtained with the structure 05-05-01. This algorithm gives

Table 9 Efficacy parameters of NIFTY50 daily forecasting using the classical GDM in testing phase

ANN architecture	05-03-01	05-05-01	05-07-01
ME	−4.63E–03	−3.39E–03	−1.99E–03
MAD	3.42E–02	3.04E–02	3.04E–02
MSE	1.79E–03	1.68E–03	1.57E–03
RMSE	4.24E–02	4.12E–02	3.97E–02
MPE%	−1.03	−0.96	−0.46
MAPE%	7.15	6.41	5.85
MPA%	92.85	93.59	94.15
R^2	0.927	0.9316	0.936

significantly larger MPA than that of the Classical GDM for all the architectures. This shows that GDM_MMD method outperforms the classical GDM.

Next, the same data as above is considered and modified backpropagation method employing the Cauchy error function (GDM_CCY), instead of conventionally used quadratic error function, is used for forecasting. These results are tabulated in Table 5.

It can be seen from Table 5 that the best MPA (94.40%), amongst all possible algorithms and architectures, for NIFTY monthly forecast could be obtained with GDM_CCY with an architecture of 05-05-01. This shows that GDM_CCY method outperforms the classical GDM.

Performance of all the classical and modified GDM algorithms, with respect to efficacy parameters MPA and MSE, of NIFTY50 monthly forecasting in testing phase, is summarized below in Table 6, for a quick glance, for all the architectures.

It can be observed from Table 6, in the testing phase, barring a few exceptions, MPAs and MSEs are in general better with the algorithms implementing the proposed variants of the optimization function as compared to the classical GDM method. Also it can be concluded that the best MPA, (94.40%), amongst all possible algorithms and architectures, for NIFTY monthly forecast could be obtained with GDM_CCY with an architecture of 05-05-01.

A comparison summary of all the MPAs obtained for various ANN architectures by using different algorithms for NIFTY50 monthly forecast, is shown graphically in the Fig. 3.

It is clear from the figure above that the best MPA is achieved by the ANN architecture 05-05-01 and deploying the GDM_CCY algorithm.

A comparison of all the MSEs obtained for various ANN architectures by using different algorithms, is shown graphically in the Fig. 4.

It can be seen from Fig. 4 that the least MSE is achieved by the ANN having the architecture 05-05-01 and deploying the algorithm GDM_CCY.

A comparison of all the R^2 values obtained for various ANN architectures by using different algorithms, is also compiled in the Fig. 5.

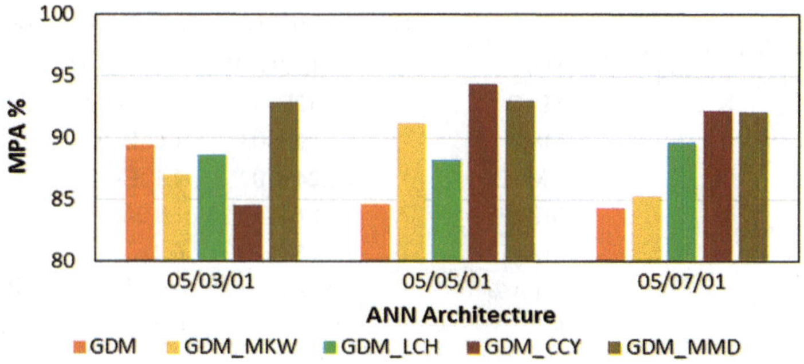

Fig. 3 MPA comparison for monthly forecasting NIFTY50 of classical GDM with modified methods during testing

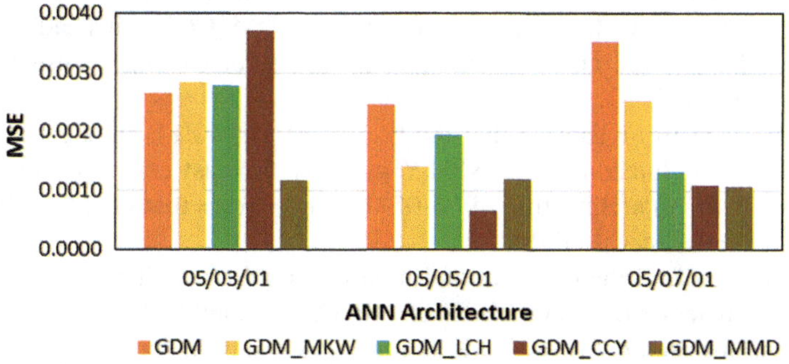

Fig. 4 MSE comparison for monthly forecasting NIFTY50 of classical GDM with modified methods during testing

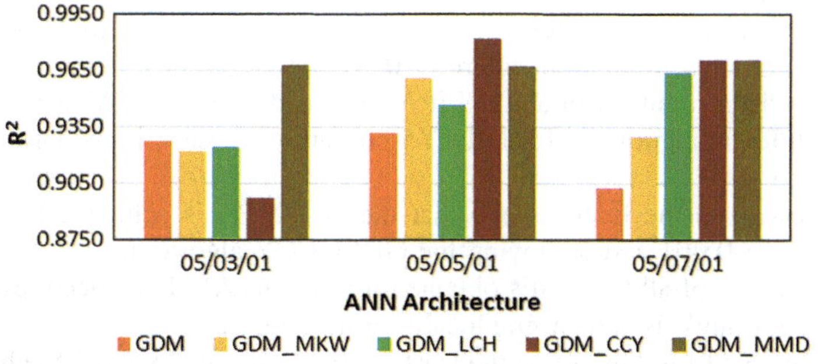

Fig. 5 R^2 comparison for monthly forecasting NIFTY50 of classical GDM with modified methods during testing

It can be seen from Fig. 5 that the best fit i.e. the highest R^2 is achieved by the 05-05-01 ANN deploying the GDM_CCY algorithm. The largest MPA obtained is 94.40% with GDM_CCY with the network architecture 05-05-01. All the efficacy parameters of this case are compared with those of the classical GDM in Table 7.

One can observe in Table 7, a significant jump of 10% in MPA while using the modified algorithm in the testing phase. Other parameters are also significantly better with the modified algorithm.

The graph in Fig. 6 plots desired versus forecasted normalized prices in the testing phase for the classical GDM with an ANN architecture 05-05-01 while as in Fig. 7 the same study is depicted employing the GDM_CCY algorithm.

It can be observed from the graphs in Figs. 6, 7 and Table 7 that the modified algorithm, GDM_CCY gives a better forecast than the classical GDM algorithm in the testing phase.

Further, the results were validated on an out-of-sample data (validation phase) also for the chronologically latest 15 months. All the efficacy parameters except R^2

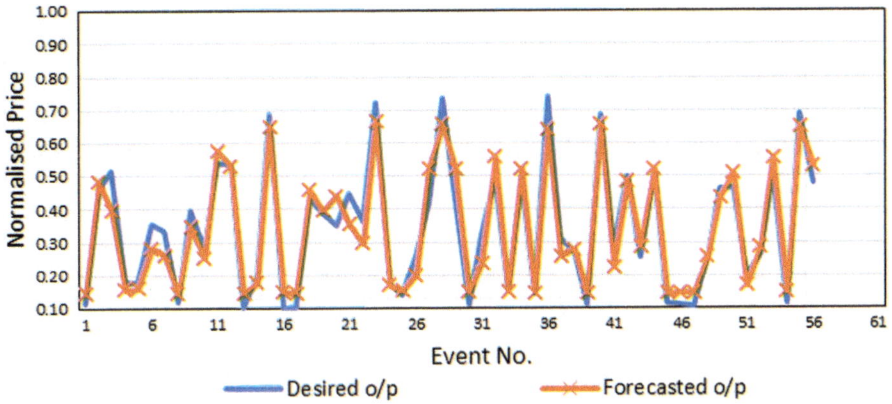

Fig. 6 Desired versus forecasted output for nifty monthly test data using classical GDM (ANN architecture 05-05-01)

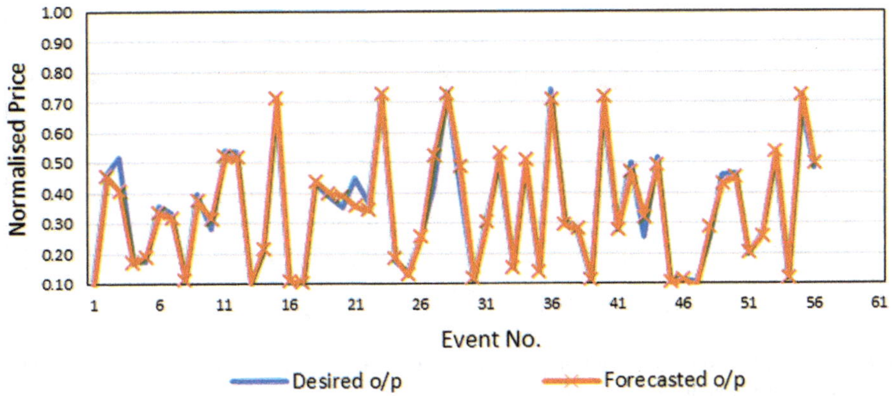

Fig. 7 Desired versus forecasted output for nifty monthly test data using GDM_CCY (ANN architecture 05-05-01)

are shown in Table 8 as R^2 is not a reliable performance indicator for the out of sample data [8].

As can be seen from Table 8 the MPA increases from 90.23% (classical GDM) to 95.66% (GDM_CCY). Other parameters are also significantly better with the modified algorithm. Graphs for these two cases are plotted below in Figs. 8 and 9.

Graphs in the Figs. 8 and 9 drawn above, plot desired versus forecasted normalized prices of the validation phase for GDM and GDM_CCY algorithms with the ANN configuration 05-05-01. It can be observed from these graphs that the modified algorithm, GDM_CCY gives a better forecast than the classical GDM algorithm in the validation phase too.

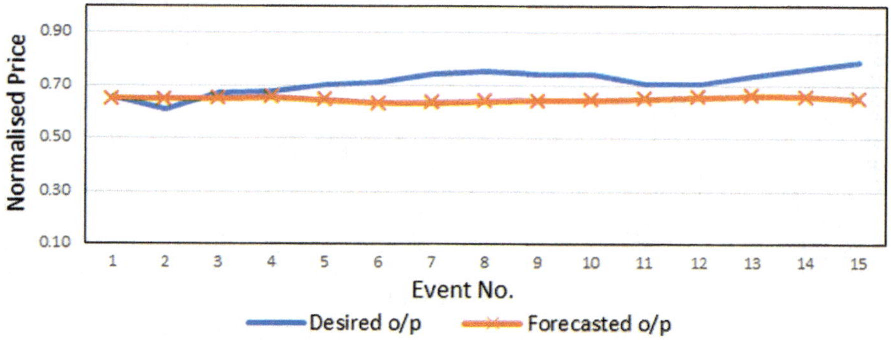

Fig. 8 Desired versus forecasted output (classical GDM) during validation for NIFTY50 monthly forecasting (ANN architecture 05-05-01)

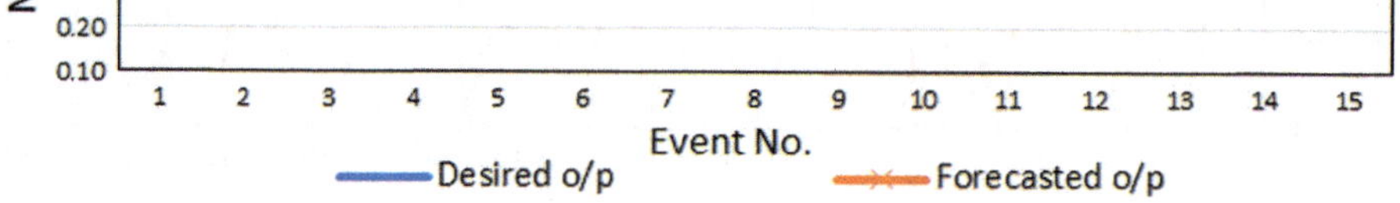

Fig. 9 Desired versus forecasted output (GDM_CCY) during validation for NIFTY50 monthly forecasting (ANN architecture 05-05-01)

5.2 *NIFTY50 Daily Forecast*

For this forecast, one year data is considered from 1 Jan 2017 to 31 Dec 2017 for NIFTY50. The data consists of the closing price of NIFTY50 for every day of the month. The prices are plotted in the Fig. 10.

One day ahead price is forecasted from the previous 5 days' prices. Efficacy parameters of daily forecasts of NIFTY50 using the classical GDM method and modified GDM methods for various ANN architectures and for testing and validation phases are summarized below in Tables 9, 10, 11, 12, 13, 14, 15 and 16. Important observations are mentioned below each table.

It can be seen from Table 9 that when the classical GDM is employed, MPA of forecast in testing phase increases as the number of hidden layer neurons is increased from 3 to 7 in the network. This parameter needs to be compared with the

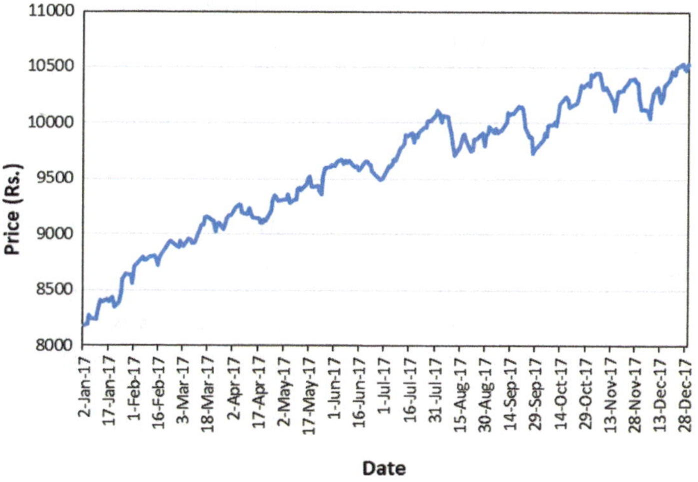

Fig. 10 Daily price variation of NIFTY50

Table 10 Efficacy parameters of NIFTY50 daily forecasting using GDM_LCH in testing phase

ANN architecture	05-03-01	05-05-01	05-07-01
ME	3.95E–03	−5.20E–03	5.64E–04
MAD	3.30E–02	2.87E–02	2.14E–02
MSE	1.74E–03	1.40E–03	7.74E–04
RMSE	4.17E–02	3.74E–02	2.78E–02
MPE%	0.9	−1.55	0.19
MAPE%	6.27	5.76	4.19
MPA%	93.73	94.24	95.81
R^2	0.9292	0.9432	0.9685

Table 11 Efficacy parameters of NIFTY50 daily forecasting using GDM_CCY in testing phase

ANN architecture	05-03-01	05-05-01	05-07-01
ME	−9.11E−03	−4.79E−03	−8.28E−03
MAD	3.54E−02	2.80E−02	2.72E−02
MSE	1.98E−03	1.22E−03	1.32E−03
RMSE	4.45E−02	3.50E−02	3.63E−02
MPE%	−1.51	−1.27	−1.53
MAPE%	6.1	5.79	4.92
MPA%	93.9	94.21	95.08
R^2	0.9194	0.9502	0.9464

Table 12 Efficacy parameters of NIFTY50 daily forecasting using GDM_MMD in testing phase

ANN architecture	05-03-01	05-05-01	05-07-01
ME	−2.39E−03	−1.52E−03	−1.74E−03
MAD	2.09E−02	2.51E−02	1.74E−02
MSE	6.73E−04	1.04E−03	5.76E−04
RMSE	2.59E−02	3.22E−02	2.40E−02
MPE%	−0.71	−0.19	−0.25
MAPE%	4.02	4.62	3.28
MPA%	95.98	95.38	96.72
R^2	0.9726	0.9577	0.9766

Table 13 Efficacy parameters of NIFTY50 daily forecasting using GDM_MKW in testing phase

ANN architecture	05-03-01	05-05-01	05-07-01
ME	1.89E−03	1.56E−03	−4.69E−03
MAD	1.75E−02	1.86E−02	2.01E−02
MSE	5.21E−04	5.78E−04	7.96E−04
RMSE	2.28E−02	2.40E−02	2.82E−02
MPE%	0.15	0.24	−0.76
MAPE%	3.22	3.67	3.51
MPA%	96.78	96.33	96.49
R^2	0.9788	0.9765	0.9676

values obtained from other methods which is demonstrated below in Tables 10, 11, 12 and 13.

It can be seen from Table 10 that both the parameters MPA and MSE are better forecasted by GDM_LCH than by the classical GDM for all the architectures. This is an advantage of the GDM_LCH method in comparison to the classical GDM.

Next, the same data as above is considered and modified backpropagation method employing the cauchy error function (GDM_CCY), instead of conventionally used quadratic error function, is used for forecasting. These results are tabulated in Table 11.

Table 14 MPA and MSE summary for NIFTY50 daily forecast in testing phase

ANN architecture	Method	MSE	MPA%
05-03-01	Classical GDM	1.79E–03	92.85
	GDM_MKW	5.20E–04	96.78
	GDM_LCH	1.74E–03	93.73
	GDM_CCY	1.98E–03	93.90
	GDM_MMD	6.70E–04	95.97
05-05-01	Classical GDM	1.68E–03	93.59
	GDM_MKW	5.80E–04	96.33
	GDM_LCH	1.39E–03	94.24
	GDM_CCY	1.23E–03	94.21
	GDM_MMD	1.04E–03	95.38
05-07-01	Classical GDM	1.87E–03	94.15
	GDM_MKW	8.00E–04	96.48
	GDM_LCH	7.70E–04	95.81
	GDM_CCY	1.32E–03	95.08
	GDM_MMD	5.80E–04	96.72

Table 15 Efficacy parameters of NIFTY50 daily forecasting using the classical GDM and GDM_MKW with 05-03-01 ANN in testing phase

Method	GDM	GDM_MKW
ME	−4.60E–03	1.90E–03
MAD	3.42E–02	1.75E–02
MSE	1.79E–03	5.20E–04
MPE%	−1.03	0.15
MAPE%	7.15	3.22
MPA%	92.85	96.78
R^2	0.927	0.9788

Table 16 Efficacy parameters of NIFTY50 daily forecasting using classical GDM and GDM_MKW in validation phase with 05-03-01 ANN

Method	GDM	GDM_MKW
ME	3.98E–02	1.36E–02
MAD	4.93E–02	2.09E–02
MSE	3.10E–03	5.90E–04
MPE%	4.57	1.59
MAPE%	5.83	2.52
MPA%	94.17	97.48

It can be seen from Table 11 that MPA of forecast by GDM_CCY is better than that by the classical GDM for all the architectures. This is an advantage of the GDM_CCY method in comparison to the classical GDM.

Next, the same data as above is considered and modified backpropagation method employing the mean median error function (GDM_MMD), instead of conventionally used quadratic error function, is used for forecasting. These results are tabulated in Table 12.

It can be seen from Table 12 that MPA of forecast is better by GDM_MMD than by the classical GDM for all the architectures. This is an advantage of the GDM_MMD method in comparison to the classical GDM.

Next, the same data as above is considered and modified backpropagation method employing the Minkowski error function (GDM_CCY), instead of conventionally used quadratic error function, is used for forecasting. These results are tabulated in Table 13.

It can be seen from Table 13 that deploying the architecture 05-03-01 and using the algorithm GDM_MKW gives the best MPA, 96.78%, out of all the algorithms and architectures considered here in the one day ahead forecast of NIFTY50 in testing phase. This is an advantage of the GDM_MKW method in comparison to the classical GDM.

A comparison of all the MPAs obtained for various ANN architectures by using different algorithms, is shown graphically in the Fig. 11.

It can be seen from the Fig. 11 that the best MPA is achieved by the 05-03-01 ANN deploying the GDM_MKW algorithm.

A comparison of all the MSEs obtained for various ANN architectures by using different algorithms, is shown graphically in the Fig. 12.

It can be seen from the Fig. 12 that the least MSE is achieved by the 05-03-01 ANN deploying the GDM_MKW algorithm.

A comparison of all the R^2 values obtained for various ANN architectures by using different algorithms, is shown graphically in the Fig. 13.

It can be seen that the best fit i.e. the highest R^2 is achieved by the 05-03-01 ANN deploying the GDM_MKW algorithm. Performance of all the algorithms, with respect to efficacy parameters MPA and MSE, in NIFTY50 daily forecasting in testing phase, is summarized below in Table 14, for a quick glance, for all the architectures.

It can be observed from Table 14 that, in the testing phase, MPAs and MSEs are better with the proposed variants of the optimization function as compared to the

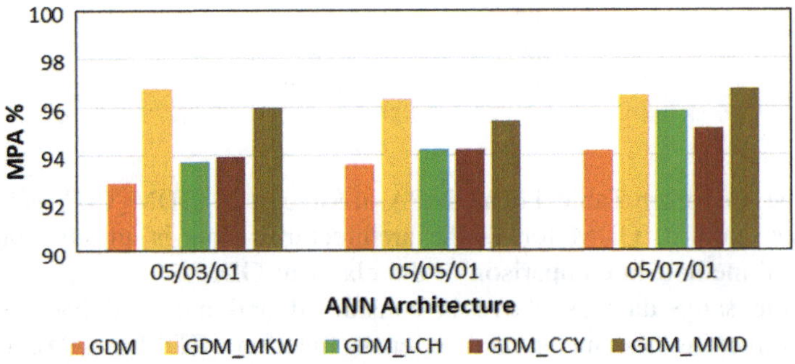

Fig. 11 Daily forecast of MPA for NIFTY50 with classical and modified GDMs during testing

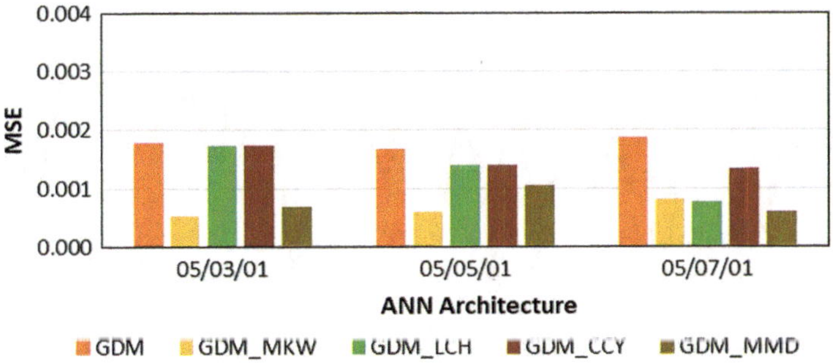

Fig. 12 Daily forecast of MSE for NIFTY50 with classical and modified GDMs during testing

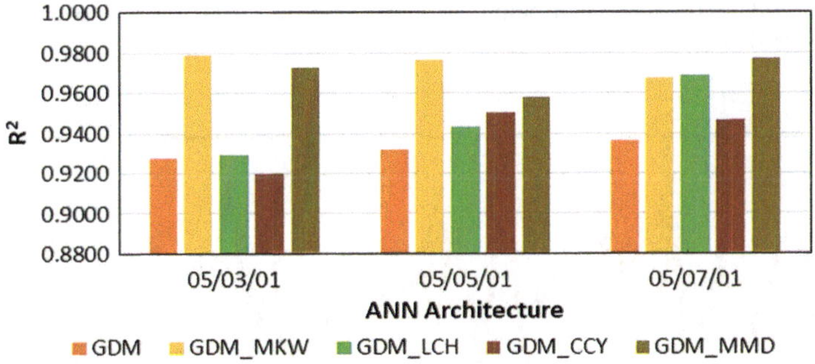

Fig. 13 Daily forecast of R^2 for NIFTY50 with classical and modified GDMs during testing

optimization function of the classical GDM method. The largest MPA obtained in the testing phase is 96.78% with GDM_MKW with the network architecture 05-03-01. All the efficacy parameters of this case are compared with those of the classical GDM in Table 15.

One can observe a significant jump of around 4% in MPA i.e. from 92.85% to 96.78% while using the modified algorithm in the testing phase as shown in the Table 15. Other parameters are also significantly better with the modified algorithm. Graphs in the Figs. 14 and 15, plot desired versus forecasted normalized prices of the testing phase for GDM and GDM_MKW algorithms with the ANN configuration 05-03-01.

It can be observed from the graphs in Figs. 14 and 15 and Table 15 that the modified algorithm, GDM_MKW gives a better forecast than the classical GDM algorithm in the testing phase.

Further, the results were validated on an out-of-sample data (validation phase) also for the chronologically latest 32 days. All the efficacy parameters except R^2 are shown in Table 16 as R^2 is not a reliable performance indicator for the out of sample data [8].

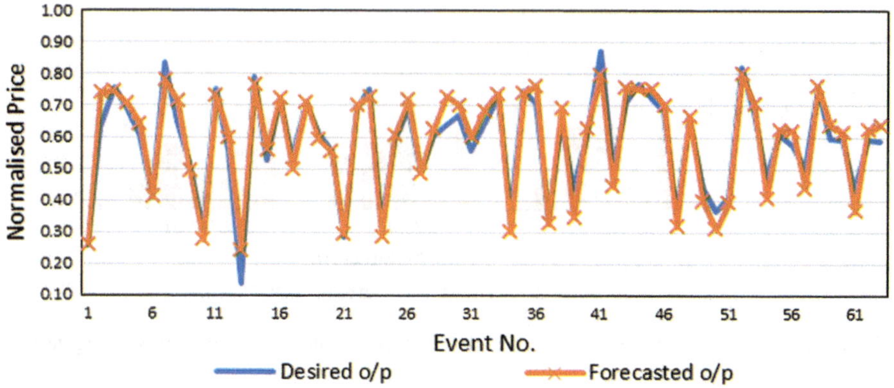

Fig. 14 Desired versus forecasted output for nifty daily test data using classical GDM (ANN architecture 05-03-01)

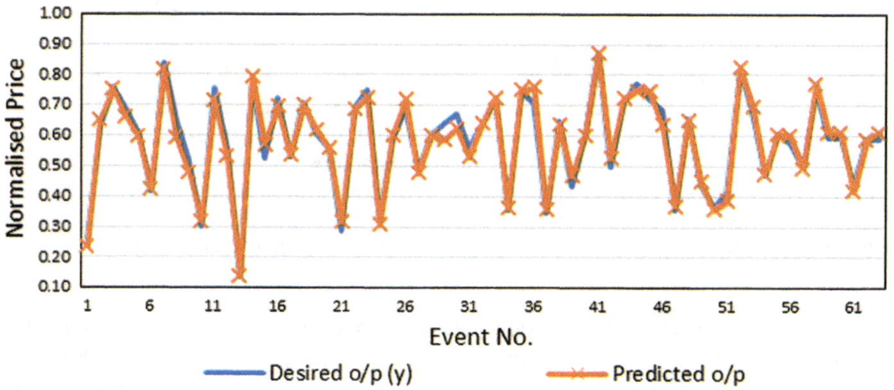

Fig. 15 Desired versus forecasted output for nifty daily test data using GDM_MKW (ANN architecture 05-03-01)

As can be seen from Table 16 the MPA increases from 94.17% (classical GDM) to 97.48% (GDM_MKW). Other parameters are also significantly better with the modified algorithm. Graphs in the Figs. 16 and 17 drawn below, plot desired versus forecasted normalized prices of the validation phase for GDM and GDM_MKW algorithms with the ANN configuration 05-03-01.

It can be observed from the graphs in Figs. 16 and 17 that the modified algorithm, GDM_MKW gives a better forecast than the classical GDM algorithm in the validation phase too.

Thus to summarize the results of the monthly and the daily NIFTY forecasting, it was found that in the monthly horizon the best results are obtained by GDM_CCY

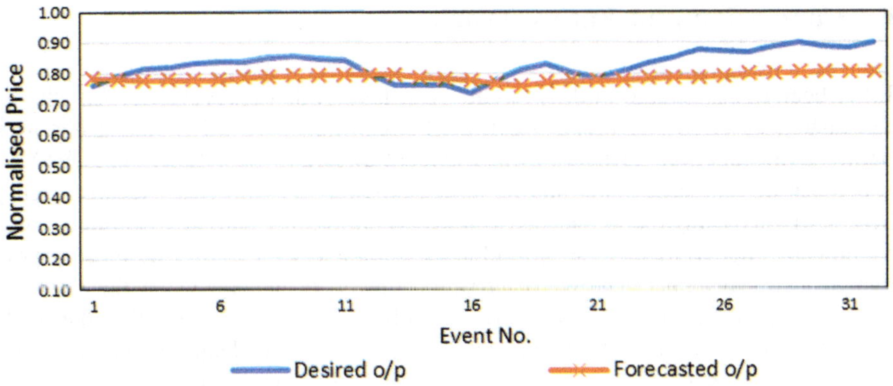

Fig. 16 Desired versus forecasted output (classical GDM) during validation for NIFTY50 daily forecasting (ANN architecture 05-03-01)

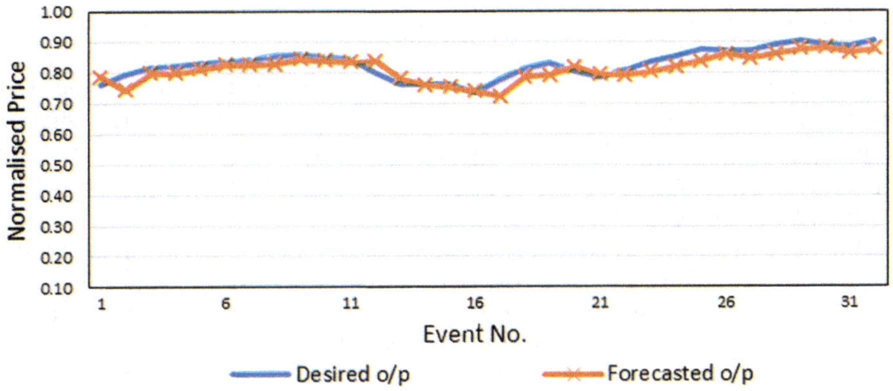

Fig. 17 Desired versus forecasted output (GDM_MKW) during validation for NIFTY50 daily forecasting (ANN architecture 05-03-01)

which is the modified algorithm using the Cauchy error function as the optimization criteria instead of the quadratic error function used in classical GDM algorithm. GDM_CCY improves the MPA by approximately 10% in the testing phase and 5% in the validation phase. In the case of daily forecasting, it was found that the best results are obtained by GDM_MKW which is the modified algorithm using the MinKowski error function as the optimization criteria instead of the quadratic error function used in classical GDM algorithm. GDM_MKW improves the MPA by approximately 4% in the testing phase and 3% in the validation phase.

6 Conclusion and Future Work

It has been demonstrated in the present work that, while deploying ANN methods, forecasting efficiency of Indian stock market can be improved by using alternate optimization functions instead of the prevalent quadratic error function. NIFTY50 was taken up as a case study and its forecasting was done in two widely differing time horizons using both the classical GDM algorithm and the modified ones. Significant improvement in several efficacy parameters including the most important one i.e. Mean Percent Accuracy was achieved in general when the alternate optimization functions were used. For monthly time horizon forecasting the modified GDM based on Cauchy Error Function gave the best Mean Percent Accuracy while for the daily time horizon forecasting the modified GDM based on Minkowski Error Function gave the highest Mean Percent Accuracy. One can therefore conclude that the choice of the error function to be employed is in principle problem dependent and error functions cannot be used as off the shelf. Also it is clear from the detailed investigation done above that one has to be particularly careful while employing an error function as there is no thumb rule which states beforehand which error function can be employed for which type of data. Thereby, the researchers have to find the error function through trial and error methods. Though, time consuming it may seem, however we strongly suggest exploiting different error functions for problem at hand to put forward the most optimized combination and the least possible error thereof, instead of always using a few conventionally postulated error minimization error functions.

Effective and robust prediction of financial markets depends upon progress put forward in technology and the readiness of the researchers to apply the tools developed. Tools that are suitable and quite effective for financial markets have now started to emerge from academia. It is therefore of utmost importance that proper exploitation of the emerging trends in AI and other fields is adhered to especially in wake of rapidly increasing academic research and falling prices of computers coupled with their ever increasing computational speeds. AI methods are therefore, clearly the technology of future for financial market exploitation and forecasting.

References

1. Jain, A., Bhatnagar, V.: Concoction of ambient intelligence and big data for better patient ministration services. Int. J. Ambient. Comput. Intell. (IJACI) **8**(4), 19–30 (2017)
2. Kamal, S., Dey, N., Nimmy, S.F., et al.: Evolutionary framework for coding area selection from cancer data. Neural Comput. Appl. **29**(4), 1015–1037 (2018)
3. Dey, N., Wagh, S., Mahalle, P., Pathan, M.: Applied Machine Learning for Smart Data Analysis. CRC Press, Boca Raton (2019). https://doi.org/10.1201/9780429440953
4. Dey, N., Ashour, A.S., Bhatt, C.: Internet of things driven connected healthcare. In: Internet of Things and Big Data Technologies for Next Generation Healthcare, pp. 3–12. Springer, Cham (2017)

5. Mo, H., Wang, J., Niu, H.: Exponent back propagation neural network forecasting for financial cross-correlation relationship. Expert Syst. Appl. **53**, 106–116 (2016)
6. Hornik, K., Stinchcombe, M., White, H.: Multilayer feedforward networks are universal approximators. Neural Netw. **2**(5), 359–366 (1989)
7. Dhar, V.K., Tickoo, A.K., Koul, R., Dubey, B.P.: Comparative performance of some popular ANN algorithms on benchmark and function approximation problems. Pramana **74**(2), 307–324 (2010)
8. Montgomery, D.C., Jennings, C.L., Kulahci, M.: Introduction to Time Series Analysis and Forecasting, 1st edn. Wiley, USA (2008)
9. Kim, K.-j., Han, I.: Genetic algorithms approach to feature discretization in artificial neural networks for the prediction of stock price index. Expert Syst. Appl. **19**, 125–132 (2000)
10. Kim, K.-j.: Financial time series forecasting using support vector machines. Neurocomputing **55**, 307–319 (2003)
11. Kim, K.-j.: Artificial neural networks with evolutionary instance selection for financial forecasting. Expert Syst. Appl. **30**, 519–526 (2006)
12. Abbasi, E., Abouec, A.: Sock price forecast by using neuro-fuzzy inference system. World Acad. Sci. Eng. Technol. **46**, 320–323 (2008)
13. Atsalakis, G.S., Valavanis, K.P.: Forecasting stock market short-term trends using a neuro-fuzzy based methodology. Expert Syst. Appl. **36**, 10696–10707 (2009)
14. Boyacioglu, M.A., Avci, D.: An adaptive network-based fuzzy inference system (ANFIS) for the prediction of stock market return: the case of the Istanbul stock exchange. Expert Syst. Appl. **37**, 7908–7912 (2010)
15. Hadavandi, E., Shavandi, H., Ghanbari, A.: Integration of genetic fuzzy systems and artificial neural networks for stock price forecasting. Knowl.-Based Syst. **23**, 800–808 (2010)
16. Kara, Y., Boyacioglu, M.A., Baykan, Ö.K.: Predicting direction of stock price index movement using artificial neural networks and support vector machines: the sample of the Istanbul stock exchange. Expert Syst. Appl. **38**, 5311–5319 (2011)
17. Ticknor, J.L.: A Bayesian regularized artificial neural network for stock market forecasting. Expert Syst. Appl. **40**, 5501–5506 (2013)
18. Moghaddam, A.H., Moghaddam, M.H., Esfandyari, M.: Stock market index prediction using artificial neural network. J. Econ. Financ. Adm. Sci. **21**, 89–93 (2016)
19. Aparna Nayak, M.M., Pai, M., Pai, R.M.: Prediction models for Indian stock market. Proc. Comput. Sci. **89**, 441–449 (2016)
20. Abdulmalik, S., Almasani, M., Finaev, V.I., Qaid, W.A.A., Tychinsky, A.V.: The decision-making model for the stock market under uncertainty. Int. J. Electr. Comput. Eng. (IJECE) **7**(5), 2782–2790 (2017)
21. Devadoss, A.V., Ligori, T.A.A.: Stock prediction using artificial neural networks. Int. J. Data Min. Tech. Appl. **2**, 283–291 (2013)
22. Yetis, Y., Kaplan, H., Jamshidi, M.: Stock market prediction by using artificial neural network. World Automation Congress, pp. 718–722. IEEE Computer Society, U.S. (2014)
23. Qiu, M., Song, Y., et al.: Application of ANN for prediction of stock market returns: Case of Japanese stock market. Chaos Solitions Fractals **85**, 1–7 (2016)
24. Kamley, S., Jaloree, S., Thakur, R.S.: Performance forecasting of share market using machine learning techniques: a review. Int. J. Electr. Comput. Eng. **6**(6), 3196–3204 (2016)
25. Rumelhart, D.E., Hinton, G.E., Williams, R.J.: Learning representations by back-propagating errors. Nature **323**, 533–536 (1986)
26. Atsalakis, G.S., Kimon, P., Valavanis, K.P.: Surveying stock market forecasting techniques—Part II: soft computing methods. Expert Syst. Appl. **36**, 5932–5941 (2009)
27. Reidmiller, M.: Advanced supervised learning in multilayer perceptrons from backpropagation to adaptive learning algorithms. Comput. Stand. Interfaces **16**, 265–278 (1994)
28. Moller, M.F.: A scaled conjugate gradient algorithm for fast supervised learning. Neural Netw. **6**, 525–533 (1993)
29. Marquardt, D.W.: An algorithm for least square estimation of non-linear parameters. J. Soc. Ind. Appl. Math. **11**(2), 431–441 (1963)

30. Lahmiri, S.: A comparative study of backpropagation algorithms in financial prediction. Int. J. Comput. Sci. Eng. Appl. **1**(4), 15–21 (2011)
31. Gangal, A.S., Kalra, P.K., Chauhan, D.S.: Performance evaluation of complex valued neural networks using various error functions. Int. J. Electr. Comput. Energ. Electron. Commun. Eng. **1**(5), 732–737 (2007)
32. www.nseindia.com. Accessed 10 April 2019

Analysis of Facial Expression Recognition of Visible, Thermal and Fused Imaginary in Indoor and Outdoor Environment

Ravindra Patil, Kiran Chaudhari, S. N. Kakarwal, R. R. Deshmukh and D. V. Kurmude

Abstract This research study performed on visible images, thermal images and fused images for facial expression recognition. Linear Discriminant Analysis has implemented for feature extraction technique and support vector machine to calculate the result. This work is implemented on a newly designed database of 20 peoples' facial expression which includes visible images, thermal images, and fused images. The extracted features of visible, thermal and fused images are utilized for classification using support vector machine. This study focuses on 5 types of facial expression. Better results are achieved on smile and anger expression. The comparative analysis of this study is done on visible, thermal and fused facial expression images. The experimental result analysis shows that fused images give better results as compared to visible images. The accuracy of smile expression is better than anger and disgust facial expression. The implementation is carried out on dataset designed in indoor and outdoor environmental setup.

Keywords Support vector machine · Linear discriminant analysis · Facial expression recognition · Visible · Thermal · Fused dataset

---

R. Patil (✉) · K. Chaudhari
Computer Science and Engineering, Marathwada Institute of Technology, Aurangabad 431001, India
e-mail: ravindra.be2004@gmail.com

K. Chaudhari
e-mail: kiran_chaudhari@rediffmail.com

S. N. Kakarwal
Computer Science and Engineering, P.E.S. College of Engineering, Aurangabad 431001, India
e-mail: s_kakarwal@yahoo.com

R. R. Deshmukh
Dept. of Computer Science and IT, Dr. Babasaheb Ambedkar Marathwada University, Aurangabad 431001, India
e-mail: ratnadeep_deshmukh@yahoo.co.in

D. V. Kurmude
Physics Department, Milind College of Science, Aurangabad 431002, India
e-mail: dvkurmude@yahoo.com

N. Dey et al. (eds.), *Internet of Things, Smart Computing and Technology: A Roadmap Ahead*, Studies in Systems, Decision and Control 266,
https://doi.org/10.1007/978-3-030-39047-1_2

1 Introduction

Human expression plays a vital role in the identification of the human state of mind. Many times, with the help of expression, a human can understand the mental state of peoples. In any situation, human expression replaces the verbal talk and gives a signal to the surrounding person so that people understand the current situation whether it is a positive one or a negative one. These types of expression recognition and analysis are important in the clinical area, criminal cases such as lie detection tests are carried out. Now a day's robotics plays a vital role in our day to day routine work. In human-computer interaction world, understanding the expression of a human being by robot or machine is an essential task. The identification of correct expression to give a proper response to that expression is an important task for that machine program. This will be beneficial for robots who are taking care of patients in intensive care units or taking care of old aged peoples at home [1–6].

Last 3 decades various methods have implemented to work on face recognition and facial expression recognition using visible images and thermal infrared images separately. Some researchers also worked on a fusion of visible and thermal images for recognition of face and their expressions [7–9].

Many researchers use only visible images like grayscale 2-D images for their experimental work. An appearance-based method, as well as geometric feature-based methods are used by these researchers to collect the various features for experimental work. Visible images performance gets vary as illumination condition changes. Many kinds of research works demonstrate that illumination condition plays a vital role during the prediction of emotion using a visible facial expression. Visible image of the same person in different illumination situation can produce varying result [10–12].

Thermal images are going to be widely used to overcome the illumination issues over the visible images. Thermal images are invariant to illumination changes as suggested by various researcher's works [13, 14]. Thermal images also work on heat patterns illuminated from human faces. This is helpful for the detection of variation in emotion using different heat patterns observed by the thermal camera [15, 16].

Each expression emits heat which normally gives varying heat patterns. These slight varying heat patterns are not detected by a normal person's eye view or even by a visible camera, but a thermal camera can easily detect the heat pattern. Many researchers use this heating pattern to capture the features for facial expression recognition. One major drawback is detected that thermal images are prone to sun-glasses so that human being wearing sunglasses can degrade the overall performance of the recognition work [17–20].

To overcome these drawbacks, few researchers works on a fusion of thermal images and visible images feature set. They have independently extracted the features from visible images and thermal images and then combine these into fused images. Such experimental works give better result as compare to only visible image dataset or thermal image dataset [21].

In many research, works demonstrate that the use of action units along with expression, use of head motion with expression gives the better result. It is observed that face recognition using thermal images have higher recognition rates as compare to visible images due to various illumination conditions. Visible images give poor performance in poor illumination condition i.e. in night. Thermal Infrared images overcome this problem but have some limitation related to occlusion to eyeglasses which degrades the performance of face recognition [22].

Heat emissivity is get captured by thermal camera. This heat pattern is useful to detect the unwanted regions of face as well as noise present in face region. These regions are not processed during feature extraction.

Separate experimentation has done on visible, thermal and fused images to measure the performance. Visible spectrum is in range 0.4–0.7 μm whereas NIR 0.7–1.0 μm. Short-wave Infrared is having range 1–3 μm and Long-wave Infrared is having range 8–14 μm.

FLIR camera have used to capture the visible, thermal and fused images. Visible images are captured under the visible spectrum useful for visible features while thermal images are acquired with thermal emissivity of face useful for identifying thermal features. Fused image is the combination of visible and thermal features [23–25].

2 Experimental Setup for Database Capturing

In development of dataset, Images are captures in indoor and outdoor locations. Figures 1, 2 and 3 show sample visible, thermal and fused images of our database.

In indoor location, the controlled environment is maintained for light intensity and temperature drift 1–2°. The camera is mounted one meter away from subject. The Height of camera from ground level is 5 feet. The wall behind the subject is painted with white color. Subjects have asked to stand up in front of camera near the wall. Subject have also asked to keep their face positions towards camera lens. Subject have asked to give different pose for smile, sadness, astonish, anger and neutral. For each pose expression, Images are clicked using FLIR camera. With this single click, Images have captured of visible, thermal and fused image for same subject. Total 20 subject participated for data capturing. In this study, 2 images per dataset per expression per environment per subject is captured. Total 1200 images are captured for 20 subjects.

Total images of 20 subjects are divided into five expression as smile, sadness, astonish, anger and neutral groups. Each group now includes 240 images out of which 120 images from indoor situation and 120 images from outdoor situation.

In indoor situation temperature is maintain at 26 °C. The Data acquisition laboratory is a normal room of size 48 m with four white color walls and one door which is closed during experimentation work. These images have captured during

9 am to 5.00 pm for 2 days' time. The age group of subjects varies from 18 to 26 years. Total 20 subjects are participated for this dataset capturing of which 5 are female and 15 are male subject. Some subject has eyeglasses which is beneficial for our dataset creation.

Before capturing the images, subjects are instructed to seat in laboratory environment for 20 min to adjust their body temperature as well as ornament temperature with data capture laboratory room temperature. Two cycles of image capturing are related to indoor controlled condition and outdoor condition respectively.

In outdoor environment, temperature was varying from 32 to 35 °C. Air impact, humidity is also varying in outdoor environment. Subjects have asked to take a walk of 1 km in 15 min for outdoor condition. Once walk is over, subject is asked to ready for data captured in outdoor condition. The distance between subject and camera, position and height of camera is kept same as indoor condition. Same set of expression are repeated for outdoor image capturing. Initially 30 subjects are shortlisted for this dataset capturing experiment. Out of which 20 subjects have successfully completed the indoor and outdoor process as per our standard condition and rules. Same experiment of data capturing has done in rainy season and in high air flow condition, but not able to get good quality images due to humidity and temperature drift. It happens due to humidity in atmosphere and loss of temperature spot signal due to air respectively.

For summer season, Experimental setup have got high quality images as temperature range is permissible for thermal image capturing. Hot day won't affect the sensor as their range is −20 to +150 °C. FLIR C2 camera is used for capturing the dataset. This camera is widely used for various engineering application where thermal image capturing is essential task [4, 21–28]. This camera is easy to use and of pocket size to carry from one location to other. The specification of camera are as follows:

Focal length: 1.54 mm
Spectral range: 7.5–14 μm
IR Sensor measurement in pixel: 80 × 60 = 4,800.
Temperature operating range: −10 to +50 °C (14–122 °F).
Visible spectrum camera resolution in pixel: 640 × 480
Digital color display measurement in pixel: 320 × 240
Accuracy: ±2 °C
Object Temperature range: –10 to +150 °C (14–302 °F).
Thermal Sensitivity: <0.10 °C (Figs. 1, 2, 3).

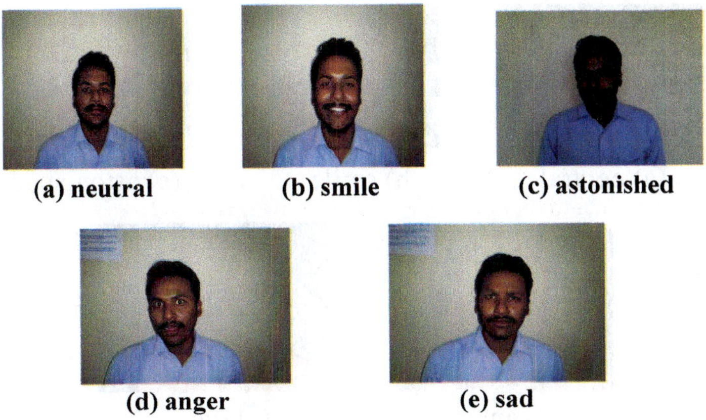

Fig. 1 Sample visible image

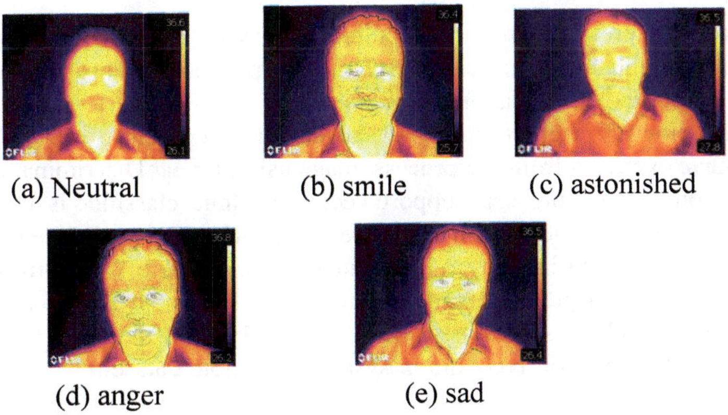

Fig. 2 Sample thermal images

3 Experiment Procedure

3.1 Experiment 1

In this experiment, only visible dataset is considered which consist of 20 subjects with 400 visible images. The images distribution per expression is as shown in Table 1.

Table 1 shows total number 400 images in visible dataset with indoor and outdoor environment. This dataset is divided into training and testing dataset such that 30 images of per expression per environment is set to training dataset while 10 images of per expression per environment is set to testing dataset. It means 300 images are assign for training dataset and 100 images are assign for testing dataset.

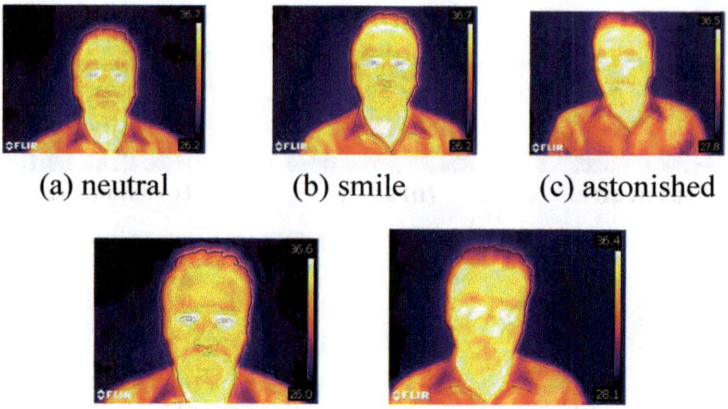

(a) neutral (b) smile (c) astonished

(d) anger (e) sad

Fig. 3 Sample fused images

In this work, pre-processed the captured images by applying rotation, scaling and translation transformation to get fixed size input image. This fixed size input image is use for experimentation purpose.

Figure 4 shows flow chart of steps to carried out experiment 1 on visible dataset. Features have extracted from pre-process image using Linear Discriminant Analysis and based on that feature set, support vector machine classifier is trained. The experiment is carried out in three different scenario as per their environment. Scenario 1 is related to indoor environment images, scenario 2 performed on outdoor environment images and scenario 3 performed on both indoor and outdoor situation images. These scenarios are useful to understand the impact of environment for facial expression recognition system on visible dataset.

Scenario 1

In this situation, indoor visible images are used for facial expression recognition. Total indoor visible images are 200 out of 150 images are assign for training dataset and 50 images are assigned for testing dataset. On trained Support Vector Machine, 50 testing images have applied to calculate the expression recognition rate. The confusion matrix of scenario 1 is shown in Table 2.

This confusion matrix shows that smile expression has achieved highest recognition rate of 63% and anger expression got lowest recognition rate of 50%. For smile expression, out of 10 images it gives correct prediction for 6 images. For anger expression have got correct prediction for 5 images. The average recognition rate for facial expression is 54%.

Scenario 2

In this situation, outdoor visible images are used for facial expression recognition. Total outdoor visible images are 200 out of 150 images are assign for training dataset and 50 images are assigned for testing dataset. On trained Support Vector

Table 1 Facial expression image distribution in training and testing dataset

Sr. No.	Dataset used	Smile		Anger		Astonished		Neutral		Sadness		Total		Total
		Indoor	Outdoor	Indoor	Outdoor	Indoor	Outdoor	Indoor	Outdoor	Indoor	Outdoor	Indoor	Outdoor	
1		40	40	40	40	40	40	40	40	40	40	200	200	400
2	Training	30	30	30	30	30	30	30	30	30	30	150	150	300
3	Testing	10	10	10	10	10	10	10	10	10	10	50	50	100

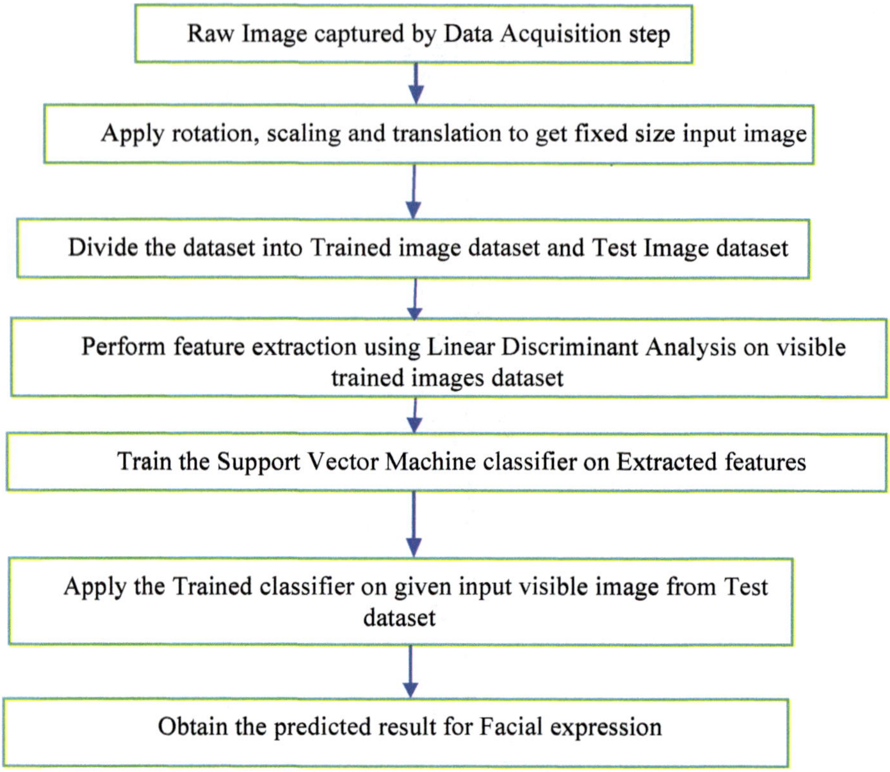

Fig. 4 Implementation of linear discriminant analysis and support vector machine for expression recognition on visible dataset

Table 2 Confusion matrix for expression recognition on indoor visible database using linear discriminant analysis and support vector machine

Expression/result % accuracy	Smile	Anger	Astonished	Neutral	Sadness
Smile	**63**	0	12	10	15
Anger	9	**50**	12	18	11
Astonished	0	16	**51**	17	16
Neutral	13	6	14	**55**	12
Sadness	3	11	11	23	**52**

Machine, 50 testing images have applied to calculate the expression recognition rate. The confusion matrix of scenario 2 is shown in Table 3.

This confusion matrix shows that smile expression have achieved highest recognition rate of 59% and astonished expression got lowest recognition rate of 44%. The average recognition rate for facial expression is 50%.

Table 3 Confusion matrix for expression recognition on outdoor visible database using linear discriminant analysis and support vector machine

Expression/% accuracy	Smile	Anger	Astonished	Neutral	Sadness
Smile	**59**	0	16	10	15
Anger	9	**45**	12	18	16
Astonished	0	16	**44**	17	23
Neutral	13	6	14	**51**	16
Sadness	3	11	11	25	**50**

Scenario 3

In this situation, indoor and outdoor visible images are used for facial expression recognition. Total indoor and outdoor visible images are 400 out of 300 images are assigned for training dataset and 100 images are assigned for testing dataset. On trained Support Vector Machine, 100 testing images have applied to calculate the expression recognition rate. The confusion matrix of scenario 3 is shown in Table 4.

This confusion matrix shows that smile expression have achieved highest recognition rate of 60% and anger expression got lowest recognition rate of 48%. The average recognition rate for facial expression is 51%.

3.2 Experiment No. 2

In this experiment, only thermal dataset is considered which consist of 20 subjects with 400 thermal images. Table 5 shows total number 400 images in thermal dataset with indoor and outdoor environment. This dataset is divided into training and testing dataset such that 30 images of per expression per environment is set to training dataset while 10 images of per expression per environment is set to testing dataset. It means 300 images are assigned for training dataset and 100 images are assigned for testing dataset. The images distribution per expression is shown in Table 5.

Table 4 Confusion matrix for expression recognition on indoor and outdoor visible database using linear discriminant analysis and support vector machine

Expression/result % accuracy	Smile	Anger	Astonished	Neutral	Sadness
Smile	**60**	4	11	10	15
Anger	9	**48**	12	20	11
Astonished	0	18	**49**	17	16
Neutral	13	6	14	**51**	16
Sadness	3	11	11	27	**48**

Table 5 Facial expression image distribution in training and testing dataset

Sr. No.	Dataset used for	Smile		Anger		Astonished		Neutral		Sadness		Total		Total
		Indoor	Outdoor	Indoor	Outdoor	Indoor	Outdoor	Indoor	Outdoor	Indoor	Outdoor	Indoor	Outdoor	
1		40	40	40	40	40	40	40	40	40	40	200	200	400
2	Training image	30	30	30	30	30	30	30	30	30	30	150	150	300
3	Testing image	10	10	10	10	10	10	10	10	10	10	50	50	100

Figure 5 shows flow chart to carried out experiment no. 2 on thermal dataset. Features have extracted from preprocessed images using Linear Discriminant Analysis and based on that feature set, Support vector machine classifier is trained. The experiment is carried out in three different scenario as per their environment. Scenario 1 is related to indoor environment images, scenario 2 performed on outdoor environment images and scenario 3 performed on both indoor and outdoor situation images. These scenarios are useful to understand the impact of environment for facial expression recognition system on thermal dataset.

Scenario 1

In this situation, indoor thermal images are used for facial expression recognition. Total indoor thermal images are 200 out of 150 images are assign for training dataset and 50 images are assigned for testing dataset. On trained Support Vector Machine, 50 testing images have applied to calculate the expression recognition rate. The confusion matrix of scenario 1 is shown in Table 6.

This confusion matrix shows that for smile expression have achieved highest recognition rate of 70% and neutral expression got lowest recognition rate of 47%. The average recognition rate for facial expression is 56%.

Scenario 2

In this situation, outdoor thermal images are used for facial expression recognition. Total outdoor thermal images are 200 out of 150 images are assign for training

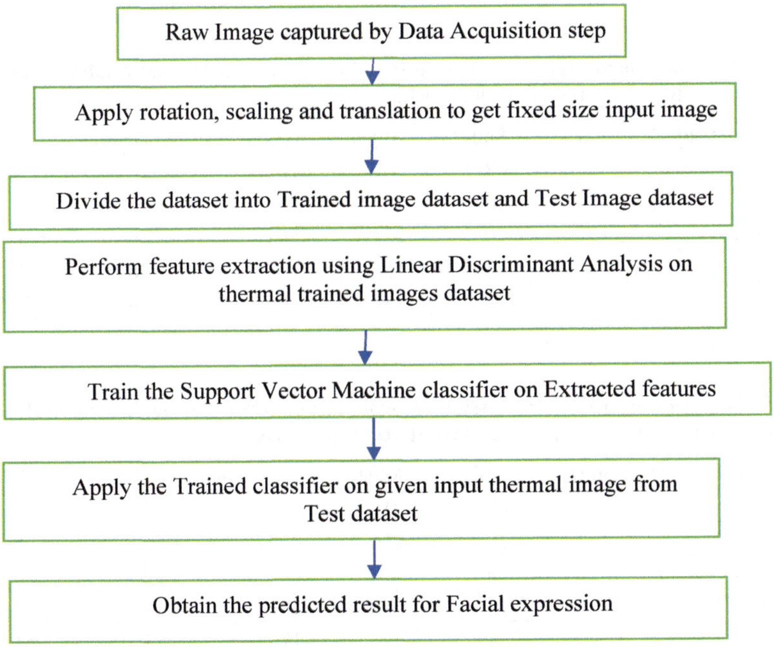

Fig. 5 Implementation of linear discriminant analysis and support vector machine on thermal database

Table 6 Confusion Matrix for expression recognition on indoor thermal database using linear discriminant analysis and support vector machine

Expression/result % accuracy	Smile	Anger	Astonished	Neutral	Sadness
Smile	**70**	0	11	10	9
Anger	10	**51**	15	17	7
Astonished	0	22	**61**	11	6
Neutral	13	16	19	**47**	5
Sadness	4	8	17	18	**53**

dataset and 50 images are assigned for testing dataset. On trained Support Vector Machine, 50 testing images have applied to calculate the expression recognition rate. The confusion matrix of scenario 2 is shown in Table 7.

This confusion matrix shows that smile expression have achieved highest recognition rate of 68% and neutral expression got lowest recognition rate of 45%. The average recognition rate for facial expression is 53%.

Scenario 3

In this situation, indoor and outdoor thermal images are used for facial expression recognition. Total indoor and outdoor thermal images are 400 out of 300 images are assigned for training dataset and 100 images are assigned for testing dataset. On trained Support Vector Machine, 100 testing images have applied to calculate the expression recognition rate. The confusion matrix of scenario 3 is shown in Table 8.

Table 7 Confusion matrix for expression recognition on outdoor thermal database using linear discriminant analysis and support vector machine

Expression/result % accuracy	Smile	Anger	Astonished	Neutral	Sadness
Smile	**68**	0	13	10	9
Anger	10	**48**	15	22	5
Astonished	0	22	**56**	16	6
Neutral	13	16	19	**45**	7
Sadness	4	8	19	18	**51**

Table 8 Confusion matrix for expression recognition on indoor and outdoor thermal database using linear discriminant analysis and support vector machine

Expression/result % accuracy	Smile	Anger	Astonished	Neutral	Sadness
Smile	**67**	0	11	13	9
Anger	10	**47**	21	17	5
Astonished	0	22	**60**	12	6
Neutral	13	16	19	**45**	7
Sadness	4	8	17	20	**51**

Table 8 confusion matrix shows smile expression have achieved highest recognition rate of 67% and neutral expression got lowest recognition rate of 45%. The average recognition rate for facial expression is 54%.

3.3 Experiment No. 3

In this experiment, only fused dataset is considered which consist of 20 subjects with 400 fused images. The image distribution per expression is as below.

Table 9 shows total number 400 images in fused dataset with indoor and outdoor environment. This dataset is divided into training and testing dataset such that 30 images of per expression per environment is set to training dataset while 10 images of per expression per environment is set to testing dataset. It means 300 images are assigned for training dataset and 100 images are assigned for testing dataset.

Figure 6 shows flow chart to carried out experiment no. 3 on fused dataset. Features have extracted from pre-process image using linear discriminant analysis and based on that feature set, support vector machine classifier is trained. The experiment is carried out in three different scenario as per their environment. Scenario 1 is related to indoor environment images, scenario 2 performed on outdoor environment images and scenario 3 performed on both indoor and outdoor situation images. These scenarios are useful to understand the impact of environment for facial expression recognition system on fused dataset.

Scenario 1

In this situation, indoor fused images are used for facial expression recognition. Total indoor fused images are 200 out of 150 images are assign for training dataset and 50 images are assigned for testing dataset. On trained Support Vector Machine, 50 testing images have applied to calculate the expression recognition rate. The confusion matrix of scenario 1 is shown in Table 10.

Table 10 confusion matrix shows that smile expression has achieved highest recognition rate of 72% and neutral expression got lowest recognition rate of 47%. The average recognition rate for facial expression is 57%.

Scenario 2

In this situation, outdoor fused images are used for facial expression recognition. Total outdoor fused images are 200 out of 150 images are assign for training dataset and 50 images are assigned for testing dataset. On trained Support Vector Machine, 50 testing images have applied to calculate the expression recognition rate. The confusion matrix of scenario 2 is shown in Table 11.

Table 11 confusion matrix shows smile expression have achieved highest recognition rate of 70% and anger expression got lowest recognition rate of 44%. The average recognition rate for facial expression is 54%.

Table 9 Facial expression fused image distribution in training and testing dataset

Sr. No.	Dataset used for	Smile		Anger		Astonished		Neutral		Sadness		Total		Total
		Indoor	Outdoor	Indoor	Outdoor	Indoor	Outdoor	Indoor	Outdoor	Indoor	Outdoor	Indoor	Outdoor	
1		40	40	40	40	40	40	40	40	40	40	200	200	400
2	Training	30	30	30	30	30	30	30	30	30	30	150	150	300
3	Testing	10	10	10	10	10	10	10	10	10	10	50	50	100

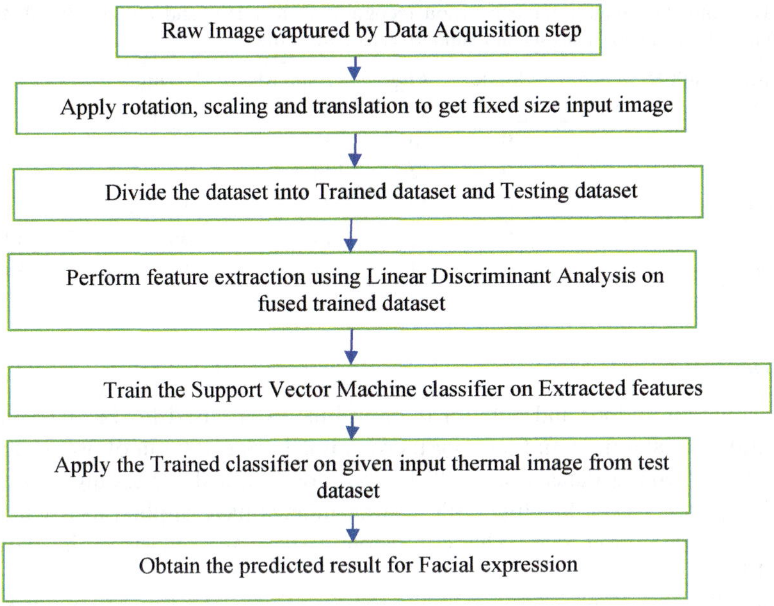

Fig. 6 Implementation of linear discriminant analysis and support vector machine on fused database

Table 10 Confusion matrix for expression recognition on indoor fused database using linear discriminant analysis and support vector machine

Expression/result % accuracy	Smile	Anger	Astonished	Neutral	Sadness
Smile	**72**	0	12	11	5
Anger	9	**55**	11	18	7
Astonished	0	22	**57**	10	11
Neutral	10	16	14	**47**	13
Sadness	6	14	8	18	**54**

Table 11 Confusion matrix for expression recognition on outdoor fused database using linear discriminant analysis and support vector machine

Expression/result % accuracy	Smile	Anger	Astonished	Neutral	Sadness
Smile	**70**	0	11	11	8
Anger	9	**53**	13	18	7
Astonished	0	12	**53**	10	25
Neutral	10	16	17	**44**	13
Sadness	6	18	8	18	**50**

Table 12 Confusion matrix for expression recognition on indoor and outdoor fused database using linear discriminant analysis and support vector machine

Expression/result % accuracy	Smile	Anger	Astonished	Neutral image	Sadness
Smile	**70**	0	15	11	4
Anger	9	**52**	14	18	7
Astonished	0	22	**55**	12	11
Neutral	10	16	14	**46**	14
Sadness	6	14	8	18	**54**

Scenario 3

In this situation, indoor and outdoor thermal images are used for facial expression recognition. Total indoor and outdoor thermal images are 400 out of 300 images are assigned for training dataset and 100 images are assigned for testing dataset. On trained Support Vector Machine, 100 testing images have applied to calculate the expression recognition rate. The confusion matrix of scenario 3 is shown in Table 12.

Table 12 confusion matrix shows smile expression have achieved highest recognition rate of 70% and neutral expression got lowest recognition rate of 46%. The average recognition rate for facial expression is 55%.

3.3.1 Experiment 4

In this experiment, visible, thermal and fused images are considered which consist of 20 subjects with 1200 images. Table 13 shows total number 1200 images with visible, thermal and fused dataset in indoor and outdoor environment. This dataset is divided into training and testing dataset such that 90 images of per expression per environment is set to training dataset while 30 images per expression per environment is set to testing dataset. It means 900 images are assigned for training dataset and 300 images are assigned for testing dataset. Figure 7 shows flow chart to be carried out for implementation of experiment 4.

The experiment is carried out in three different scenarios as per their environment. Scenario 1 is related to indoor environment images, scenario 2 performed on outdoor environment images and scenario 3 performed on both indoor and outdoor situation images. These scenarios are useful to understand the impact of environment on combine features of visible and fused facial expression recognition system.

Scenario 1

In this situation, indoor visible, thermal with fused images are used for facial expression recognition. Total indoor visible, thermal with fused images are 600 out of 450 images are assign for training dataset and 150 images are assigned for testing dataset. On trained Support Vector Machine, 150 testing images have applied to

Table 13 Facial expression visible, thermal and fused image distribution in training and testing dataset

Dataset used for	Smile						Anger						Astonished					
	Indoor			Outdoor			Indoor			Outdoor			Indoor			Outdoor		
	V	F	T	V	F	T	V	F	T	V	F	T	V	F	T	V	F	T
Total	40	40	40	40	40	40	40	40	40	40	40	40	40	40	40	40	40	40
Training image	30	30	30	30	30	30	30	30	30	30	30	30	30	30	30	30	30	30
Testing image	10	10	10	10	10	10	10	10	10	10	10	10	10	10	10	10	10	10

	Neutral						Sadness					
	Indoor			Outdoor			Indoor			Outdoor		
	V	F	T	V	F	T	V	F	T	V	F	T
Total	40	40	40	40	40	40	40	40	40	40	40	40
Training image	30	30	30	30	30	30	30	30	30	30	30	30
Testing image	10	10	10	10	10	10	10	10	10	10	10	10

Total						Total
Indoor			Outdoor			
V	F	T	V	F	T	
200	200	200	200	200	200	1200
150	150	150	150	150	150	900
50	50	50	50	50	50	300

Table 14 Confusion matrix for expression recognition on indoor visible, thermal with fused database using linear discriminant analysis and support vector machine

Expression/result % accuracy	Smile	Anger	Astonished	Neutral image	Sadness
Smile	**75**	0	14	11	0
Anger	0	**62**	19	15	4
Astonished	0	16	**67**	13	4
Neutral	11	6	10	**68**	5
Sadness	0	15	8	12	**65**

calculate the expression recognition rate. The confusion matrix of scenario1 is shown in Table 14.

This confusion matrix shows that smile expression has achieved highest recognition rate of 75% and anger expression got lowest recognition rate of 62%. The average recognition rate for facial expression is 67%.

Scenario 2

In this situation, outdoor visible, thermal with fused images are used for facial expression recognition. Total outdoor visible, thermal with fused images are 600 out of 450 images are assign for training dataset and 150 images are assigned for

Table 15 Confusion matrix for expression recognition on outdoor visible, thermal with fused database using linear discriminant analysis and support vector machine

Expression/result % accuracy	Smile	Anger	Astonished	Neutral image	Sadness
Smile	**72**	0	12	10	6
Anger	0	**62**	19	15	4
Astonished	0	16	**65**	15	4
Neutral	11	6	8	**65**	10
Sadness	0	13	8	17	**62**

testing dataset. On trained Support Vector Machine, 150 testing images have applied to calculate the expression recognition rate. The confusion matrix of scenario 2 is shown in Table 15. This confusion matrix shows that smile expression have achieved highest recognition rate of 72% while anger and sadness expression got lowest recognition rate of 62%. The average recognition rate for facial expression is 65%.

Scenario 3

In this situation, indoor and outdoor visible, thermal with fused images are used for facial expression recognition. Total indoor and outdoor visible, thermal with fused images are 1200 out of 900 images are assigned for training dataset and 300 images are assigned for testing dataset. On trained Support Vector Machine, 300 testing images have applied to calculate the expression recognition rate. The confusion matrix of scenario 3 is shown in Table 16.

This confusion matrix shows that smile expression have achieved highest recognition rate of 73% while anger expression got lowest recognition rate of 62. The average recognition rate for facial expression is 67%.

Table 16 Confusion matrix for expression recognition on indoor visible, thermal with fused database using linear discriminant analysis and support vector machine

Expression/result % accuracy	Smile	Anger	Astonished	Neutral image	Sadness
Smile	**73**	0	14	10	0
Anger	9	**62**	12	15	2
Astonished	0	16	**67**	13	4
Neutral	11	6	10	**67**	6
Sadness	0	15	8	12	**65**

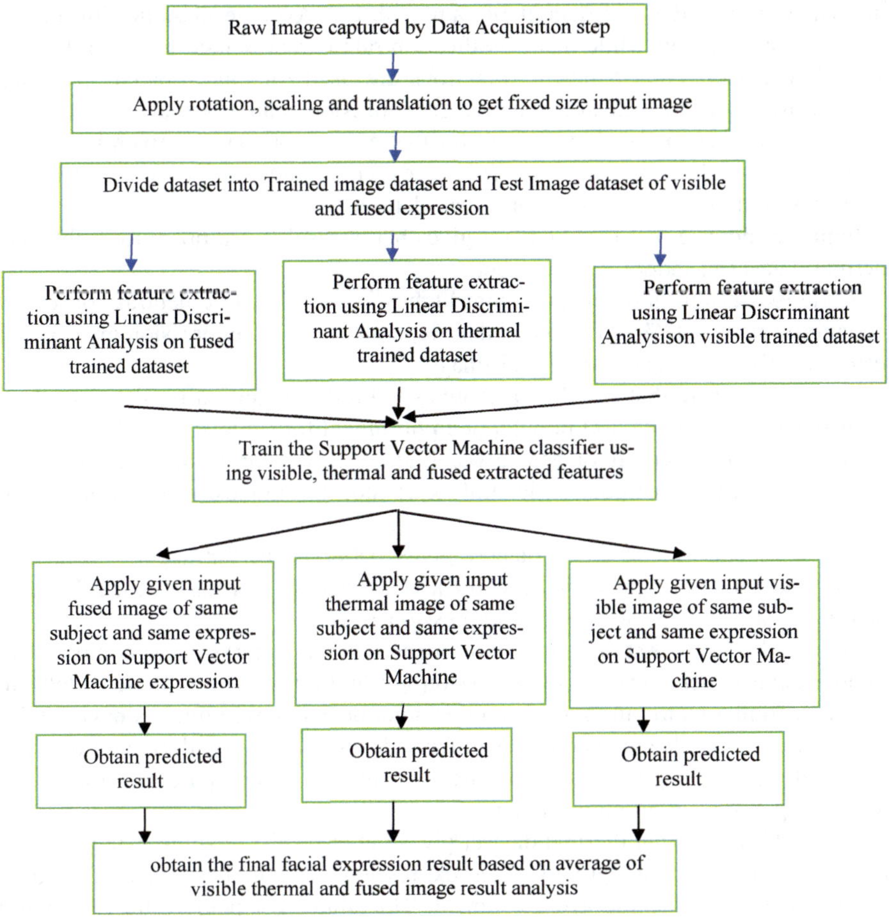

Fig. 7 Implementation of linear discriminant analysis and support vector machine on visible, thermal and fused database

3.4 Results and Discussion

Based on above experimental results, performance analysis is calculated based on average expression recognition. This recognition accuracy is useful to find out best method and poor method in different scenario.

Recognition accuracy (Percentage recognition) is defined as follows:

Recognition accuracy = (number of correct matches)/Total number of matches * 100.

The best recognition system gives highest result. This study calculates count of expressions correctly recognized against total number of expressions used. This work is going to improve accuracy for the indoor as well outdoor situation using thermal, visible and fused dataset of same subject. Average classification rate is given in diagonal part while mis-classification rate is represented in rest of field of confusion matrix. Total 5 label expressions are used for this research are smile, anger, astonished, neutral and sadness. In confusion matrix actual class label is given in row wise and predicted result for that expression class is given in column wise per expression for that row. This study compared the indoor environment result with outdoor environment on given dataset.

Figure 8 indicate that fused dataset gives better result as compare to visible and thermal dataset in indoor environment.

Smile expression is recognized with highest 72% rate while neutral got lowest recognition rate of 47% in fused and thermal dataset. Anger expression gives better result of 55% as compare to neutral image.

Figure 9 indicate that fused dataset gives highest recognition result as compare to visible and thermal dataset in outdoor environment for smile 70% and anger 53% expression while thermal dataset gives highest result for astonished 56% and sadness 51%. In visible dataset, neutral expression got highest recognition rate of 56%.

Figure 10 shows that fused dataset gives better result for anger, astonished, neutral and sadness expression as compared to visible, thermal dataset. Thermal dataset gives better result for smile expression is 70%.

Figure 11 shows that indoor environment gives better result as compared to outdoor environment. The smile expression got highest recognition rate of 68% in indoor as compared to outdoor and indoor-outdoor both environments having 65% each. This method also gives poor result for sadness (54%) in outdoor environment. This study achieved good result for anger expression as compare to sadness and neutral expression in all environment.

Figure 12 shows that thermal dataset gives better result for smile expression for all environment as compare to other expression. The highest result for smile expression is 70% in indoor environment. This study got poor result for neutral

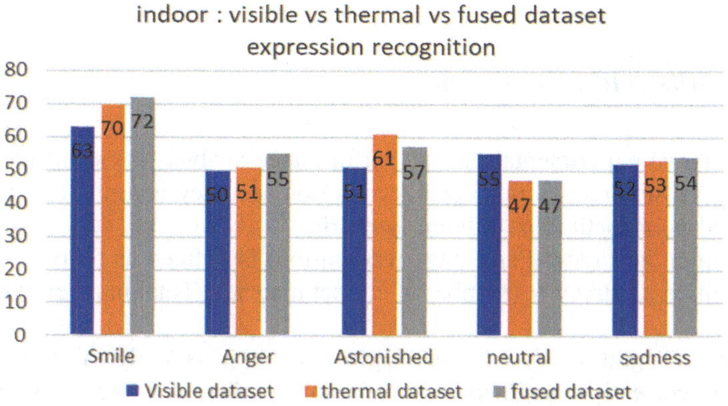

Fig. 8 Comparison analysis of visible versus thermal versus fused in indoor

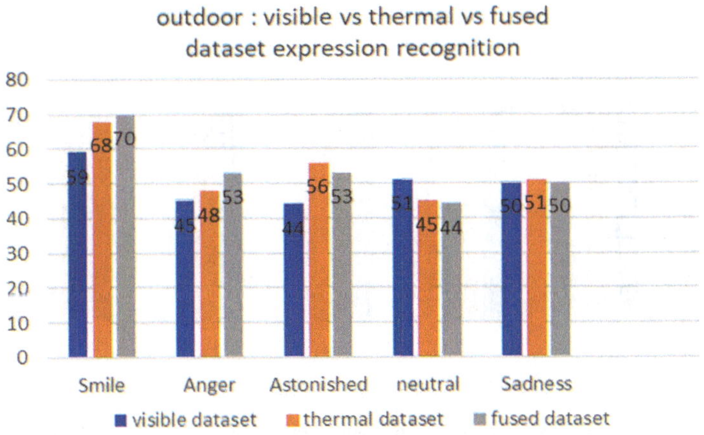

Fig. 9 Comparison analysis of visible versus thermal versus fused in outdoor

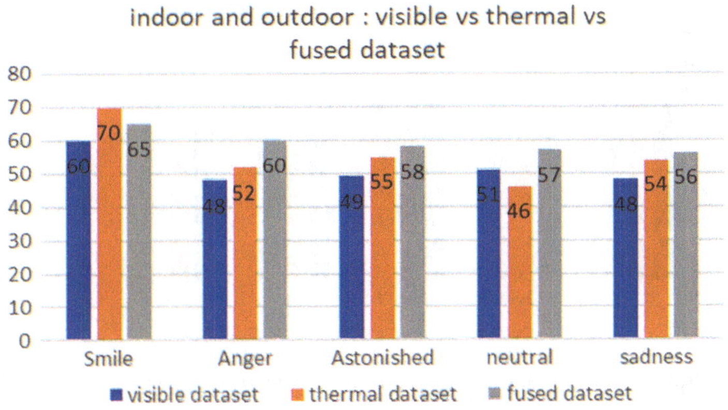

Fig. 10 Comparison analysis in both indoor, outdoor environment

expression in all three types of environment. The lowest expression recognition rate is 45% on neutral in outdoor and indoor-outdoor environment.

Figure 13 shows performance analysis of fused dataset in all environment. This study achieved highest result on smile expression as compare to other expression in all environment scenario. Neutral expression got lowest result for all environment. The lowest recognition result is 44% on neutral expression in outdoor environment [29–37].

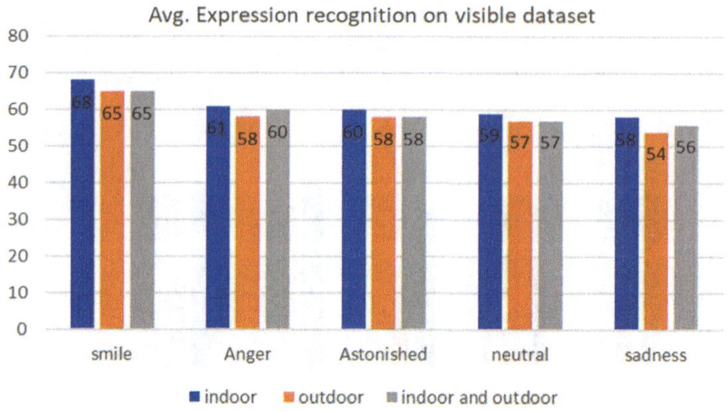

Fig. 11 Performance analysis for visible in indoor versus outdoor versus indoor and outdoor

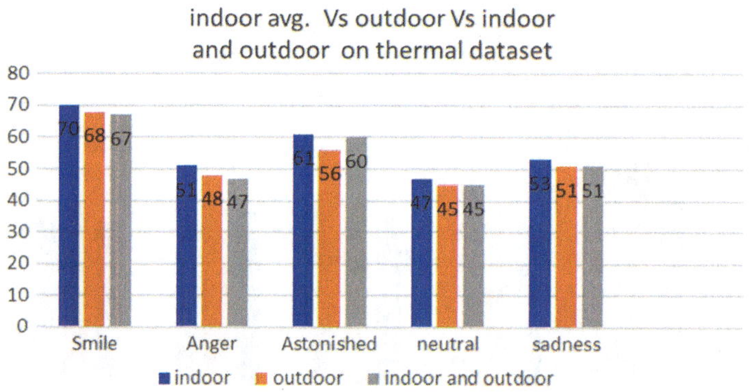

Fig. 12 Performance analysis for thermal in indoor versus outdoor versus indoor and outdoor

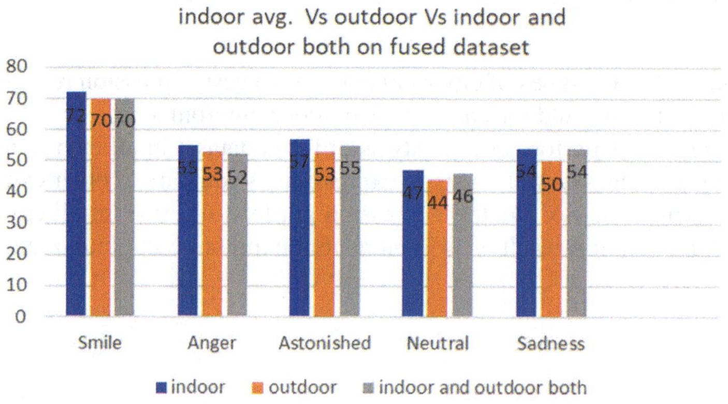

Fig. 13 Performance analysis for fused in indoor versus outdoor versus indoor and outdoor

4 Conclusion

This study has done the performance analysis of facial expression recognition on visible, thermal and fused dataset captured on Indian faces with age group 19–22 years in indoor and outdoor environment. This study conducted 7 types of experimentation with indoor, outdoor and indoor – outdoor environment situation. Total 21 ways of experimentation have done on given dataset. Subjects have also asked to do warm up before going for outdoor image capturing. This warm-up scenario is useful to verify that is daily routing work, warm up and physical work plays an important role for facial expression recognition or not? Result achieved through this method proved that during warm up condition fused dataset gives better result as compare to visible and thermal dataset.

In this study, Fused dataset got highest recognition rate for all five expression as compared to visible and thermal expression in both indoor and outdoor environment.

Overall fused dataset smile expression got highest recognition rate of 72%, 70% and 65% in indoor, outdoor and indoor-outdoor environment respectively.

Fused neutral expression got lowest recognition accuracy of 44% in outdoor environment.

Fused sadness expression achieved lowest recognition as 50% for fused image and visible images as compared to 51% recognition accuracy in thermal images.

Total five facial expression are considered for this research in which fused dataset achieved better result as compared to visible and thermal dataset. Smile expression is got highest recognition rate (72%).

References

1. Zhang, Z.: Feature-based facial expression recognition: sensitivity analysis and experiments with a multi-layer perceptron. Int. J. Pattern Recognit. Artif. Intell. **13**(6), 893–911 (1999)
2. Chibelushi, C.C., Bourel, F.: Facial expression recognition: a brief tutorial overview (2002)
3. Wang, S. (IEEE Member), Liu, Z., Lv, S., Lv, Y., Wu, G., Peng, P., Chen, F., Wang, X.: A natural visible and infrared facial expression database for expression recognition and emotion inference. https://doi.org/10.1109/TMM.2010.2060716
4. Nguyen, H., Chen, F., Kotani, K.: Fusion of visible images and thermal image sequences for automated facial emotion estimation. J. Mob. Multimed. **10**(3&4), 294–308 (2014). Rinton Press
5. Bebis, G., Gyaourova, A., Singh, S., Pavlidis, I.: Face recognition by fusing thermal infrared and visible imagery. Image Vis. Comput. **24**, 727–742 (2006). Elsevier
6. Shen, P., Wang, S., et al.: Facial expression recognition from thermal video. In: Intelligent Autonomous System 12. AISC, vol. 194, pp. 323–333. Springer, Heidelberg (2013)
7. Wang, S., He, S.: Spontaneous facial expression recognition by fusing thermal infrared and visible images. In: Intelligent Autonomous System 12. AISC, vol. 194, pp. 323–333. Springer, Berlin (2013)
8. Ross, A., Jain, A.: Information fusion in biometrics. Pattern Recognit. Lett. **24**, 2115–2125 (2003)

9. Monwar, Md.M., Gavrilova, M.L.: Multimodal biometric system using rank-level fusion approach. IEEE Trans. Syst. Man Cybern.—Part B: Cybernet. **39**(4), 867–878 (2009)
10. Jain, A.K., Ross, A., Prabhakar, S.: An introduction to biometric recognition. IEEE Trans. Circuits Syst. Video Technol. **14**, 420 (2004)
11. Jain, A.K., Ross, A.: Multibiometric systems. IEEE Trans. **47**, 34–40 (2004)
12. Sheena, S., Mathu, S.: A study of multimodal biometric system. IJRET **3**, 93–97 (2014)
13. Deriche, M.: Trends and challenges in mono and multi biometrics. In: IEEE Image Processing Theory, Tools and Applications, pp. 1–9 (2008)
14. Ives, R.W.: A multidisciplinary approach to biometrics. IEEE Trans. Educ. **48**, 462–472 (2005)
15. Prabhakar, S., Kittler, J., Maltoni, D., OGorman, L., Tan, T.: Introduction to the special issue on biometrics: progress and directions. IEEE Trans. Pattern Anal. Mach. Intell. **29**, 513–516 (2007)
16. Karande, K.J., Talbar, S.N.: Simplified and modified approach for face recognition using PCA. In: IET-UK ICTES 2007, Dr. M.G.R. University, Chennai, Tamil Nadu, India, pp. 523–526 (2007)
17. Zhao, H., Yuen, P.C., Member, I.E.E.E., Kwok, J.T.: A novel incremental principal component analysis and its application for face recognition. IEEE Trans. Syst. Man Cybernet. Part B: Cybernet. **36**(4), 873–887 (2006)
18. Hermosilla, G., Ruiz-del-Solar, J., Verschae, R., Correa, M.: A comparative study of thermal face recognition methods in unconstrained environments in Elsevier. Pattern Recognit. **45**, 452–459 (2012)
19. Kumar, M.G.S., Saravanan, D.: A novel approach to face recognition based on thermal imaging. IJRET **03**(03), 141–145 (2014)
20. Bhowmik, M.K., Saha, K., Majumder, S., Majumder, G., Saha, A., Sarma, A.N.: Thermal infrared face recognition a biometric identification technique for robust security system. In: Intech, pp. 113–139
21. https://en.wikipedia.org/wiki/File:AirportThermographicCamera.jpg
22. https://www.thermowork.com/emissivitytable
23. https://www.flir.com/product/c2
24. https://www.flir.com/globalassets/imported-assets/document/flir-ets320-usermanual.pdf, pp. 69–70
25. Ruggieri, S.: Efficient C4.5. IEEE Trans. Knowl. Data Eng. **14**(2), 438–444 (2002)
26. Kwon, O., Sim, J.M.: Effects of data set features on the performances of classification algorithms. Expert Syst. Appl. **40**, 1847–1857 (2013)
27. Socolinsky, D., Selinger, A., Neuheisel, J.: Face recognition with visible and thermal infrared imagery. Comput. Vis. Image Underst. **91**, 72–114 (2003)
28. Turk, M., Pentland, A.: Eigenfaces for recognition. J. Cogn. Neurosci. **3**(1), 71–96 (1991)
29. Kakarwal, S.N.: Development of feature extraction techniques for face recognition. Ph.D. thesis, Dr. Babasaheb Ambedkar Marathwada University, Aurangabad, Maharashtra, India (2012)
30. Uddin, Md.Z. (IEEE Senior Member), Khaksar, W., Torresen, J. (IEEE Senior Member): Facial expression recognition using salient features and convolutional neural networking. In: IEEE (2017). https://doi.org/10.1109/access.2017.2777003
31. Ghimire, D., Lee, J., Li, Z.-N., Jeong, S.: Recognition of facial expressions based on salient geometric features and support vector machines. Multimed. Tools Appl. **2017**(76), 7921–7946 (2017)
32. Lajevardi, S.M., Hussain, Z.M.: Automatic facial expression recognition: feature extraction and selection. SIViP **6**, 159–169 (2012). Springer
33. Samad, R., Sawada, H.: Extraction of the minimum number of Gabor wavelet parameters for the recognition of natural facial expressions. Artif. Life Robot. **16**, 21–31 (2011). https://doi.org/10.1007/s10015-011-0871-6

34. Tsai, H.-H., Chang, Y.-C.: Facial expression recognition using a combination of multiple facial features and support vector machine. Soft Comput. (2017). https://doi.org/10.1007/s00500-017-2634-3. Springer, Berlin
35. Patil, R.A., Sahula, V., Mandal, A.S.: Features classification using geometrical deformation feature vector of support vector machine and active appearance algorithm for automatic facial expression recognition. Mach. Vis. Appl. **25**, 747–761 (2014)
36. Belarbi, M.A., Mahmoudi, S., Belalem, G.: PCA as dimensionality reduction for large-scale image retrieval systems. Int. J. Ambient. Comput. Intell. (IJACI) **8**(4), 45–58 (2017)
37. Dey, N., Wagh, S., Mahalle, P., Pathan, M. (ed.): Applied Machine Learning for Smart Data Analysis. CRC Press, Boca Raton. https://doi.org/10.1201/9780429440953

AI—Assisted Chatbot for E-Commerce to Address Selection of Products from Multiple Products

Pathan Mohd Shafi, Gauri S. Jawalkar, Manasi A. Kadam, Rachana R. Ambawale and Supriya V. Bankar

Abstract Artificial Intelligence has been used nowadays increasingly to develop and advance numerous fields and industries, including finance, healthcare, education, transportation, and more. Machine Learning is a subset of AI techniques which includes various algorithms that gives machines the ability to learn from data or while interacting with the world without being explicitly programmed. E-commerce websites are trending nowadays due to online shopping makes customer's life easier. Similar to this, Chatter Robots i.e. Chatbots are providing better customer service through Internet. A chatbot is a computer or software program for simulating intelligent conversations with human using rules or artificial intelligence for machine. Users can interact with the chatbot program via conversational interface through written or spoken text as input. With the help of E-commerce website sellers can reach to larger audience and with the help of chatbots, sales can be increased by personal interaction with the users. Chatbots will welcome a user to conversation, guide to customer to make purchase which will reduce customer's struggle. Chatbots will ask customers all the relevant questions to find the perfect fit, style, and color for them. Chatbots are the future of marketing and customer support. Chatbots are one such means of technology which helps humans in a lot of ways, by helping them in e-commerce to increase sales whilst providing great customer satisfaction.

P. M. Shafi (✉) · G. S. Jawalkar · M. A. Kadam · R. R. Ambawale · S. V. Bankar
Department of Computer Engineering, Smt. Kashibai Navale College of Engineering, Pune, India
e-mail: shafipathan@gmail.com

G. S. Jawalkar
e-mail: gaurisjawalkar@gmail.com

M. A. Kadam
e-mail: manasi.kadam84.mk@gmail.com

R. R. Ambawale
e-mail: rachanaambawale98@gmail.com

S. V. Bankar
e-mail: supriyabankar70@gmail.com

N. Dey et al. (eds.), *Internet of Things, Smart Computing and Technology: A Roadmap Ahead*, Studies in Systems, Decision and Control 266,
https://doi.org/10.1007/978-3-030-39047-1_3

Keywords Online shopping · E-commerce · Chatbot · Customers · Machine learning · Artificial intelligence · Natural language processing · Supervised learning · Neural networks

1 Introduction

With the increasing development of internet technology, internet service plays an increasingly individual role in people's everyday life. People are expecting that they can get the satisfied services or goods in a convenient manner and within very short time. Hence, at this moment the electronic commerce that is E-Commerce plays a very evaluative and important part. Also, it is very agreeable for people to look at the goods available online and it also reduces people's time required for shopping. Along with this, for the enterprises and organizations, it diminishes intermediate links, and it can lessen the geographic restrictions and compress the merchandise inventory constraint, hence, it can greatly retain business operating cost. E-Commerce is fast achieving ground as associate degree acquired and used business prototype. Many more additional business homes area unit implementing websites providing practicality for acting industrial transactions over the net. It is reasonable to mention that the method of searching on the net is turning into common place.

An online shopping store may be a virtually available store over the internet wherever customers will look around the catalogue and choose merchandise which is fascinating. The preferred items could also be gathered in a shopping cart. At checkout, the items in the shopping cart will be presented as an order for purchase. For this purpose, more information regarding customer and account details will be needed to complete the transaction. Usually, the customer or client are going to be asked to provide a billing details, a shipping address, any shipping related option, and payment details such as credit card number. An e- mail notification for purchase order is sent to the customer or client as presently because the purchase order is placed.

A chatbot also known as a computer code or software program for simulating intelligent conversations with human using rules of artificial intelligence or machine learning. Users interact or talk with the chatbot via conversational interface through written or spoken text as input. Chatbots will welcome a user to conversation, guide to customer to make purchase which will reduce customer's struggle. Chatbots are the future of marketing field and customer support for services. Chatbots are the means of technology which helps humans or users in a lot of ways, by helping them increase sales whilst providing great customer satisfaction. With the help of E-commerce website sellers can reach to larger audience and with the help of chatbots, sales can be increased by personal interaction with the users.

Digitization, the rise of the web and versatile gadgets has changed the way individuals associated with each other and with companies. The internet has boosted ecommerce and the development of remote systems and portable gadgets by clients or clients has driven to the advancement of e-commerce websites. Artificially intelligent chatbots or conversational specialists can be utilized to computerize the interaction between a company and client. Chatbots are a computer program that communicate with its customers or clients by using natural human language and engages in a conversation with its customer by generating human language response as output.

Chatbots have been around in online web based situations for very a few times and are commonly utilized to encourage client benefit. Chatbots can react with messages, proposals or call-to-action buttons and clients can shop for items by going through a item carousel, all within the flag-bearer interface. A chatbot will recognize the buyer's aim and refine offerings backed the buyer's choices and inclinations. It can at that point encourage the deal, arrange, and delivery process.

1.1 Motivation

There are various systems available with chatbots which are currently in use. The available systems are only related to few categories, such as starbucks is related to food category i.e. coffee and snacks, Sephora is related is associated with makeup material and makeup tutorials. Due to destitute memory, Chatbots are not able to memorize the past discussion which powers the client to sort the same thing once more and once more. This could be cumbersome for customers or clients and bother them because of efforts required for product purchase.

Due to known or fixed programs, chatbots may not be able to respond if an unsaved query is asked to them. This may lead to client dissatisfaction and effect in loss. Also it is additionally the numerous messaging that can be demanding for users and worsen the overall general experience of user of the website.

1.2 Problem Definition

E-Commerce becomes one of the well-liked ways in which of searching as they fancy their online searching as they enjoy their online because of its easiness and convenience. The combination of e-commerce site with AI- assisted chatbots will provide better customer service to make profitable sales by personalized marketing. Design and implementation of E-Commerce website including various categories such as Clothing, Accessories, Books, Electronic Appliances, etc. with Chatbots to offer suggestions to help customer to make a purchase decision is the Proposed System. It will be easy to handle customers, provide better customer support by interacting with the help of Chatbot. Chatbot will help customers to browse or search through E-commerce website and purchase products or add them to the Shopping Cart.

The key reasons or necessity to develop the proposed system against existing systems are:

- To combine the multiple categories of products available in e-commerce.
- To allow chatbot system to memorize the past conversation to improve the communication.
- To reduce the efforts of customers required for product purchase using previous purchase history.
- To allow chatbot to improve itself along with the conversation with the customers.

2 Literature Review

There are various systems available with chatbots which are currently in use. Though these chatbots assisted systems are in use there are some limitations associated with it. Chatbot programs are installed for motivation to accelerate the reply and improve chatbot—client interaction. However, because of limited availability of data and time needed for self-improving, this process requires more time consuming and more expensive. Therefore, instead of addressing many customers or clients simultaneously, chatbots may get confused about communicating with people simultaneously.

2.1 Related Work

Anwesh Marwade, Shubham Mundada, Nakul Kumar and Jagannath Aghav have published a paper "Augmenting E-Commerce Product Recommendations by Analyzing Customer Personality" in 2017 [1] in which they had focused on customer or client specific personalization. The e-commerce industry predominantly makes use of several machine learning algorithms for product suggestions and analyzing a customer's operating patterns. With the help of e commerce based conversational bot, the personality understanding to develop a unique product recommendation system can be utilized based on purchase order history and conversation data that the bot-application would collect over time from users.

Adhitya Bhawiyuga, Eko Sakti Pramukantoro, M. Ali Fauzi, Widhi Yahya have published a paper "Design of E-Commerce Chat Robot for Automatically Answering Customer Question" in 2017 [2] in which they had focused on design and application of e-commerce chatbot interaction system which provides automatic output response to the customer-to-seller query or question. The proposed system includes two main agents: communication and intelligent agent which can give automatic answer in less than few seconds with relatively better matching accuracy.

S. J. du Preez, S. Sinha, M. Lall Lall have published "An Intelligent Web-Based Voice Chatbot" in 2009 [3] in which they are presenting the design and development of intelligent web-based voice recognition chat bot. The paper for web based chatbot represents a technology manifestation to verify a proposed structure required to support such intelligent bot for a web based service. By introducing an artificial brain through this intelligent bot, the web-based intelligent bot generates customized client responses for particular user, aligned to the required character.

Bayu Setiaji, Ferry Wahyu Wibowo have published a paper, "Chatbot Using A Knowledge in Database", in 2017 [4] in which they describe about a conversation bot or chatbot aims to make a interaction between both human client and machine. The computer machine which will interact with human has been embedded knowledge to recognize the sentences given by human and making a decision itself as output reply to answer a question. The response principle is used for matching provided input sentence given from customer to generate a response. From given input sentence, it will be scored to get the matching similarity of sentences, the higher score gained the more similar of reference sentences.

Godson Michael D'silva, Sharddha More, Sanket Thakare and Jeril Kuriakose have published a paper "Real World Smart Chatbot for Customer Care using a Software as a Service (SaaS) Architecture" in 2017 [5] in which they proposed a system architecture which will try to defeat the above shortcoming by analyzing messages of each ejabberd users to check if it's actionable or not. If it's actionable then an automated interaction Chatbot will initiates conversation with that client and help the customer to resolve the issue by providing a human like interactions using LUIS and cognitive services. To provide a highly robust, extensible and scalable architecture, this system is implemented on AWS public cloud.

Cyril Joe Baby, Swathi J. N., Faizan Ayyub Khan have published a paper "Home Automation using IoT and a Chatbot using Natural Language Processing" in 2017 [6] in which they focused on a web based application using which the lights, fans and other electrical home appliances can be controlled over the Internet. The most important features of web based application are that firstly, we have a conversational chatbot algorithm such that the user can send text information to control the functioning of the electronic appliances at home. The messages sent using the conversation chatbot is processed using Natural Language processing (NLP) techniques. Secondly, any gadget connected to the local area network of the home can control the gadgets and other electronic appliances in the house. Thirdly, the web based application used to enable home automation also has a security feature that only enables specific users to access the application. And finally, it additionally has a functionality feature of sending an email alert notification when attacker or intruder is detected using motion sensors (Table 1).

Table 1 Literature review analysis

Title of paper	Author	Publication year	Key points
Real world smart Chatbot for customer care using a software as a service (SaaS) architecture	Godson Michael D'silva, Sharddha More, Sanket Thakare and Jeril Kuriakose	2017	1. Respond to actionable messages 2. Initiate conversation and help to solve issue 3. Implemented on AWS public cloud
Chatbot using a knowledge in database	Bayu Setiaji, Ferry Wahyu Wibowo	2017	1. Conversation between human and machine 2. Use of response principle 3. Machine identify sentence & make decision as response
Home automation using IoT and a chatbot using natural language processing	Cyril Joe Baby, Swathi J. N, Faizan Ayyub Khan	2017	1. Web application to control home appliances 2. Message sent to chatbot for control security and network connectivity
Design of E-Commerce chat robot for automatically answering customer question	Adhitya Bhawiyuga, Eko Sakti Pramukantoro, M. Ali Fauzi, Widhi Yahya	2017	1. Provides automatic response 2. Use of communication and intelligent agent 3. Good pattern matching accuracy
Augmenting E-Commerce product recommendations by analysing customer personality	Anwesh Marwade, Shubham Mundada, Nakul Kumar and Jagannath Aghav	2017	1. Use of ML models analyzes customers' behavioral pattern 2. Utilizes personality insights (order history)
An intelligent web-based voice chatbot	S. J. du Preez1, M. Lall, S. Sinha	2009	1. Proposed framework to support web based bot 2. Use of artificial brain to generate customized user responses

2.2 Existing Systems

There are various systems available with chatbots which are currently in use. Though these chatbots assisted systems are in use there are some limitations associated with it. The limitations could be,

- The available systems are only related to few categories, such as starbucks is related to food category i.e. coffee and snacks, Sephora is related is associated with makeup material and makeup tutorials.
- Poor memory can be a disadvantage to the system. Due to minimum amount memory, Chatbots may not able to memorize the past discussion which forces the user to ask the same queries repeatedly. This could be cumbersome for customers or clients and bother them because of efforts required for product purchase.
- Due to known programs, chatbots may not be able to respond if an unsaved query is asked to them. This may lead to client dissatisfaction and effect in loss. Also it is additionally the numerous messaging that can be demanding for users and worsen the overall general experience of user of the website.

Chatbot programs are installed for motivation to accelerate the reply and improve chatbot—client interaction. However, because of limited availability of data and time needed for self-improving, this process requires more time consuming and more expensive. Therefore, instead of addressing many customers or clients simultaneously, chatbots may get confused about communicating with people simultaneously.

- Starbucks
  Starbucks has developed an Android and iOS application to place an order for favorite drink or snacks. The order can be placed with help of voice commands or text messaging.
- Spotify
  Spotify chatbots allow users to search for, listen to their favorite music. Also it allow users to share music.
- Whole Foods
  Whole Foods is related to groceries and food material. It allow to search for grocery items to shop for. Also it provides interesting recipes for users to try.
- Sephora
  Sephora is associated with makeup material such as foundation, face primer, concealer, blush, highlighter, etc. Sephora chatbots also suggest for makeup tutorials for which user is interested.
- Pizza Hut
  Pizza Hut chatbot can help customer to order pizza with favorite toppings and carryout delivery. A customer can reorder favorite pizza based on previous orders and can ask questions about current deals.

- SnapTravel
  SnapTravel helps users to book hotels according to their convenient location and timings. A customer can also get to know about current deals available at various hotels and resorts.
- 1–800 Flower
  1–800 Flower will help customers to gift flowers and gifts to someone for events like birthday, anniversary or any special occasion. It also offer suggestions for gifts to customers.

These available chatbots are related to only few categories. To combine all the categories together at single place to integrate with chatbots for customer service.

3 Proposed System

The Proposed System E-commerce with Chatbots will permit consolidation of customer login, browse and purchase products available, manage orders and payments, engaging customer with personalized marketing, qualifying recommendations based on history. The main users of the project are customers who want to shop various products and services online. There are essentially three classes of users of the proposed system: the general users, the customers and the administrators. The combination of e-commerce site with AI- assisted chatbots will provide better customer service to make profitable sales by personalized marketing.

3.1 Project Scope

The Proposed System E-commerce with Chatbots will permit consolidation of customer login, browse and purchase products available, manage orders and payments, engaging customer with personalized marketing, qualifying recommendations based on history. The main users of the project are customers who want to shop various products and services online.

From end-user perspective, the proposed system consists of functional elements: login module to access online products and services, browse and search products, purchase and pay for products, communicate to chatbot for better product and offer recommendation.

According the back end logic, Natural Language Processing (NLP) will be used to understand messages sent by user through messaging platform. The Chatbot will launch an action as answer with real time information based on machine algorithms such supervised and unsupervised learning. The bot will improve with increasing number of messages received.

The important features of the system are handling thousands of customers simultaneously which will provide better satisfaction to customers. Also it will be a virtual but personal assistant for each customer. Similarly like chatbots, e-Commerce has become one of the preferred and well-liked ways of shopping as they enjoy their online because of its easiness and convenience. The combination of e-commerce site with AI- assisted chatbots will provide better customer service to make profitable sales by personalized marketing.

The risks associated such as privacy issues can be handled with the help of authentication and Authorization to provide strong access control measures. Intellectual property related risks can be avoided by proper instructions to upload data with restrictions. Online security is the most important risk to be considered while developing the system regarding to customers' credentials, online products and services available. Data storage could be a risk associated with chatbots as they store information to interact with the users. The best solution in this situation to avoid such risks is to store the sensitive data in a secure place for a certain amount of time and to delete it after that.

3.2 User Classes and Characteristics

There are essentially three classes of users of the proposed system: the general users, the customers and the administrators. General users will be able to see and browse through products available to purchase, but they cannot buy the products and services. Customers are the users of the E-commerce System who will be able to browse, purchase, pay and add products and services to the cart with available functionality. Chatbots will help them to make a purchase decision based on various criteria and suggestions by chatbot algorithms. Also customers can write reviews or feedbacks on products and services they purchased.

The administrators will be having advanced functionality to add, edit, update and delete products available in inventory. Also administrator will be able to authorize and authenticate the users logged into the system. Administrator will able to see daily sales and details about deliveries. He will be able to see the feedbacks or reviews given by the customers.

3.3 System Architecture

Systems design is the procedure of defining the architecture, layout, components, modules, interfaces, and data for a system to satisfy specified requirements. Systems design can be seen as the application of systems functionality theory to product development. There is some overlap with the disciplines or processes of systems analysis, systems architecture and systems engineering.

System architecture includes the modules used in the project and relationships between them based on data flow and processing. AI—Assisted Chatbots For E-Commerce System consists of following components:

- General User
- Customer
- Administrator.

The System Architecture shows the main elements of Proposed System. There are three main user classes of system such as General User, Customer and Administrator. Along with these users the important components of the system are mentioned in diagram such as E-Commerce Website Home Page, Product Categories, Inventory, Sales and Marketing, Shopping Cart, Purchase and Invoice Generation and Order Tracking Shipment.

A General User is the basic component of the Proposed System who will be able to browse and search through the filters. General User will be directed to the E-Commerce Website Home Page with various products of different categories, deals of the specific products, Menu option to search various categories and log in to the system. Also Chatbot will interact with the customer with some basic standards of interaction.

A General User can be Administrator with help of Seller Portal of the E-Commerce Website. Administrator will be the authorized person who will be able to handle products in inventory, sales of products, refunds to the customers, marketing of products, purchase and transactions made by customers, shipment of products ordered, records of invoices, inventory and customers, etc. Administrator must be authenticated as per well-defined rules and standards with his/her personal information, contact information, product manufacturing information and other required information.

Administrator is responsible to manage quantity of products available in inventory, deals and discounts associated with specific products, manage marketing of products on regional basis, manage pricing of the products, advertising of products to make them as sponsored products to increase sales, etc.

A Customer will be a General User who logged into the E-Commerce System. A customer need to login to the Proposed System with help of login credentials such as username and password, along with Name, Address, Phone No., E-Mail, Credit Card Details. A customer can search products with filters as per need, purchase products, add products to shopping cart, track order shipment online, track record of invoice, write reviews to purchased products. A General User can log into the system to make use of more features and functionalities of the system. A User can be Customer after logging into the system that not only can add products to the shopping cart but also can purchase the products.

The proposed system will verify and authorize person by verifying Phone No and E-mail ID. Also it will verify Credit Card Details if provided by user. User Authentication allows user to search and purchase products available in inventory with provided address, date of delivery, shipping type, etc. Also a shopping cart is available for each individual customer to add products from inventories which are to be purchased. Chatbot will use search and purchase history if user is authenticated by system for product recommendations (Fig. 1).

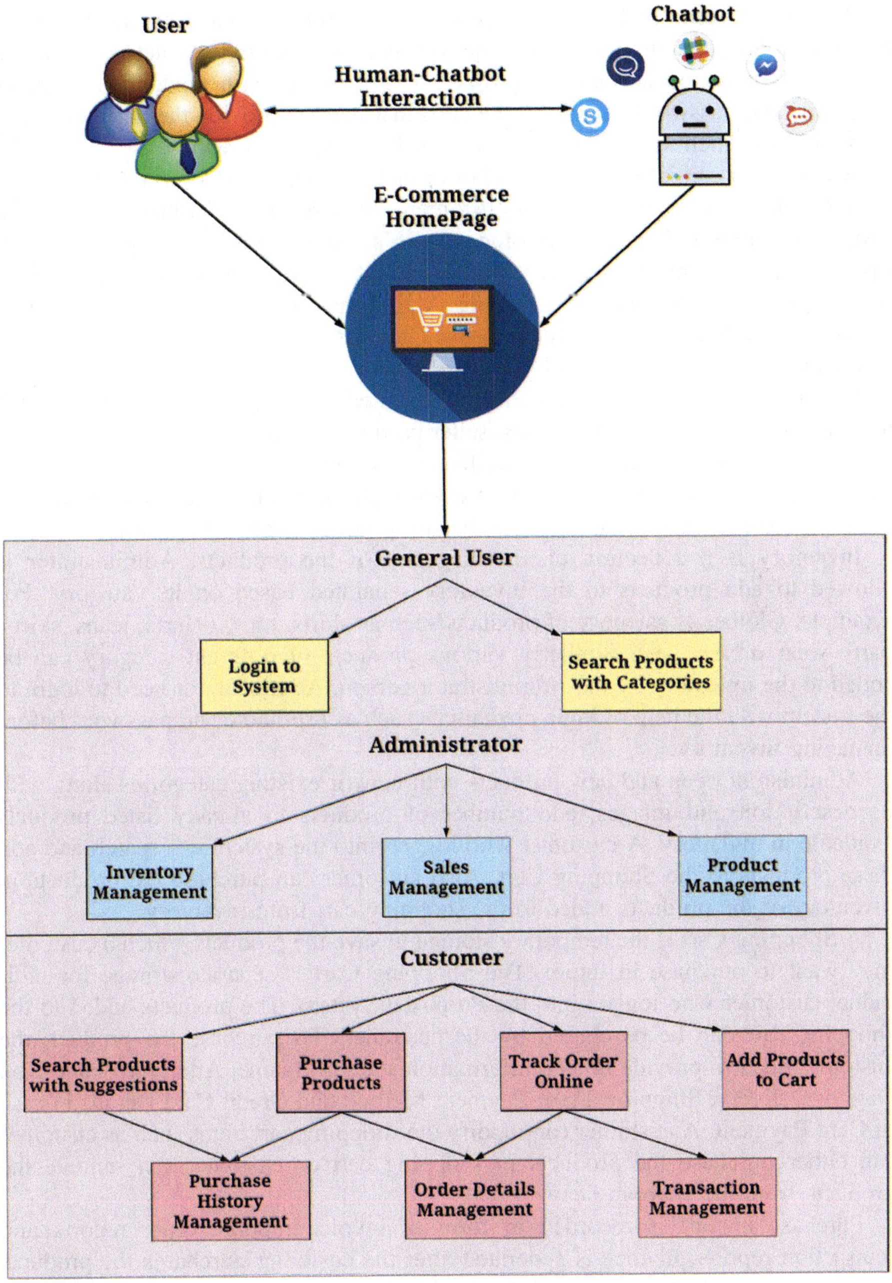

Fig. 1 System architecture

A customer can modify his/her profile or account created into the Proposed System. Before updating the profile the user need to authenticate that he/she is the original user of the account or profile by providing login credentials such as username and password. After this the customer need to provide the attributes need to be updated such address, phone no, mail ID, credit card details, etc.

A Customer can interact with chatbot to make a purchase decision. Chatbot will interact with a customer based on customer's browse or search history, purchase history. Chatbot will make use of customer's record to suggest products from inventory to customer to purchase. Initially Chatbots will interact with basic set of rules designed with Machine Learning Algorithms. With more interaction with customer, chatbots will also get improved to recommend products to customers based on customer's search and purchase history.

E-Commerce Website's Home Page is designed with the important features such as deals of the specific products, best seller products, discounts of products, various categories of the products to choose, feature to login to the system and a way to interact with chatbot. A user can browse through these categories to view various products such as clothes, accessories, beauty products, shoes, bags, etc.

Inventory is a collection of all categories of the products. Administrator is allowed to add products to the inventory separated based on its category. For Example, Clothes is category of products such as shirts, tops, t-shirts, jeans, skirts, party wear dresses, etc. Similarly various products of different category can be added to the inventory by the Administrator person. Administrator need to login to the system with the help of login credentials such as username and password before managing inventory.

Administrator can add new products with new or existing categories along with its description and images, add number of products to already listed products available in inventory. A customer who logged into the system can search and add these products to the Shopping Cart. Also customer can purchase the products in inventory or the products added to the shopping cart from inventory.

A Shopping Cart is the temporary storage to save the products which a customer may want to purchase in future. The Shopping Cart is separate storage for individual customer who logged into the Proposed System. The products added to the shopping cart can be purchased by the customer. To purchase the product, the customer need to provide related information such as, Name, Address, Phone No., Date of Delivery, Shipping Type, Payment Method and Credit Card Details in case of Card Payment. A customer can modify the shopping cart items such as customer can either purchase the products in shopping cart or customer can remove the products from the shopping cart.

Purchase history is recorded in form of invoice reports, order reports and transaction reports. Invoice is generated after the customer purchases the products from inventory which includes all the details of purchase and transactions made by the customer. It includes the details of product purchased such as price, quantity, product ID, product category along with customer details such as name, address of delivery, shipping type, date of delivery, phone no and payment method. All the details regarding purchase history are used by chatbot to interact with the customer based on his/her history to suggest or recommend products. A customer can track the shipment of order based on invoices recorded or transactions saved to his/her profile. The online tracking of order can help to customer to locate his/her product.

Sales and Marketing involves the techniques to suggest a customer to purchase particular products with advertisement. It is done based on the keywords searched by the customers for the products he/she wants to purchase. Advertising of products is a way of Marketing to increase Sales of products. All these things are managed by the Administrator to make maximum sales of products with the help of marketing of products.

3.4 Overview of System Modules

- **Sign Up and Login**

A Customer can log into the system to make use of more features and functionalities of the system. A General User need to provide login credentials to log into the system. The login credentials are Username, Password, Name, Address, and Phone No. E-Mail ID, Credit Card Details (optional), Country, etc. the proposed system will verify and authorize the person by verifying the Phone No and E-mail ID. Also, it will verify the Credit Card Details if provided by the user. All details of each individual customer will be stored in database with unique ID.

User Authentication allows user to search and purchase the products available in inventory with provided address, date of delivery, shipping type, etc. Also, a shopping cart is available for each individual customer to add products from inventories which are to be purchased. Chatbot will use the search and purchase history if the user is authenticated by the system for product suggestions and recommendations.

- **Product Browse/Search and Selection**

A customer can search products with filters as per need, purchase products, add products to shopping cart, track order shipment online, track record of invoice, write reviews to purchased products. A General User can log into the system to make use of more features and functionalities of the system. A User can be Customer after logging into the system that not only can add products to the shopping cart but also can purchase the products.

- **Product Shopping Cart**

A Shopping Cart is the temporary storage to save the products which a customer may want to purchase in future. The Shopping Cart is separate storage for individual customer who logged into the Proposed System. The products added to the shopping cart can be purchased by the customer. To purchase the product, the customer needs to provide related information such as, Name, Address, Phone No.,

Date of Delivery, Shipping Type, Payment Method and Credit Card Details in case of Card Payment. A customer can modify the shopping cart items such as customer can either purchase the products in shopping cart or customer can remove the products from the shopping cart.

- **Product Purchase and Checkout**

The products added to the shopping cart can be purchased by the customer. To purchase the product, the customer needs to provide related information such as, Name, Address, Phone No., Date of Delivery, Shipping Type, Payment Method and Credit Card Details in case of Card Payment.

Invoice is generated after the customer purchases the products from inventory which includes all the details of purchase and transactions made by the customer. It includes the details of product purchased such as price, quantity, product ID, product category along with customer details such as name, address of delivery, shipping type, date of delivery, phone no and payment method. All the details regarding purchase history are used by chatbot to interact with the customer based on his/her history to suggest or recommend products. A customer can track the shipment of order based on invoices recorded or transactions saved to his/her profile. The online tracking of order can help to customer to locate his/her product.

- **Invocation of Chatbot**

A General customer can interact with the chatbot using basic queries and questions. Before interacting with the chatbot, it is necessary to invoke the chatbot. The chatbot is integrated with the Google Assistant to reach out more number of users. While developing chatbot, particular phrase is specified to invoke the Chatbot. This phrase must be used by end user or customer to invoke the chatbot.

- **Human-Machine Conversation**

A Customer can interact with chatbot to make a purchase decision. Chatbot will interact with a customer based on customer's browse or search history, purchase history. Chatbot will make use of customer's record to suggest products from inventory to customer to purchase. Initially Chatbots will interact with basic set of rules designed with Machine Learning Algorithms. With more interaction with customer, chatbots will also get improved to recommend products to customers based on customer's search and purchase history.

- **Product Recommendation**

Chatbot will interact with a customer based on customer's browse or search history, purchase history. Chatbot will make use of customer's record to suggest products from inventory to customer to purchase. With more interaction with customer, chatbots will also get improved to recommend products to customers based on customer's search and purchase history.

3.5 Algorithm Details

I. Natural Language Processing

Natural Language Processing (NLP) the implementation of computational techniques to study and synthesis of natural human language—is objective of AI. It might sound theoretical and technical as per the concept, but it's all around us which is we are using today: in spell check for documents; spam filters for emails; data mining for search engines; translation software. Chatbot takes some union of steps to transform the customer's input message or speech input into structured knowledge that's accustomed choose the connected response or output. Following are the Natural Language Processing elements are:

- Sentiment Analysis: Chatbot try to understand if the user is having a better experience or if required after some point interaction should be handled to the human.
- Tokenization: Natural Language Processing separates a string of input words into small chunks or pieces that are linguistically symbolic which can be useful for the application.
- Named Entity Recognition: Chatbot program used for interaction recognizes categories of input words called as entities, such as the name of shopping product, the client's name or shipping address, whichever details is required.
- Normalization: Chatbot program processes the message or input of user to find spelling mistakes or errors through which user may intent to convey input as query. This provides additional impact like human of Chatbot to clients.
- Dependency Parsing: The Chatbot searches for subjects and objects such as verbs, adjectives, nouns and phrases in the user's input to find interdependent and related phrases in input that users may be try to provide through input [7–10].

The above NLP elements are used in algorithm of NLP steps to understand the human language:

- Sentence Segmentation
- Word Tockenization
- Identify parts of speech
- Finding text forms or lemmas
- Identify stop words
- Dependency Parsing
- Named Entity Recognition

II. Machine Learning—Supervised Learning

Machine learning may be a technique of knowledge research that automates logical model building. It's a branch of AI supported the concept that systems will learn from knowledge, determine patterns and build selections with smallest human interference [11]. Although there is a wide range of various approaches and techniques enclosed within the machine learning, most of the techniques are statistics.

This allows algorithms to create forecast or predictions supported extracting matching patterns from incomplete and infrequently abuzz input file. Supervised learning is when you have the given data set and also the result, all you have to do is concoct a relationship between the input (given data set) and the output (the given result). There are two types of supervised learning:

- Regression:
  In Regression Learning technique, all we need is a continuous output to the user that is the output should not be in discrete terms.
- Classification:
  In this type the output should be in discrete terms that is it should be either yes (1) or no(0).In some cases the options May increase to more than two.

Using these two types of supervised learning following algorithm steps are performed:

- Prepare the information required
- Select the type of algorithm
- Fit a model using fitting function
- Examine accuracy of fitted model
- Update and examine model until satisfied
- Use accurate model for generating prediction [12–17]

III. **Artificial Neural Networks**

Artificial Neural Networks (ANN) are a method of predicting and calculating the output from the provided input using weighted connections which can be calculated from repeatedly performed iterations while training initial data. Each action through coaching knowledge changes the weights leading to the output response with accuracy [18, 19].

Each input sentence is counteracted into totally variant words and every input word then is employed as input required for the neural networks. The weighted connections are then obtained by different repeated iterations through the training data numerous times. Every time increase weightage to creating it correct. The final trained data of neural network is a comparable algorithm of steps more and less code. When there is comparably tiny example, whereas training data of sentences have 300 different words with 20 classes, then it would be a matrix of 300×20. But the matrix size will increase by n times additional step by step and might cause a large variety of errors. In this reasonably things, processing speed should be considerably high.

Using these two types of supervised learning following algorithm steps are performed:

- Assign random weights to all links
- Using inputs find activation rate for hidden layer nodes
- Using these activation rates find activation rate of output layer nodes
- Find error rate at output layer node

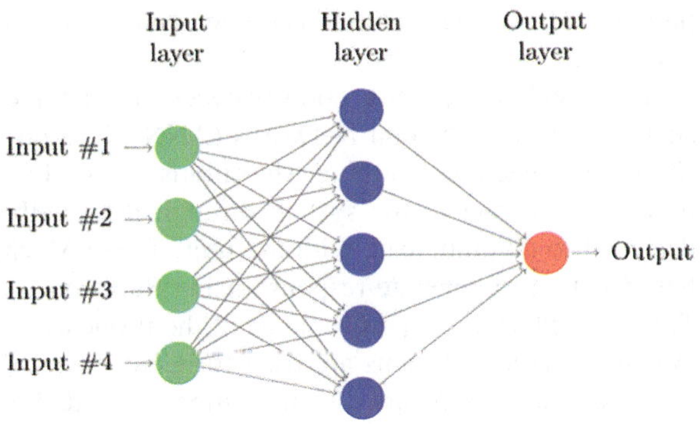

Fig. 2 Artificial neural networks

- Using the error recalibrate the wights from output nodes to hidden nodes and back to input nodes
- Repeat process until convergence criteria is met
- Using final weights score activation rate of output layer nodes (Fig. 2).

4 Results

The Proposed System E-commerce with Chatbots will provide of customer login, browse and purchase products available, manage orders and payments, engaging customer with personalized marketing, qualifying recommendations based on history. The main users of the project are customers who want to shop various products and services online. There are essentially three classes of users of the proposed system: the general users, the customers and the administrators. The combination of e-commerce site with AI- assisted chatbots will provide better customer service to make profitable sales by personalized marketing.

4.1 Outcomes and Screenshots

The outcome of the proposed system is E-commerce website with multiple categories of products such as Clothing, Footwear, Accessories, Electronics, Appliances, Mobiles and Computers. Along with the website a Chatbot is deployed as result using Google Assistant Simulation to interact with the customers. A customer can communicate with the chatbot to provide the specific criteria for product search. The chatbot will use these criteria and will provide a link recommendation to the customer for required products. The proposed system is efficient and effective for E-commerce website which are having products of multiple categories. As the available systems are providing chatbots only for single category products. Hence the proposed is implemented to provide chatbot for all categories at one place.

To provide better interaction between Human and Machine, the concepts of AI such as, Artificial Neural Networks (ANN), Natural Language Processing concept (NLP) and Machine Learning Algorithms are used in the Proposed System. Human users will interact to the system through the Chatbot which a Software program designed to communicate with the user. These Machine Learning Algorithms will help Chatbot to generate response using Supervised and Unsupervised algorithms. This will enhance the performance of the proposed system as compared to existing system as fixed programs are used for Chatbot in existing system.

Recommendations are the suggestions provided to the user based on the search or browse history and purchase history of the particular user. These recommendations provided by Chatbot can be in the form of Product recommendations with links, updates for latest products. To facilitate the customer service and support recommendations will play more important role by providing personalized suggestions. It will improve the performance of the proposed system due to personal recommendations.

AI concepts such as Artificial Neural Networks (ANN), Natural Language Processing (NLP), and Machine Learning Algorithms (ML) are used in the proposed system. Machine learning algorithms such as Supervised and Unsupervised Algorithms will improve the performance of the proposed system. Linear Regression are capable for prediction Modelling and minimizing risk of failure. Linear Regression is used for predicting responses with better accuracy. It makes use of relationship between input values and output values. Naïve Bayes Algorithm is used for large data set for ranking or indexing purpose. It will help to rank the products based on customer reviews. Semi-supervised algorithms will help to handle the combination of both labelled and unlabelled data. NLP is useful to understand the human understandable language by machine and generate the response in human language. It will make use of elements of Named Entity Recognition, Speech Recognition, Sentiment analysis and OCR. All these concepts will help to enhance the performance of the proposed system (Figs. 3, 4, 5, 6).

4.2 Advantages

The advantages of the proposed system can be given as:

- The system is very convenient due to online shopping of products will reduce the customers' efforts.
- Chatbot will be more useful to order products instead of browsing through the whole website.

Fig. 3 Home page 1

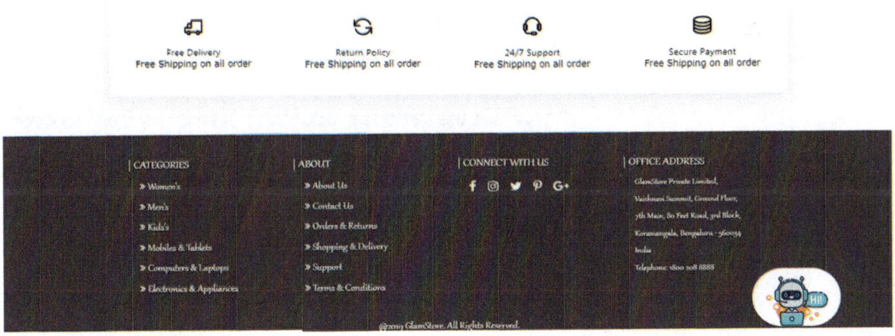

Fig. 4 Home page 2

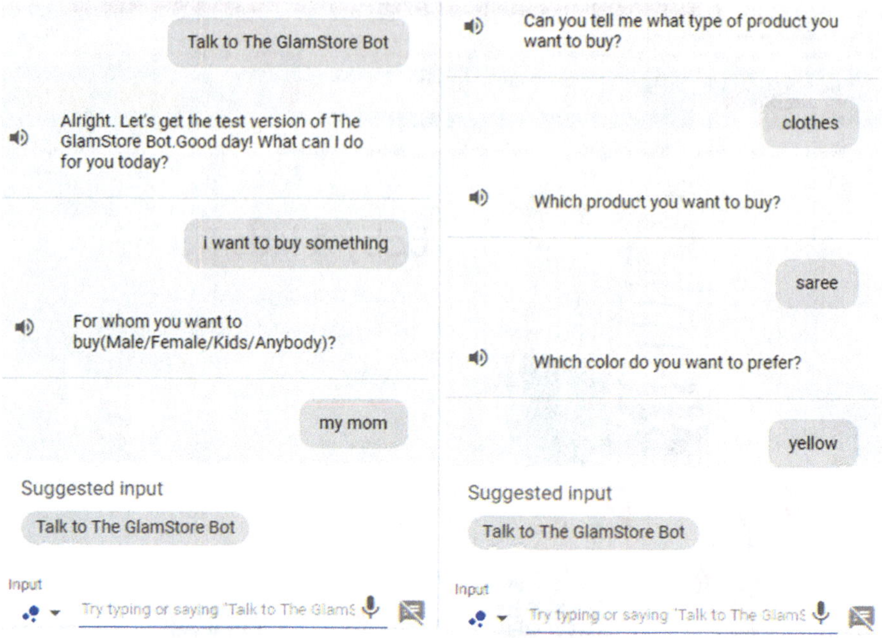

Fig. 5 Chatbot invocation and conversation

Fig. 6 Chatbot recommendation through link

- Description of products in inventory will help both customers and chatbots to search products and purchase them.
- Customers are able to quickly find out discounts and deals for items or services provided by different vendors.

4.3 Applications

The Proposed System can be used in various ways such as:

- Chatbot Interaction System for individual category of products listed in E-Commerce System for product based companies.
- Customer Services based applications can use the chatbot interaction technique to handle the maximum customers to provide better customer support.
- E-Commerce system can be implemented for more creative and new products which are yet to be discovered by rest of the E-commerce Systems.
- An Android application can be developed for purchasing products from E-Commerce System with the help of Chatbot Assistance.

4.4 Performance Analysis Comparison with Parameters

The performance of the proposed system can be analyzed based on few parameters. These parameters can be used to measure the performance of the system in comparison with the existing system. The parameters can be used for this analysis are,

- Human Machine Interaction
  To provide better interaction between Human and Machine, the concepts of AI such as, Artificial Neural Network (ANN), Natural Language Processing (NLP) and Machine Learning Algorithms are used in the Proposed System. Human users will interact to the system through the Chatbot which a Software program designed to communicate with the user. These Machine Learning Algorithms will help Chatbot to generate response using Supervised and Unsupervised algorithms. This will enhance the performance of the proposed system as compared to existing system as fixed programs are used for Chatbot in existing system.
- Better Recommendations to User
  Recommendations are the suggestions provided to the user based on the search or browse history and purchase history of the particular user. These recommendations provided by Chatbot can be in the form of Product recommendations with links, updates for latest products. To facilitate the customer service and support recommendations will play more important role by providing personalized suggestions. This will help customers to make a purchase decision which will increase profitable sales by personalized marketing. It will improve the performance of the proposed system due to personal recommendations.
- Use of AI Concepts
  AI concepts such as Artificial Neural Networks (ANN), Natural Language Processing (NLP), and Machine Learning (ML) Algorithms are used in the proposed system. Machine learning algorithms such as Supervised and Unsupervised Algorithms will improve the performance of the proposed system.

Linear Regression are capable for prediction Modelling and minimizing risk of failure. Linear Regression are used for predicting responses with better accuracy. It makes use of relationship between input values and output values. Naïve Bayes Algorithm is used for large data set for ranking or indexing purpose. It will help to rank the products based on customer reviews. Semi-supervised algorithms will help to handle the combination of both labelled and unlabelled data. NLP is useful to understand the human understandable language by machine and generate the response in human language. It will make use of elements of Named Entity Recognition, Speech Recognition, Sentiment analysis and OCR. All these concepts will help to enhance the performance of the proposed system.

4.5 Limitations

The limitations of the proposed system can be given as:

- The proposed system is not accessible to worldwide as online hosting is required.
- Load balancer integration is not efficient to distribute loads of the system.
- Data security and system security need to be improved.
- More improvement in chatbot interaction for customer support is required.
- Backup mechanism for taking backup of database is not available which is important on regular basis on different servers.

5 Conclusion

The Internet Network has become a serious asset in trendy business, hence electronic shopping or online purchase has obtained significance from the entrepreneur's as well as from the customer's perspective. For the start-up businesses or entrepreneur, electronic or online shopping generates new business circumstances and for the customer, it makes shopping comparatively possible. As per survey, most customers of online stores are spontaneous and usually make a decision to stay on shopping website within the first few seconds. Hence we've designed and developed the project system to produce the client with simple navigation, easy retrieval of data and required feedback as much as possible.

As we have seen in this proposed project, the process of developing a user-friendly and straightforward customer platform is one with complexity. From understanding customer requirements to complete system design and finally proposed system prototype and finalization, every action needs in detail knowledge and commitment towards achieving the objective of proposed project.

So this is an efficient and effective way for the customers to purchase products online with the help of Chatbot within a few steps.. With the help of E-commerce website sellers can reach to larger audience and with the help of chatbots, sales can be increased by personal interaction with the users. In this way, this application provides an optimized solution with better availability, maintainability and usability.

6 Future Enhancement

This project was developed to meet user requirement; but there square measure several scope to boost the performance of the projected System within the space of interface, database performance, and query processing time Etc.

So there square measure several things for future improvement of this project. The future enhancements that square measure doable within the project square measure as follows.

- Host the platform on online servers to make it accessible to world wide
- Integrate multiple load balancers to distribute loads of the system.
- Improvements in the area of data security and system security can be made in future.
- Provide more online tips and help with help of chatbot interaction.
- Implement the backup mechanism for taking backup of database on regular basis on different servers.

Reference

1. Marwade, A., Kumar, N., Mundada, S., Aghav, J.: Augmenting E-Commerce Product Recommendations by Analyzing Customer Personality (2017)
2. Bhawiyuga, A., Ali Fauzi, M., Pramukantoro, E.S., Yahya, W.: Design of E-Commerce chat robot for automatically answering customer question. University of Brawijaya Malang, Republic of Indonesia (2017)
3. du Preez, S.J., Lall, M., Sinha, S.: An intelligent web-based voice chatbot (2009)
4. Setiaji, B., Wibowo, F.W.: Chatbot using a knowledge in database (2017)
5. D'silva, G.M., Thakare, S., More, S., Kuriakose, J.: Real world smart chatbot for customer care using a software as a service (SaaS) architecture (2017)
6. Baby, C.J., Khan, F.A., Swathi, J.N.: Home automation using IoT and a chatbot using natural language processing (2017)
7. Pratt, E.: Artificial intelligence and chatbots in technical communication (2017)
8. Shawar, B.A.: Arab Open University, Information Technology Department, Jordan: Integrating computer assisted learning language systems with chatbots as conversational partners (2017)
9. Deshpande, A., Shahane, A., Gadre, D., Deshpande, M., Joshi, P.M.: A survey of various Chatbot implementation techniques. Int. J. Comput. Eng. Appl. **XI** (2017)

10. Abdul-Kader, S.A., Woods, J.: Survey on chatbot design techniques in speech conversation systems. Int. J. Adv. Comput. Sci. Appl. (IJACSA) **6** (2015)
11. Pereira, M.J., Coheur, L.: Just. Chat-a platform for processing information to be used in chatbots (2013)
12. Lokman, A.S., Zain, J.M.: One-match and all-match categories for keywords matching in chatbot. Am. J. Appl. Sci. **7** (2010)
13. Ghose, S., Barua, J.J.: Toward the implementation of a topic specific dialogue based natural language chatbot as an undergraduate advisor. In: Proceedings of IEEE of 2013 International Conference on Informatics, Electronics & Vision (ICIEV) (2013)
14. Kar, R., Haldar, R.: Applying Chatbots to the Internet of Things: opportunities and architectural elements
15. McTear, M., Callejas, Z., Griol, D.: Creating a Conversational Interface Using Chatbot Technology. Springer International Publishing, Cham (2016)
16. Yusril, M., Setyawan, H., Awangga, R.M., Efendi, S.R.: Comparison of multinomial Naive Bayes algorithm and logistic regression for intent classification in chatbot. In: 2018 International Conference on Applied Engineering (ICAE) (2018)
17. Hossain, M.Y., Hossain, I., Banik, M., Hossain, M.I.A., Chakrabarty, A.: Embedded system based Bangla intelligent social virtual robot with sentiment analysis. In: 2018 Joint 7th International Conference on Informatics, Electronics & Vision (ICIEV) and 2018 2nd International Conference on Imaging, Vision & Pattern Recognition (icIVPR) (2018)
18. Saratha Devi, G., Vinothini, G., Janani Devi, C., Jeyachandrakala, C.: An enhanced voice based chatbot using artificial intelligence. Int. J. Eng. Technol. (2017)
19. Elhayatmy, G., Dey, N., Ashour, A.S.: Internet of Things based wireless body area network in healthcare. In: Internet of Things and Big Data Analytics Toward Next-Generation Intelligence, pp. 3–20. Springer, Cham (2018)

A Novel Approach to Detect Microcalcification for Accurate Detection for Diagnosis of Breast Cancer

P. P. Patil and S. Kotrappa

Abstract Breast Cancer is one in all the top cancers inflicting death of assorted woman's across the globe. It is really dangerous but at the same can be cured completely in case of early detection of it happens. Worldwide radiologist and doctors are working on various techniques and they found that mammographic identification of breast cancer images can be important in early detection of breast tissue cancer. Primary and main indication of the breast cancer is presenting the clusters of microcalcification. Although it becomes very difficult and also time consuming to make the arrangement of microcalcifications by the radiologists as benign or malignant. Also, the interpretation of mammograms isn't straightforward thanks to tiny variations in densities of various tissues at intervals the image. It's absolutely matching case for dense breast. This work presents novel approach that can perform classification of the clusters which is part of micro classification presents in the mammograms. In case of multiscale morphology, it observed the inter-connectivity of each microcalcification. This work is different one from the existing which mainly has attainment on the morphology of the microcalcification and the cluster feature for representing the cluster structure of microcalcification it generates a graphs. Theoretical features of a graph are extracted to maintain topological features used to make classification of the clusters of microcalcification using the k-nearest neighbour algorithm. This projected work evaluated with the assistance of 2 most accepted digitized dataset referred to as MIAS and DDSM and capable digital dataset. This work shows the different topology modeling is extremely necessary for accurate estimation of the microcalcification. Microcalcification thanks to the improved accuracy of microcalcification and conjointly the topological measures that is in a position to couple with clinical understanding.

P. P. Patil (✉)
Computer Science and Engineering, VTU, Belgaum, India
e-mail: Pramodkumar.patil@gmail.com

S. Kotrappa
Computer Engineering, VTU, Dr. MSS Sheshgiri College of Engineering and Technology, Belgaum, India
e-mail: kotrappa06@gmail.com

N. Dey et al. (eds.), *Internet of Things, Smart Computing and Technology: A Roadmap Ahead*, Studies in Systems, Decision and Control 266,
https://doi.org/10.1007/978-3-030-39047-1_4

Keywords Classification · Graphs · Rough set · Mammography · Microcalcification · Topology

1 Introduction

Nowadays most of women's are affected by the breast cancer worldwide. As per research in Europe, cancer is the leading reason of mortality among Men's and Women's. Among all types of cancers more no of deaths by the breast cancer in women's. In the U.S 12.15% of women's are suffering from this cancer during her lifetime. In solution to that mammography is most regular solution for detecting breast cancer which helps to detect in early stages. Most of countries trying to implement population based program on mammography screening because of that they encourage women's to take a part in breast examination regularly which may help to examine the cancer in early stages. In U.S mostly annual mammographic screening for the women at the normal risk in the age 40 and in the U.K women's are in the age 50–70 recommended for screening program which is performed in every three years.

Microcalcifications square measure nothing however little deposits of the Calcium salt within the breast tissue that square measurebright spots in mammograms. It'll acquireclusters of the microcalcification within the primary stage of carcinoma. The radiological definition of the cluster is that is the area of 1 cm^2 which is not fewer than the three microcalcifications and the spatial high resolution of the mammography mostly in range between 40 and 100 Âµm per pixel. Because of that we can easily identify the breast cancer in early stage. It's not necessary that all the clusters are the part of cancer; the cancer clusters are slight different which has high probability of malignancy.

As per clinical prospective it's terribly tedious method to classify the malignant and benign microcalcifications. Such types of the results shows very high rate of unnecessary biopsy examinations. Because of that need to improve the accuracy of the interpreted micro-calcifications by the radiologist in mammograms. Mostly the CAD (Computer-aided diagnosis) used for the analysis and to reduce the false positive rate at the time of maintaining sensitivity [1, 2]. There are so many CAD techniques are proposed. Some different types of the features used to characterize the microcalcification after that make classifications of the abnormalities which falls into one malignant or benign based on the cluster shape, morphological, intensity and also the texture features [1, 3]. The existing research shows that morphological microcalcifications used to differentiate malignant or benign cases [4]. Morphological characteristics extracted from the particular microcalcifications to describe the morphological features just like size, roughness and shape [5–9]. By adopting the morphological options for form and cluster that makes focus on the worldwide properties of the microcalcification clusters [2, 8, 10–12]. Some of them are used for making prediction of the morphology of clusters like cluster diameter, cluster area, circularity, elongation and the eccentricity.

The digital mammogram is low radiation test for early stage breast cancer prediction and it is mostly used worldwide tool for first sight investigation tool. Basically the cancerous and non-cancerous cells will be easily identified with the presence of microcalcification in mammogram test and afterword's the oncologist will suggest for biopsies or related treatments.

The sample mammogram scan for cancerous and non cancerous cells looks like the image shown in Fig. 1.

The figure shows the cancerous (Malignant) and Non cancerous (Benign) tumor samples. This is a challenging work to accurate classification of Benign and Malignant tumor accurately sometimes due to dense breast and ductal particals its being a challenging task to accurate classification of image. various techniques for CAD to the breast cancer are largely developed within the a part of decades, a number of them have obtained the nice sensitivity for the actual abnormality to mechanically classify the actual microcalcification as malignant or benign that is remaining the challenge owing to their nature, several of the present work having their own specific disadvantages. First the work based on the morphology/shape of the certain microcalcifications [5, 6], informative characteristics is not able to achieved when the microcalcifications are very small which means that it is occupying only few pixels so that it is seen to be pointless to examine the morphological properties/shape of the such type of small objects. Secondly microcalcifications has an awfully a lot of low distinction by considering with the encompassing tissue specifically once microcalcifications is produce at intervals the dense tissue that has high and uniform intensity. Although the deficiency of the helpful texture data at intervals the background region that has effects on the potential of the works that's counting on the intensity differentiation and also the texture options [10, 13]. Additionally, this is often showing the spatial distribution of

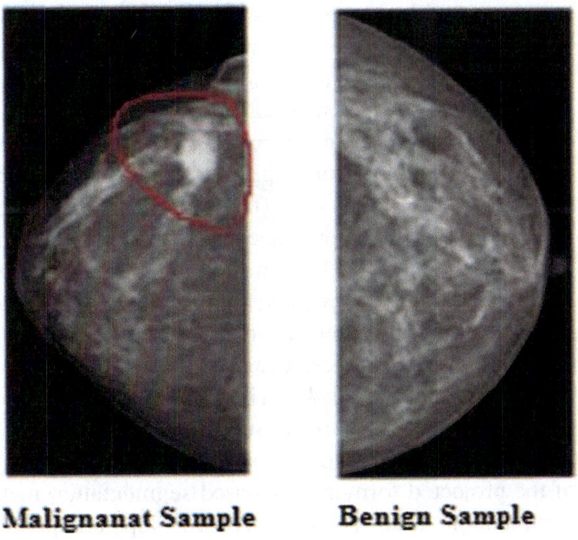

Fig. 1 Cancerous and non-cancerous samples

the microcalcifications among the cluster, the world cluster options were represented on the premise of mounted resolution and therefore the distance primarily based options depend upon the initial spatial resolution of the diagnostic procedure. this is often giving lead to the deficiency of lustiness and therefore the adaptiveness to the another completely different spatial resolutions of the mammograms within the specific screen-film mammograms that is earned by the numerous of various digitizers.

2 Literature Review

In this work, we have been proposed Cancer detection on the Medical field using Image processing schemes. Multiple techniques and schemes are already available for cancer detection to have been discussed earlier. There are many intelligent schemes and treatment is used to detect cancer of different categories, we provide details of the same. With the help of cancer datasets clearly informed that Breast, Liver, and Brain tumor are three major cancers occurs in people. FPGA and 3D-USCT with the image reconstruction algorithm are the basic method and algorithm used to detect cancer. Available numerous states of strategies in cancer detection and analysis of every technique with comparison square measure mentioned [14].

For affective and efficient extraction of information medical image analysis science is used for analyzing or finding medical issues with totally different analysis techniques. It has turned up as a topmost research area in engineering and medicine field. In recent years machine learning algorithms are rapidly used in medical image analysis. To extract compact info for the performance of medical image analysis system machine learning techniques square measure used, as compared with ancient ways that use extraction of handcrafted options. Deep learning is Associate in nursing invasion in machine learning techniques that have to be compelled to adjure the sector of pattern recognition and pc vision analysis by providing the state of the result. Different machine learning algorithms provided by Deep learning that has high-level information abstraction and not rely upon handcrafted options. to supply auspicious result deep learning strategies distribution deep convolution neural network that are applied to medical image analysis. The appliance space contains the total spectrum of medical image analysis which has detection, classification, segmentation, and computer-aided designation. This paper presents a review of the progressive convolutional neural network primarily based techniques used for medical image analysis.

In this, various cancer detection methods and algorithms have been discussing. Cancer detection is considered in define with three various compound sensible methods give out in the literature. Using cancer datasets the 3 most cancer issues that area unit the brain, breast and liver area unit analyzed and thanks to the inherent similarity of the projected formula CA-based segmentation methodology is economical as its computation time is low. To develop compound wise algorithms for image analysis and supply best resolution for analysis and diagnosing and study the time period cancer pictures area unit enclosed within the future work.

The overall studies prove that at the time of defining the conventional pixel brightness threshold the mammographic compactness is developed risk factor. We can detect existing breast tumors using a mammographic image which has features that can suspect both risks of breast cancers and the possibility of the tumor which is missed during the masking. To detect breast cancer this is important for the future. This paper describes the breast cancer risk indicators that show a distinct mode of detection. There's sturdy proof that such cancers square measure foreseen best once the density of mammographic was outlined because the bright areas usually. As compared to traditional mammographic technique or density outlined by white space a lot of risk analysis info can be gathered from a mammogram. A measure of mammographic density which is percentage is applied to the interval of cancer and the masking phenomenon is confirmed [15]. The high pixel brightness represents the analysis space that is healthier tissues of the breast are vital for the disciplines like to spot the particular markers of tissues and tailored screening utilized in scientific research.

Areas of high picture element brightness represents the higher tissue with an informal role in carcinoma is vital for several disciplines, that square measure scientific research that establish specific tissue markers, genetic analysis that try and notice pathways concerned in unwellness risk and public health initiatives, by distinguishing girls at raised risk of masking and/or of the unwellness itself. Risk-predicting info is obtained from a X-ray than from the construct of mammographic, or breast, density outlined by the white or bright areas. The traditional live of mammographic density in percent applies to the interval of cancers and therefore the masking development have confirmed.

3 Proposed System

The proposed work is to identify and detect the breast cancer on basis of microcalcification clusters, there are different methods and different algorithms is available from them we can apply on the mammograms i.e. the set of breast images or the dataset of the mammograms.

The objectives of this work are as follows:

(1) The first objective is to increase the accuracy in ROI detection by applying the segmentation mechanism.
(2) The second and the most important part is the classification and to detecting accurate clusters of microcalcification from all segmented in mammographic images.
(3) Reduce the processing time with accuracy improvement.

From the segmentation we can easily divide an image into its different multiple non-overlapping especially part depends on the particular specific benchmark or we can say that the condition or criteria such as the set of pixels of images or some inherent attributes like texture, contras and color. Mostly the key feature of image segmentation is to represent the image within the explicit pregnant kind that's it's in a position during a manner that's appropriate utilized and resolved. This mechanism is extremely attention-grabbing that is employed to scale back the search space within the image which suggests by dividing the first image into the 2 categories like as object and background.

The main aimed of segmentation is to make an identification of the possible parts of interest (ROI). Segmentation method is used for detecting the abnormal masses which is being exist in the mammographic image. Segmentation can be easily done with the help of a different view or the single unique view of the breast.

There are the number of classifier were used to make the difference malicious from the benign masses in the mammographic image dataset, but in this proposed work we are using the K nearest neighbors. These KNN classifiers are main used in this work for the classification. To classify the microcalcification clusters into malignant and benign. The standard KNN classifier is appraise as a popular and relating to intuitive instance-based learning approach. Maximum number of the choice is working which is used to attempt for direct some of inherent shortcomings of the classical KNN. This KNN classification is relaying on the simple majority vote.

Mainly to showing the differentiation in between the classification algorithms results that is relaying on manually and we can say that automatically consisting of or divided into segments microcalcifications and also for inspecting the robustness of the proposed mechanism to make the microcalcification segmentation differentiation, This will be easily tested with the help of both manual annotations and automatic detection results for the given MIAS dataset.

3.1 System Architecture

The system methodologies are discussed in detail below,

Figure 2 depicts the proposed system architecture on which the entire system works basically it starts processing with input image as Digital Mammogram and apply undergone through the preprocessing such as contrast enhancement through Contrast Limited Adaptive Histogram Equalization (CLAHE), Noise removal, Segmentation for generating exact Region of Interest (RoI). Then afterword's feature extraction and feature selection will be applied to the image.

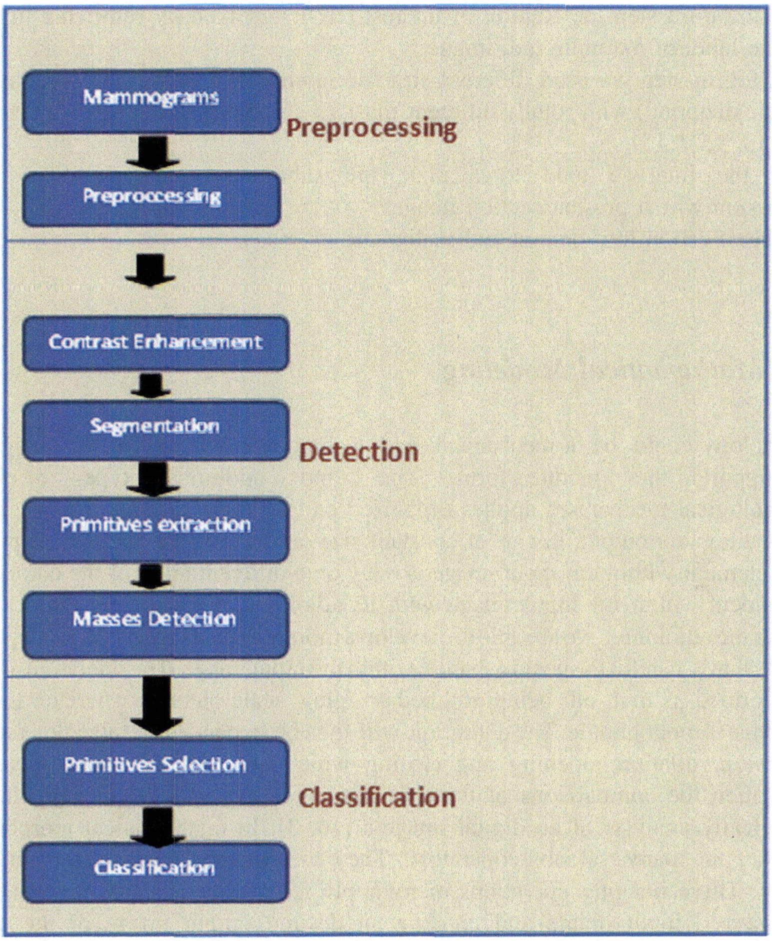

Fig. 2 Proposed system architecture

3.2 *Proposed Algorithm*

The proposed algorithm for classification of Digital Mammogram in Benign or Malignant.

(1) The first step is to browse the X ray pictures, which had taken from hospital or from the datasets we used. The format of those pictures is "jpg", "png" with size 256 * 256.
(2) The second step is to convert supply image to grey scale that is mean every pixel has one color that is grey scale, that is, carries solely the powerful and powerful data.

(3) In the third step the Region of interest (ROI) resolved by removing time and date label of mammogram image.
(4) In fourth step we used different structure component (disk, line, square, periodic diamond) with totally different mask (3 * 3, 5 * 5, 7 * 7, 9 * 9, 11 * 11, 13 * 13, 15 * 15),
(5) In the final we tend to calculate the subtraction between dilation and erosion which present section image.
(6) Classification into benign and malignant.

3.3 *Mathematical Modeling*

Morphology could be a mechanism and technique employed in the image process supported the pictures form, size and additionally type of objects. Morphological mechanism applies organized part to associate input image, associated making an output image at constant size and form. the purpose or price of every element within the input image is relay on a differentiation of the corresponding element within the input image with its adjoining. By selecting the scale and form of the adjoining, we are able to develop a morphological operation (i.e. open and closed) that's careful to specific shapes within the input image. The operation we tend to performed is first off being outlined on gray scale pictures wherever the supply image is monophonic. The definition will then be expanded to full color pictures.

Erosion, dilation, opening and closing which are morphological operations. More often the comparisons of these operations are considered to perform morphologically analysis of an digital image's [16, 3]. In mathematical morphology, there are so many valuable operators. They are dilation, erosion, opening and closing. These imaging operations many apply for restructuring of elements to an prepossessed input image and creating an desired output image of the similar features. Irrespective of the nature of the structuring element, the origin is marked at its center. Morphological opening of the image is as,

$$\gamma_{\mu\beta}(f)(x) \tag{1}$$

and respectively Morphological closing is as,

$$\phi_{\mu\beta}(f)(x) \tag{2}$$

$$\gamma_{\mu\beta}(f)(x) = \delta_{\mu b}(\varepsilon_{\mu\beta}(f))(x)$$

$$\phi_{\mu b}(f)(x) = \varepsilon_{\mu\beta}(\phi_{\mu\beta}(f))(x)$$

where μ a homothetic parameter, of size is as μ It means a respective square of (2 1) (2 1) μ μ + × + pixels. B is the structuring element of size 3 × 3 (here μ = 1).

4 Dilation

Dilation is another root operation in morphological. Dilation is additionally originally outlined for the binary pictures. Dilation is look that produces the image that's identical form because the original image however could be a totally different size. Extension of Dilation (maximize) or shrinks the first. Dilation will increase the valleys picks and enlarges the breadth of maximum regions of the input image, thus it will makes a negative impact on noises however do very minimum on positives ones. The dilation of A by associative structuring elements of B is estimated by:

$$A \oplus B = \bigcup_{b\varepsilon\beta} A_b \tag{3}$$

If B is centered, as before, the dilation of A to B can therefore be interpreted as a curve of the points covered by B when the center of B turns inside into A. Dilation process the of image 'f' by reframing the elements is given by f ⊕ s. The well framed element s is well positioned to its origin point at (x, y) and respective new pixel value is determined using the rule:

$$g(x) = \begin{cases} 1 & \textit{if shits f} \\ 0 & \textit{otherwise} \end{cases} \tag{4}$$

5 Erosion

Erosion is one among the 2 root operation in morphological. Erosion is originally outlined for the binary pictures. And it's essentially outlined as '⊕' essentially it's accustomed scale back the objects within the pictures and noted that its minimizes the associated peaks and increases the widths of the regions, that's why it will take away the positive noises however result negative spontaneous noises very little.

Erosion of the navy sq. leading to lightweight blue sq. As erosion of the digital image A by reframing the image component of B is outlined by:

$$A!B = \{\ z\varepsilon E \setminus B \in A\ \} \tag{5}$$

From datasets the images, if anyone pixels from the image are set to 0, the respective output pixel is set to 0 as therefore.

Erosion of the digital image 'f' by the corresponding structuring element s is obtained by f ⊕ s. In the view of this the structuring element s is positioned with its starting point at (x, y) and the new pixel value is determined by the rule:

$$g(x) = f(x) = \begin{cases} 1 & \textit{if shits f} \\ 0 & \textit{otherwise} \end{cases} \tag{6}$$

6 Image Segmentation

Any of the gray scale images is to observed as a topographic region of interest in which the maximum intensity denoted peaks and hills denoted low intensity denotes valleys. We can started by filling each of isolated valleys like local minima with the use of various colored water like labels. If the water increases based on the peaks such as gradients near to the water from the various valleys obviously with various colors will started to combine. To ignore this, easily created the obstacles in the locations where as water combines. Continuing of filling water to the areas and also building barriers up to when all the peaks covered by the water around. After that the barriers created at giving the readings of segmentation. This is main philosophy behind the watershed.

But it is also giving you over segmented result because of noise or any other irregularities in the given image. Opencv is implementing a marker-based watershed algorithm where as it identifies which of all valley points are to be combined and which are to be not. It is the interactive image segmentation. Here, have to give; label to a region which have to be confirmed by being the front or the targeted object with the use of any color or the intensity also used to assigning labels to the region which have to be confirmed of being background or non-object with the help of another color and finally the region which we are not sure of anything, label with the 0. This is known as a marker and after that applying the watershed algorithm. Then marker is well modified with the help of assigned labels and also the targeted boundaries of the targeted objects having a value as −1.

7 Feature Extraction

The Parameters (Features) is known as a function of primary estimation of parameters or attributes which are specifying any of quantifying observations of an object and it is also most beneficial for the accurate classification as well as pattern recognition. It is very domain specific task to obtain a good data representation which is related to the available measurements. Many of the different features employed that can be categorized into low quality and high level characteristics. In particular, the most complicated processing is required quantitatively and qualitatively to derive high-level characteristics from an picture compared to low-level characteristics. Low-level characteristics are the basic characteristics to be obtained straight from an picture without any description of the item, while high-level extraction characteristics involve finding forms and items in computer pictures and are based on low-level characteristics [17]. The characteristics of the low level can be classified as follows: general characteristics, global features and specific characteristics of the domain. Application independent features such as color, texture and shape are general characteristics. Depending on the abstraction level, these characteristics can be further split into: pixel-level features, which are the features calculated at each pixel (e.g. color, place, etc.) and local characteristics, which are features calculated over a subdivision of the picture bands resulting from segmentation of the image or edge detection. Global characteristics are characteristics that are evaluated in the general image or only frequent sub-area. Application-dependent characteristics such as human faces, fingerprints, character recognition and conceptual characteristics are the domain-specific characteristics. Obviously, this classification is not that sharp, there exist overlaps between them. Also It should be regarded to be explicit as well as meaningful (associate with important scene components) and detectable (location algorithm must be present) in some apps as computer vision apps.

Human expertise, which is often required to convert “raw” data into a set of useful features, can be escorted by automatic feature extraction methods. Feature extraction is the method of transforming raw data into more informative signatures or systems features that represent the information most efficient or effective for evaluation and classification.. In pattern recognition and in image processing, feature extraction is a special form of dimensionality reduction. Too broad are input data to process an algorithm and it is thought to be redundant (many data, but not a lot of information) so that the input information is converted into a feature set of decreased representations (vector functions). In order to execute the required job with this decreased representation instead of the complete size input, it is anticipated that the characteristics vector will extract the appropriate information from the input data.

Graph theoretical methods are very much useful in microcalcification cluster selection as it has many advantages to work with clusters and also with complex nature clusters. Here microcalcification cluster always having invariant nature in growing so its somewhat hard to identify at the initial level so that there may be delay in analysis of breast cancer at early stage.

8 Feature Selection

Sampling feature is also called selection variable or attributes selection. The most important thing for your prediction modeling issue is the automatic choice of characteristics in your information (e.g. columns in tabular information). Selection of features differs from a reduction in the total of dimensionality. Methods aim to reduce the number of attributes in the dataset, but they do so by creating new attribute combinations, while feature selection methods include and exclude data attributes without altering them. Strategies of feature selection help you generate a precise computational model in your task. Help you identify features that give you better or even more accurate information and require less feature selection methods may be used to identify and delete data attributes which are pointless, irrelevant and redundant, and that do not contribute to the predictive model’s accuracy or actually decrease the model’s accuracy. Fewer attributes is desirable because it reduces the complexity of the model, and a simpler model is simpler to understand and explain.

$$A_{grab=}\{a_{1,grab1} + a_{2,grab2} + \cdots + a_{ngrabn}\} \tag{7}$$

8.1 Feature Selection Algorithms

Three widely used variants of feature selection algorithms: The filter methods, the wrapper methods and embedded methods,

Filter Methods

Filter methods used in feature selection can applies statistical measure to allocate weighted scores to each feature. Selected features grouped by score and in case of selected features or possibly to remove the features from the dataset. The methods are very much effective and useful for the selection of the features or can be correlated with the selected independent variable. Chi squared test, information gain and correlation coefficient scores are some of the filtering methods useful for the case.

Wrapper Methods

Wrapper methods preferably uses set of features as a searching criteria for feature selection where more no of combinations are primarily selected, prepared, evaluated and compared to other combinations of parameters. A primary predictive analysis model is used to evaluate combination of features and along with the score based model for accuracy calculation. The best fit methods could be a main search criterion in this cases, there may be possibility of stochastic approach like hill climbing algorithms or may be some heuristic forward and backward for feature addition or removal.

Embedded Methods

Embedded methods are such type of methods who can able to contribute to the accuracy of the system model while in the developing phase of the model. The most frequently used feature selection approach is regularization approach. It has been also known as penalization method that may has some additional rules for the optimization problem of predictive algorithms (Regression models) that model is going to defend towards other lover complexity (i.e. fewer coefficients). Some examples are as, LASSSO, Elastic net and Ridge regression.

9 Result and Conclusion

In the proposed system we are currently taking the cancer affected image and used the erosion and dilation morphological transformations on it with the help of the open cv. Using erosion and dilation we can do a processing on the images. In this proposed system we presented a mechanism for classification of microcalcification clusters in mammograms that relay on morphological topology analysis. This viewpoint is used to analyze microcalcifications in the form of topology and connectivity for discriminating malignant from friendly or we can say that the benign clusters. Most features i.e. shape/morphological features in previous publications that are extracted at a single unit of scale, the more than one scale characteristics which are wrapping a representation of microcalcification clusters was developed in this work. By the help of multi-scale the morphology was analyzing by the topology/connectivity of microcalcification clusters. For narrating the topological structure of microcalcification clusters was developed by the set of microcalcification graphs at multiple scales. We are going to extracted eight graph metrics from the microcalcification graphs which are generated at multiple scales. At a time of analyzing the topology of microcalcifications, that are number of sub-graphs, which are resulting feature of eight graph sets were aggregated and constituted the multi-scale topological feature vector that has been used to classify clusters of microcalcification into malicious and friendly or benign.

The proposed method has to be assess the three well known data sets: DDSM, MIAS and Digital, Four k-nearer neighbors'-based algorithms KNN, FNN, FRNN, and VQNN that is been used or the classification task. For each and every data and the datasets the superior classification results have been obtained. A set 'S' values for number of multi-scales by investigating and with a range of k values for the classifier, the best classification accuracy was obtained 95% for MIAS with the manual annotation, 95% for MIAS with automatic detections, 96% for Digital, 85.2 ± 5.7% for DDSM depends on ten-fold cross-validation, 86% for DDSM using LOOCV and the largest area under the ROC curve was 0.96, 0.96, 0.90 and 0.91 ± 0.05, respectively.

Future Scope

Although we had tried the image processing and image segmentation of mammographic images obtained through mammogram images for breast cancer detection, we had classified and obtained the feature selection process through the dataset, but there is still a scope In future development like we need to work on the density of the breast cancer tissues so that the prediction and accuracy can be increased and this will help to identify the exact breast tissues to take the preventive and measurable action during treatment.

Also for complex microcalcification clusters detection we may use directed as well as undirected graph methods any also different variants to detect complex invariant nature clusters so definitely it will be beneficial for predicting breast cancer at early stage.

Acknowledgements I would like to express my deep thanks to Dr. S. Kotrappa sir for timely valuable guidance and motivation. The way we carried out the entire survey and started working together your valuable inputs leads me too learn from the failures and experiments. The way you made resources available was helped me a lot. Thank you for your cooperation and contribution.

References

1. Cheng, H.D., et al.: Computer-aided detection and classification of microcalcificationsin mammograms: A survey. Pattern Recognit. **36**(12), 2967–2991 (2003)
2. Papadopoulos, A., et al.: Characterization of clustered microcalcifications in digitized mammograms using neural networks and support vector machines. Artif. Intell. Med. **34**(2), 141–150 (2005)
3. Elter, M., Horsch, A.: CADx of mammographic masses and clustered microcalcifications: A review. Med. Phys. **36**(6), 2052–2068 (2009)
4. Lanyi, M.: Morphologic analysis of microcalcifications. In: Zander, J., Baltzer, J. (eds.) Early Breast Cancer, pp. 113–135. Springer, Berlin (1985)
5. Shen, L., et al.: Application of shape analysis to mammographic calcifications. IEEE Trans. Med. Imaging **13**(2), 263–274 (1994)
6. Ma, Y., et al.: A novel shape feature to classify microcalcifications. In: Proceedings of 17th IEEE International Conference on Image Processing, pp. 2265–2268 (2010)
7. Dengler J., et al.: Segmentation of microcalcifications in mammograms. IEEE Trans. Med. Imaging **12**(4), 634–642 (1993)
8. Betal, D., et al.: Segmentation and numerical analysis of microcalcifications on mammograms using mathematical morphology. Brit. J. Radiol. **70**(837), 903–917 (1997)
9. Chan, H.P., et al.: Computerized analysis of mammographic microcalcifications in morphological and texture feature spaces. Med. Phys. **25**(10), 2007–2019 (1998)
10. Dhawan, A.P., et al.: Analysis of mammographic microcalcifications using gray-level image structure features. IEEE Trans. Med. Imaging **15**(3), 246–259 (1996)
11. Jiang, Y., et al.: Computerized classification of malignant and benign clustered microcalcifications in mammograms. In: Proceedings of 19th Annual International Conference on IEEE Engineering in Medicine and Biology Society, pp. 521–523 (1997)
12. Shao, Y.Z., et al.: Characterizing the clustered microcalcifications on mammograms to predict the pathological classification and grading: a mathematical modeling approach. J. Dig. Imaging **24**, 764–771 (2011)
13. Chan, H.P., et al.: Computerized classification of malignant and benign microcalcifications on mammograms: texture analysis using an artificial neural network. Phys. Med. Biol. **42**, 549–567 (1997)
14. Velusamy, P.D., Karandharaj, P.: Medical image processing schemes for cancer detection: a survey
15. Nguyen, T.L., Aung, Y.K., Li, S., Trinh, N.H., Evans, C.F.: Predicting interval and screen-detected breast cancers from mammographic density defined by different brightness thresholds
16. Breast Cancer Facts & Figures 2011–2012, pp. 11–15. American Cancer Society, Atlanta (2011)
17. Aguado, A.S.: Feature extraction and image processing for computer vision. Elsevier Science Publishing Company (2012)

Big Data

Ensemble Classifier for Praise or Complaint Classification and Visualization from Big Data

Sujata Khedkar and Subhash Shinde

Abstract With the advent in Big Data Analytics, IoT and Machine Learning newer opportunities are created for Business organizations to analyze, monitor and mine user-generated contents in real time for business intelligence using cognitive IoT. Customers share their opinions online through social media platforms like review sites, Twitter and Facebook, etc. Sentiment analysis combined with real-time reporting can provide precise valuable contextual insights enabling more improved decision making. The existing sentiment analysis techniques identify only positive, negative or neutral sentiments and do not consider informativeness of reviews while analyzing the sentiments. The extreme opinions like praise and complaint sentences are informative subsets of positive and negative sentences and are very difficult to find. This chapter proposes the Ensemble classifier using linguistic features for praise or complaint classification from big customer review datasets and visualization of it. The Praise and Complaint sentences are further classified based on aspect and analysis at aspect level is presented from business intelligence point of view. The performance of the four different supervised machine learning classifiers, namely Random forest, SVC, KNeighbours, MLP with linguistic hybrid features and Ensemble of above algorithms is evaluated on Hotel and Amazon product reviews dataset using parameters Accuracy, Precision, Recall, and F1-score. The proposed approach has given excellent 99.7% Accuracy and 99.6% F1-Measure and outperforms existing approaches.

Keywords Big data · Ensemble algorithm · Machine learning · Text analytics · Praises · Complaints · Clustering · LDA

S. Khedkar (✉)
Department of Computer Engineering, VESIT, Mumbai University, Mumbai, India
e-mail: sujata.khedkar@ves.ac.in

S. Shinde
Department of Computer Engineering, LTCE, Mumbai University, Mumbai, India

N. Dey et al. (eds.), *Internet of Things, Smart Computing and Technology: A Roadmap Ahead*, Studies in Systems, Decision and Control 266,
https://doi.org/10.1007/978-3-030-39047-1_5

1 Introduction

With the various social media sites, customers can easily share their opinion on the social platform w.r.t product/service. It is important for business organizations to collect data from various channels, analyze it and measure customer satisfaction, understand customer behavior, recognize customer needs through reviews or social media channels. Many Business Intelligence applications across various domains use opinion mining as the key enabling technology for their Business Intelligence [1–3] from big review data. The system is built by aggregating the opinions from various channels like blogs, comments, reviews or tweets. The major challenge is to find the extreme opinions in terms of praises and complaints to understand customer true opinions w.r.t product/service rather than just analyzing positive or negative sentiments [4].

Sentiment analysis is a field that analyses people's opinions from unstructured text. Sentiment analysis works are three different levels of granularity namely Document level, Sentence level and Aspect level [5, 6]. Most early research on sentiment analysis was focused on document level and is used in the domains like Hotels, Restaurants, E-commerce products and Politics etc. These online reviews have a significant impact on customer's purchase decisions [7].

The sentiment classification classifies the text into three predefined categories as positive, negative or neutral based of a set of features using a classification method [2, 8]. However, it is possible to make other types of classifications by considering informativeness of reviews and to detect and classify reviews as praise or complaint. Praises are a subset of positive reviews and complaints are a subset of negative reviews These Praises and Complaints can be called as extreme opinions.

Praises express the best views or judgments and complaints express worst views or judgments w.r.t product or service. The customer always wants to know about the best and worst aspects of product or service while making online purchase decisions. These very positive praise reviews and very negative complaint reviews have a very strong impact on product sales [9].

Main motivation of this work is to develop new approach to identify praise and complaint sentences from review corpus and analyze them from Business Intelligence point of view. As these extreme opinion's expresses customers opinions more precisely as compared to plain positive and plain negative sentences.

Existing approaches works on whole review corpus, here we will be filtering out noninformative plain positive and plain negative sentences and working only with informative sentences so we can reduce the volume of big review data. Another motivation for the identification of praises and complaints is the current use of social media monitoring, branding and chat bot application's which helps to sell products or to know more about likes and dislikes of the customers. The performance and speed of these applications will be improved significantly by only analyzing extreme opinions.

The objective of the paper is to examine the effectiveness and limitations of different linguistic features based on linguistic properties of praise and complaint sentences and Ensemble classifier to identify praise and complaint sentences from Hotel and Amazon Product domains reviews. Our main contribution is to analyze the effectiveness of linguistic features of praises and complaints in the supervised classification task. We performed various experiments and compare some supervised classifiers, namely Random Forest (RF), KNeighbors, SVC, MLP and proposed Ensemble Classifier for binary classification tasks of identifying praise vs complaint sentences. Lexicon based Multiclass single label classification method and analysis of Praises and Complaints is described for Hotel domain review sentences. The rest of paper is organized as follows. In the following Sect. 2, we discuss the related work. Then Sect. 3, describes the proposed methodology. Section 4, describes the experiments, evaluations, and results. Conclusion and future work are described in Sect. 5.

2 Related Work

2.1 Modeling Customer Satisfaction

In order to be successful and to survive in today's competitive business environment its essential for organizations to provide exceptional services/products to customers. With the various social media sites, customers can easily share their opinion on the social platform w.r.t product/service. It is important for business organizations to collect data from various channels, analyze it and measure customer satisfaction, understand customer behavior, recognize customer needs through reviews or social media channels. Many Business Intelligence applications across various domains use opinion mining as the key enabling technology for their Business Intelligence [1–3, 10, 11] from big review data. The system is built by aggregating the opinions from various channels like blogs, comments, reviews or tweets. Customer satisfaction is defined as interaction between customer's expectation and post purchase evaluation. With lot of reviews available online the challenge is to find the extreme opinions in terms of praises and complaints to understand customer true opinions w.r.t product/service rather than just analyzing positive or negative sentiments [4].

2.2 Sentiment Analysis for Business Intelligence

User generated contents like reviews or tweets provides a rich source of data to extract the dimensions of customer satisfaction. Most the literature largely ignores online reviews and considers only online numerical ratings. Online reviews and twitter messages are text based often comprise of large information repositories beyond the analytical capabilities of traditional statistical methods. Sentiment analysis can be used to investigate online reviews to explore fully the potential of the data to measure customer satisfaction. We have empirically developed and identified the dimensions of customer satisfaction based on praises and complaints shared by users through big review data. These praises and complaints can be used for automatically generating Pro and Con of product/service. It can also used for generating automatic description of product/service.

We found two main approaches to find the sentiment at the sentence level. First, machine learning technique's based on training corpus annotated with polarity information and, second strategies based on lexicons.

The classification performance depends on the identification of quality features and classification methods. Various types of features have been used in classification tasks such as lexical features, stylistic features, content features, sentiment features, semantic features [12]. The influence of these features for classification task has been evaluated and analyzed by some sentiment analysis studies [9, 13, 14]. But some reviews are more informative (praise/complaints) as compare to plain positive/negative text. These informative reviews are very important from the Business Intelligence point of view for decision making. With an abundant amount of opinionated text, methods for automatic extraction of praises or complaints from a given text is not there in existence.

Customer reviews reflect the buyer's subjective perspective and include personal opinions, experiences, and complaints. Each review contains multiple sentences and its practically impossible to consume the whole corpus to know about customer's opinions. Customers often read only a few reviews and cannot get useful information from it. There are many noninformative sentences in the reviews. Most of the research previously was based on quantitative ratings given by the customer on online websites. The extreme opinions like Praises and Complaints are very challenging to identify from review corpus. Only 5–10% reviews contain extreme opinions [15].

Ganesan and Zhou [8] studied the properties of praise and complaint sentences where Praise is a subset of Positive only reviews and Complaint is a subset of negative only reviews. The praise sentence contains more Adjectives, Intensifiers, Nouns, the Length of praise is longer than avg length of sentence. On the contrary Complaint, the sentence has fewer Adjectives and Nouns, more Past Tense and conjunctions. The existing sentiment analysis algorithms can be improved by only considering the analysis of praises and complaints sentences from big customer reviews data (Table 1).

Table 1 Shows examples of different types of sentences

Sentence	Type
This hotel is not good	Negative only
This hotel is costly, rooms not clean and bad service	Complaint
I like this hotel because it's very cheap, nice food, good service, and good location	Praise
This is a good hotel	Positive only

Krishnamoorthy [16] has developed an opinion summarization system based on linguistic properties of review sentences and further classified as a direct, comparative or superlative opinion.

Moschitti and Basili [1] had developed Review Ranking system based on helpfulness score. The system classifies low quality and high-quality reviews based on hybrid features like nouns, adjectives, verb, difficult words, wrong words, entropy, review ranking, etc. and random-forest classifier. Low-quality reviews are ignored and only high-quality reviews are displayed.

Liu et al. [2] had developed a novel method of automatic product defect identification from online forums using contextual features derived from replies and uses a multi-view ensemble learning in the automotive domain.

Reddy et al. [17], discovered product defect using UGC in both the automotive and the consumer electronics domains. They have used principal component analysis and logistic regression to produce a multivariate explanatory analysis relating defects to quantitative measures derived from the text.

Hu et al. [4], have used a novel structural topic model text analysis method for analyzing negative reviews of customers to understand customer's opinions about hotels in New York City

Almatarneh and Gamallo [18, 19] have used Linguistic features to identify extreme opinions, they have used a bag of words, word embeddings, polarity lexicon and set of textual features to identify Complaint reviews and from a set of reviews.

Customer reviews contents are spontaneous and insightful feedback of customers widely available free and easily accessible from anywhere, anytime. There does not exist a standard structured procedure that prescribes how organizations should acquire and analyze User-generated content in order to generate Managerial information for decision making. So, the proposed approach identifies the dimensions of customer satisfaction insights based on praises and complaints shared by users through big data from unstructured textual Reviews of three different domains Hotel, Amazon product reviews of Home Theatre and Vacuum cleaner.

3 Proposed Approach

The proposed approach involves following steps as shown below in Fig. 1a.

3.1 Data Preprocessing

Hotel Domain reviews and Amazon product dataset is used in the analysis. We have used Python and NLTK library for data preprocessing. Data Preprocessing step cleans the extracted data before analysis. The reviews are converted into sentences. Non-textual sentences and sentences that are irrelevant for the analysis are

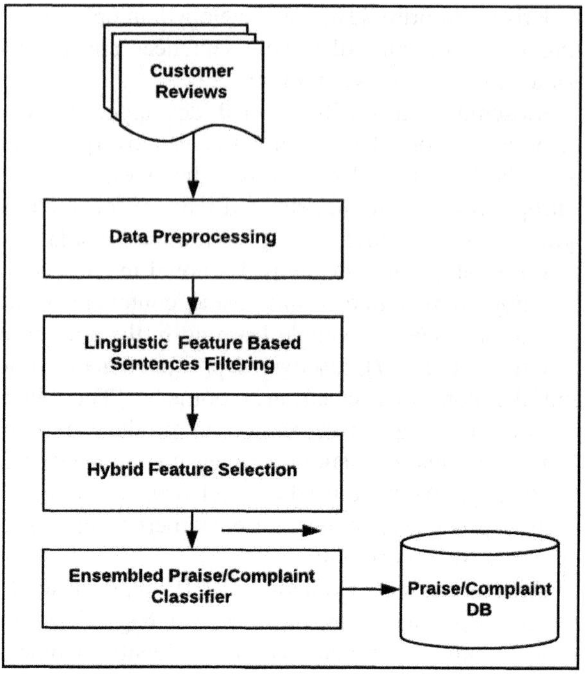

Fig. 1 Proposed approach for praise or complaint classification

identified and eliminated. Data preprocessing techniques such as stop word removal, stemming etc are used for preprocessing. A review might contain positive as well as negative sentences so the same numerical rating cannot be applied to all sentences. The Affin Dictionary is used to compute the sentiment score of each sentence The sentence's which are neutral sentences are eliminated from further analysis.

3.2 Linguistic Feature Based Sentence Filtering

In this step the extracted sentences are linguistically evaluated by using linguistic properties of sentence As the praise sentence contains more Adjectives, Intensifiers, Nouns, the Length of praise is longer than average length of sentence and on the contrary Complaint sentence has fewer Adjectives and Nouns, more Past Tense and conjunctions. The subjective sentences are divided into Praise_set and Complaint_set based on Affin sentiment score. Then the final Praise and Complaint Score of a sentence is calculated using formula (2) and (4). Sentences with subjective expression (complaints or Praises) are retained and sentences which are not subjective (facts, factual information) are discarded

$$Pscore = 0.1 * Length(Sentence) + 0.2 * no(Nouns) + 0.3 * no(Adjectives) + 0.4 * no(Intensifiers) \quad (1)$$

$$Praise\,score = Pscore(sentence) * Sentiment\_score(sentence)/10 \quad (2)$$

$$cscore = -\,0.1 * Length(Sentence) + 0.1 * no(Nouns) + 0.2 * no(Conjection) + 0.3 * no(Adverb) + 0.3 * no(Verb) \quad (3)$$

$$complaint\,score = cscore(sentence) * sentiment\,score(sentence)/10 \quad (4)$$

All Neutral sentences having zero sentence score are eliminated from a corpus

3.3 Hybrid Feature Selection

This Module selects a subset of relevant features. Hybrid Feature selection Method uses various features like

- **Meta Features**—The length of the sentence, no of non-stop words, average word length.
- **Synthetic features**—The number of nouns, number of adjectives, number of verbs, number of intensifiers, number of adjectives, number of adverbs etc
- **Content Features**—Words which are Verbs, Nouns, Adjectives, Adverbs, Verb Phrases etc.
- **Semantic Features**—Positive or negative rating and Praise/Complaint score of the sentence from the Sentence filtering module.

Table 2 shows features that are considered for evaluation.

Various experiments with a combination of the different feature set are performed on the selected dataset.

Table 2 Description of features

Features	Descriptions
FS1	Unigram TF-IDF (1-gram)
FS2	Bigram TF-IDF (2-gram)
FS3	Trigram TF-IDF (3- gram)
FS4	FS1 (Unigram TF-IDF) + FS4 (Hybrid − (Meta + Synthetic + content + Semantic))
FS5	FS2 (Bigram TF-IDF) + FS4 (Hybrid Features)
FS6	FS3 (Trigram TF-IDF) + FS4 (Hybrid Features)

3.4 Ensembled Praise or Complaint Classifier Module

Many researchers have focused on the use of traditional classifiers such as Naïve Bayes, SVM, Maximum Entropy, etc. to solve classification Problems In addition to this traditional implementation, there is also the combination of multiple classifiers to generate a single Ensemble classifier. We have studied the usefulness of various classification algorithms such as SVC, Random Forest, MLP, KNeighbors algorithms for this problem.

3.4.1 Learning Algorithms

- **SVC**—It is discriminative classifier defined by a separating hyperplane. Given a labeled training data algorithm outputs an optimal hyperplane which categories new examples. SVC uses linear kernel. It gives good accuracy, faster than Naïve Bays and uses less memory.
- **Random Forest (RF)**—It is an ensemble algorithm. Random Forest creates set of decision trees from randomly selected subset of training set. It aggregates votes from different decision trees to decide final classification of test object.
- **Multilayer Perceptron (MLP)**—It is a class of feedforward neural network. It consist of three layers of nodes: an input, hidden and output layer. It uses supervised learning technique called backpropagation for training.
- **KNeighbors**—The algorithm classify the object by majority vote of its neighbors, with the object being assigned to the class most common among its k nearest neighbor's.

The proposed Ensembled classifier uses above Machine Learning Algorithms and Hard voting where a model is selected from an ensemble to make the final prediction by a simple majority vote for accuracy.

4 Experiments

4.1 Data Preprocessing

In order to extract praises and complaints, we require to analyze customer reviews with opinion ratings and extract those documents associated with the lowest and highest scale. But as the numerical rating cannot be applied equally to all sentences that are part of the review, we have developed filtering method based on linguistic properties of reviews and Affin dictionary is used to compute the score of the sentence. We obtained the dataset of Amazon product reviews dataset (ACL15-Chan-Dataset) of and analyzed reviews of product's Vacuum Cleaner and

Table 3 Datasets with a total number of review sentences in each class

Datasets	#n of review sentences	Negative only	Complaints	Positive only	Praises sentences
Hotel	**1,00,012**	14,491	**23,545**	26,896	**35,080**
Home theater	**5117**	687	**706**	1200	**2524**
Vacuum cleaner	**4637**	579	**702**	1025	**2331**

Home Theater and Hotel reviews dataset (https://www.kaggle.com/harmanpreet93/hotelreviews/activity).

The datasets are pre-processed using the following steps:

(1) Neutral reviews (3-star rating) are deleted since they do not convey any views.
(2) Reviews are converted into sentences and stop words, digits, special characters are removed.
(3) Review sentences having short lengths i.e. Sentence length < 30 characters are eliminated.
(4) Sentiment score of each sentence is calculated using Affin Dictionary.
(5) The neutral sentence's having Affin sentiment score zero are eliminated.
(6) Linguistic properties of each sentence are computed and Praise_score computed using Eqs. 1 and 2/Compaint_score of each sentence is computed based on Eqs. 3 and 4. Table 3 shows the number of reviews in each class.
(7) Dataset is divided into Praise-set and Complaint -set based on their sentiment score as positive or negative

Table 3 shows the number of Plain _Positive, Praise, Plain _Negative and Complaint sentences in Hotel and Amazon product datasets. As shown in Fig. 2 there are 35.1% Praise and 23.5% Complaint sentences as extreme opinions in Hotel Reviews. The noninformative Plain Positive (26.9%) and Plain Negative (14.5%) sentences are eliminated from the analysis. From Amazon Product dataset, for Vacuum Cleaner there are 50.3% Praise and 15.1% Complaint, for Home Theater there are 49.3% Praise and 13.8% Complaint review sentences.

4.2 *Training and Testing*

We have developed ensembled classifier using Random Forest, Kneighbours Classifier, SVC, and Multilayer Perceptron. We have compared, summarized and discuss the behaviour of these base classifier and proposed ensembled classifier with the linguistic hybrid features as discussed above. The dataset was randomly partitioned into training (80%) and testing (20%). We have used stratified sampling and used 5_fold_cross_validation. The output is evaluated using four parameters Precision, Recall, F1-Score, and Accuracy. Macro Avg. of all parameters are calculated. We have used Python and Scikit-Learn library for machine learning to train Praise or Complaint classification model.

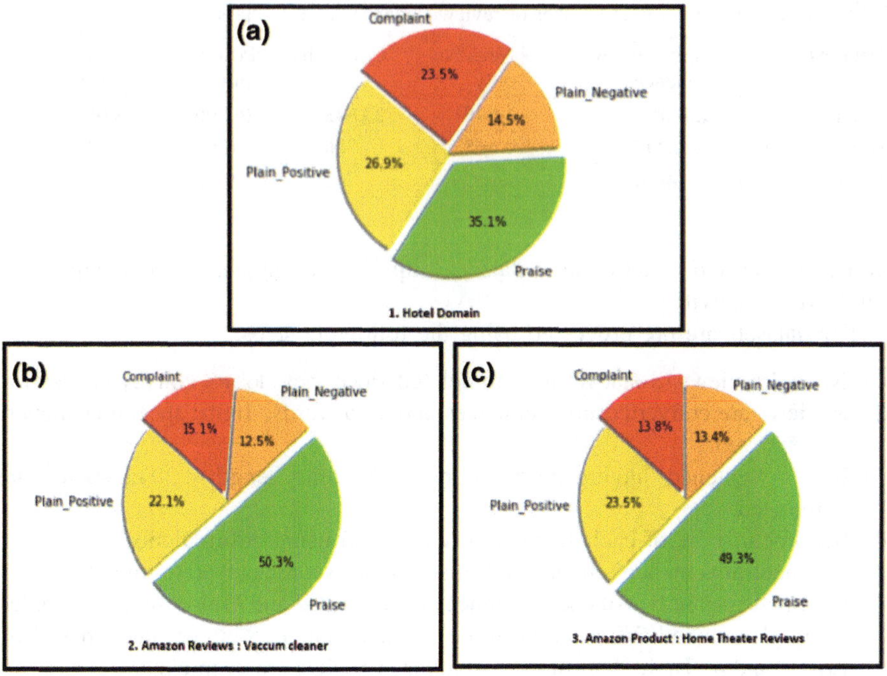

Fig. 2 Exploratory data analysis of customer reviews: **a** Hotel domain. **b**, **c** Amazon product reviews

4.3 Results

Tables 4, 5, and 6 show the classification results obtained by base classifier Random Forest, Kneighbours, SVC, Multilayer Perceptron and Ensemble classifier for all datasets. These classifiers are evaluated based on base features and linguistic hybrid features and their combinations. The final scores were computed using Precision (P), Recall (R), F1 scores and Accuracy.

For Hotel domain reviews as shown in Fig. 3, the proposed Ensemble algorithm with FS4 (Unigram tfidf and Hybrid Features) gives F1-Measuere of 0.998 and outperforms other classifiers with Accuracy 99.6% (see Table 4). For Vacuum cleaner reviews as shown in Fig. 4, the proposed Ensemble classifier with FS4 (Unigram tfidf and Hybrid Features) outperform other classifiers with F1-Measure of 0.993 and Accuracy 99% (see Table 6).

For Vacuum cleaner reviews as shown in Fig. 4, the proposed Ensembled classifier with FS4 (Unigram tfidf and Hybrid Features) outperform other classifiers with F1-Measure of 0.993 and Accuracy 99% (see Table 6). For Amazon Home theatre product reviews as shown in Fig. 5, the proposed Ensembled algorithm with FS4 (Unigram tfidf and Hybrid Features) outperform other classifier with F1-Measure of 0.991 and Accuracy 98.6% (see Table 5).

Table 4 Classification results for hotel domain reviews

Hotel domain reviews	Random forest			SVC			Kneighbor			MLP			Proposed ensembled classifier		
	P	R	F1	P	R	F1	P	R	F1	P	R	F1	P	R	F1
FS1	0.877	0.820	0.844	0.420	0.50	0.456	0.809	0.705	0.740	0.948	0.928	0.983	0.994	0.989	0.966
FS2	0.870	0.839	0.925	0.420	0.50	0.456	0.788	0.677	0.710	0.924	0.898	0.910	0.935	0.982	0.958
FS3	0.864	0.838	0.851	0.420	0.50	0.456	0.793	0.652	0.687	0.923	0.897	0.909	0.943	0.981	0.962
FS4	0.998	0.997	**0.997**	0.841	1.0	**0.913**	0.946	0.990	**0.967**	0.997	0.998	**0.997**	**0.995**	**0.999**	**0.997**
FS5	0.996	0.999	0.997	0.841	1.0	0.913	0.945	0.989	0.967	0.995	0.998	0.994	0.995	0.998	0.997
FS6	0.998	0.994	0.996	0.841	1.0	0.913	0.945	0.987	0.966	0.995	0.998	0.994	0.996	0.998	0.996

Table 5 Classification results for amazon product reviews: vacuum cleaner

Vacuum cleaner	Random Forest			SVC			Kneighbor			MLP			Proposed ensembled classifier		
	P	R	F1	P	R	F1	P	R	F1	P	R	F1	P	R	F1
FS1	0.554	0.428	0.444	0.769	0.80	0.783	0.538	0.403	0.473	0.453	0.505	0.473	0.877	0.989	0.930
FS2	0.549	0.424	0.439	0.800	0.769	0.783	0.543	0.405	0.425	0.563	0.443	0.459	0.858	0.980	0.915
FS3	0.545	0.421	0.433	0.800	0.769	0.783	0.533	0.399	0.418	0.564	0.446	0.465	0.852	0.984	0.914
FS4	0.995	0.992	**0.994**	0.771	1.0	**0.870**	0.932	0.983	**0.957**	0.980	0.991	**0.985**	0.989	0.997	0.993
FS5	0.992	0.990	0.991	0.769	1.0	0.869	0.935	0.985	0.959	0.964	0.992	0.978	0.976	0.997	0.987
FS6	0.985	0.983	0.984	0.769	1.0	0.869	0.936	0.985	0.960	0.965	0.991	0.978	0.976	1.00	**0.998**

Table 6 Classification results for amazon product reviews-home theater

Home theatre	Random forest			SVC			Kneighbor			MLP			Proposed ensembled classifier		
	P	R	F1	P	R	F1	P	R	F1	P	R	F1	P	R	F1
FS1	0.549	0.523	0.437	0.782	0.60	0.791	0.532	0.513	0.425	0.559	0.554	0.464	0.879	0.984	0.928
FS2	0.542	0.524	0.433	0.782	0.60	0.791	0.534	0.508	0.423	0.551	0..542	0.450	0.847	0.978	0.908
FS3	0.535	0.510	0.423	0.782	0.60	0.791	0.528	0.507	0.421	0.545	0.533	0.444	0.853	0.990	0.916
FS4	0.986	0.983	0.985	0.783	1.0	0.870	0.931	0.985	**0.957**	0.976	0.989	**0.983**	**0.984**	**0.998**	**0.991**
FS5	0.984	0.988	0.986	0.782	1.0	**0.877**	0.929	0.984	0.956	0.960	0.989	0.974	0.982	0.996	0.989
FS6	0.992	0.984	**0.988**	0.782	1.0	0.877	0.929	0.984	0.955	0.958	0.985	0.971	0.982	0.998	0.990

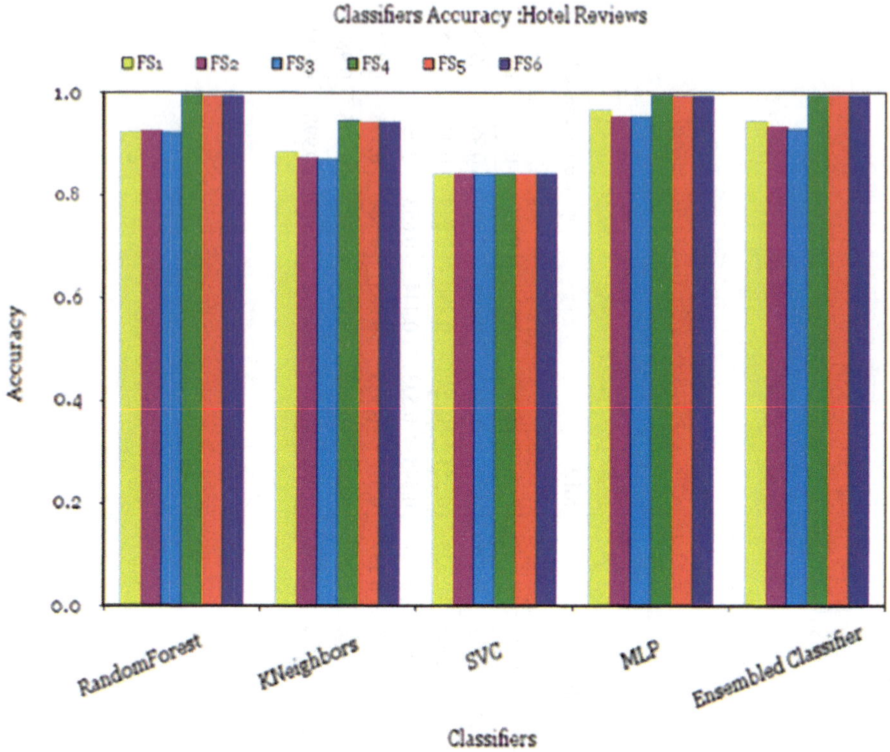

Fig. 3 Classifiers accuracy for hotel reviews

Base features combined with hybrid features have given better results as compared to simple base features.

4.4 Visualization and Reporting for Hotel Domain Reviews

The Praise and complaint review sentences are further analyzed using K-means clustering, Topic Modeling and Uniform Manifold Approximation and Projection (UMAP), dimension reduction technique is used for topic visualization. Figure 6 shows praises visualization and Fig. 7 shows complaints visualization.

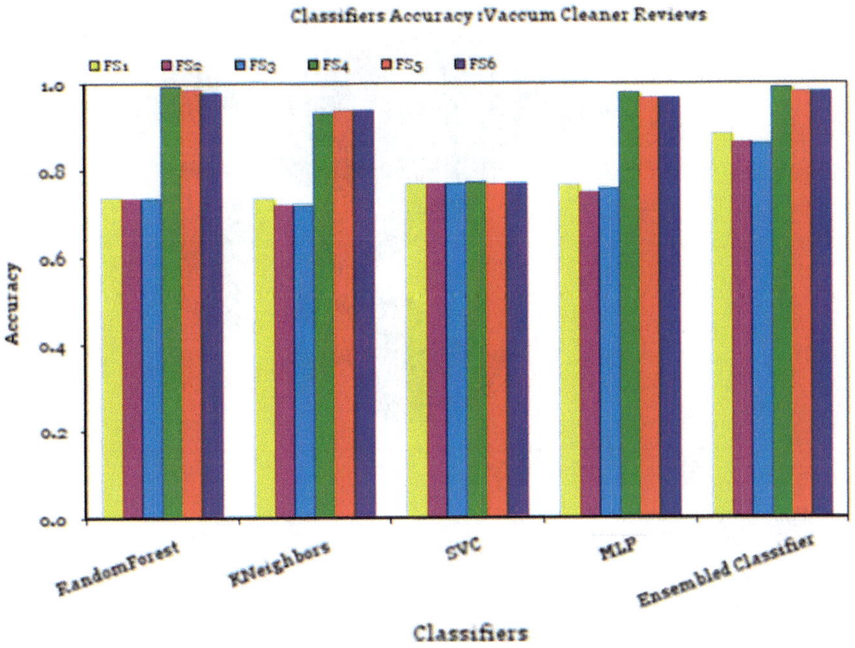

Fig. 4 Classifiers accuracy for amazon vacuum cleaner reviews

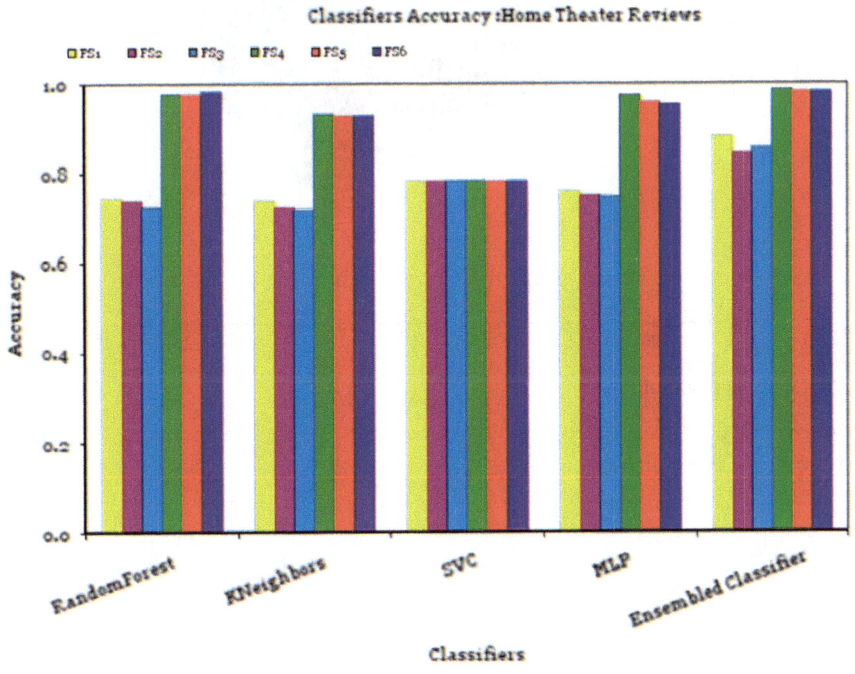

Fig. 5 Classifiers amazon home theater reviews

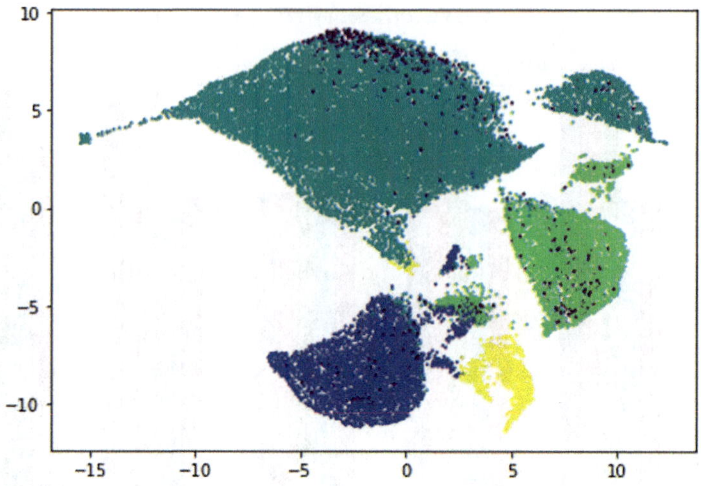

Fig. 6 Praises visualization

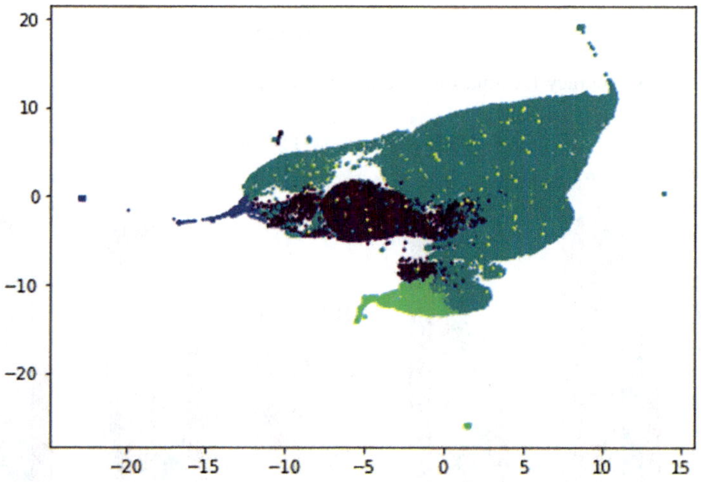

Fig. 7 Complaints visualization

Praise Topics using LDA model:

Topic #0: staff hotel good friendly breakfast helpful great really nice quite suite located extremely stay food lovely coffee enjoyed free station excellent extra choice shop location wonderful lobby home morning absolutely town included travel especially make thought fruit overall fresh metro

==

Topic #1: room clean comfortable nice small bathroom large pool area dont free feel spacious size quiet shower center modern able space lobby lot like little huge complimentary pretty Wi-Fi floor king issue deal amenity beautiful internet certainly people recently market appointed

==

Topic #2:room service stayed night desk time best check view review arrived long just hour think shuttle fantastic day early guest staff late year amazing went checked trip asked better airport pleasant light nicely booked checkin floor help came left weekend

==

Topic #3:street restaurant room door bed great walking noise experience view floor area sure didnt like distance elevator just nice little really nearby water look outside loved concierge window place wine drink actually reception open right plenty thing evening make hear

==

Topic #4:hotel stay location just great time definitely walk place close recommend away block night square price parking right minute city restaurant downtown staying need problem park perfect york easy rate want visit come area business subway work really highly shopping

The complaints are analyzed and visualized as shown in Fig. 7.

```
Complaint Topics using LDA model:
Topic #0:hotel good stay really location great place stayed breakfast lobby area staff restaurant want nice just price walk food coffee away better night sure right money friendly city close free recommend like pretty felt overall walking thing best getting doe
==================================================
Topic #1:desk hotel room service check called guest told staff time rate minute experience took hour wanted housekeeping internet night come arrived wait wouldn just long asked finally stay change ready line phone clerk stayed checked waiting drink half wasn reservation
==================================================
Topic #2:room hotel door said make people morning time booked night think desk just didn manager elevator review day came asked service outside checked staying told year open like going hear later customer card stay gave thought early problem couple tell
==================================================
Topic #3:room night went floor parking noise left business feel paid suite check wasn sleep view given street extra park probably free offered working valet time king just hotel stay luggage available space different charge double able instead fine upgrade like
==================================================
Topic #4:room bathroom clean didn small shower work nice pool floor like look wall bed window comfortable water looked looking carpet dirty property light large need week hotel towel thing tried quite little hard rude weekend sheet area noisy place staff
==================================================
```

The big customer review data can be analyzed and visualized for getting meaningful insights from managers point of view using visualization and reporting module.

4.5 Lexicon Based Approach for Aspect Level Classification of Praise and Complaint Sentences

Praise and Complaint sentences are further classified into six predefined aspect categories by manually analyzing the data. These categories represent the aspect frequently occurring in Hotel review dataset. The Hotel review dataset, following aspects and frequent recurring words associated with the aspect were defined (Table 7).

Aspect based Classifier:

1. Begin Procedure
2. Create list of aspects and frequently occurring keywords.
3. Repeat unless all praise and complaint sentences are processed
    a. Repeat for each of the praise and complaint sentence
        i. Repeat for all words in praise and complaint sentence
            1. Compare the word with list of aspect words from aspect list
            2. Count number of matching words per aspect and find respective aspect count for each type of aspect.
        ii. Find out which aspect have maximum count
        iii. Assign the sentence to the that aspect.
4. End Procedure

The above algorithm creates the labeled praise and complaint dataset with Aspect label for each sentence. The sentences where multiple aspect counts are encountered with two or more same aspect count are labeled as Multi aspect Sentences. Figure 8 shows the count of each type of aspect sentence in praise and complaint set.

It can be observed from the Fig. 4 there are more praises for staff, service, location and facilities aspect and more complaints for Room and Value. From Multi aspect sentences it can be observed that number of praises are more as compared to complaints.

Table 7 Manually identified aspects and their frequent words

Aspect	Aspect words
Staff	Good, polite, helpful, useful, friendly, coordination, quick, clear, rude
Room	Room, bed, waiting, time, dirty, bathroom, bad, fridge
Location	View, station, Airport, far, distance, close, train
Facilities	Pool, spa, Wi-Fi, gym, internet
Value	Value, price, amount, rate, cheap, worth, low, money, free, expensive
Service	Service, clean, cleanliness, facilities, desk, check -in, check-out, fast

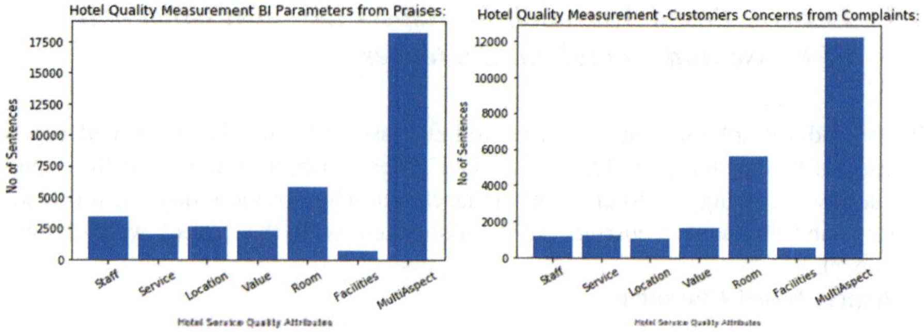

Fig. 8 Hotel quality measurement BI parameters for praise and complaint sentences

The LDA algorithm is applied on Aspect based sentences for identifying five topics about each aspect. Most frequently used aspect words per topic is shown to the user. Word cloud visualization for each type of aspect category for praise and complaint sentences is displayed from business intelligence point of view (Fig. 9).

It can be observed from the above Praise sentences word cloud that hotel provides fast clean service, value wise this is good deal, location is great and close to airport. Hotel provides nice facilities like pool, gym, spa, some of the services are free. Staff is friendly and nice.

After Analyzing Complaint sentences, it can be observed that Hotel rooms are expensive, service and facilities are good, location of hotel is far away.

5 Conclusions

In this paper, to deal with big review data, we have proposed the approach for filtering noninformative review sentences from the review corpus based on linguistic properties of extreme opinions (Praise/Complaint).

We have used properties of extreme opinions such as praise and complaint sentences and proposed hybrid features for praise and complaint classification. We have examined the performance of these features with supervised learning classifiers such as Random Forest, SVC, Kneighbours to identify extreme opinions like praises or complaints from Hotel Reviews and Amazon Product Reviews. The existing sentiment analysis techniques can be improved by only considering informative sentences in the sentiment analysis process. The linguistic features have high impact in increasing classification accuracy w.r.t extreme opinions.

The Ensembled Classifier and Random Forest classifier outperforms all other classifiers with (unigram+ hybrid features) by giving 99.6% Accuracy. The lexicon based approach for aspect level classification of praise or complaint sentences is presented. The Aspect level visualization of praise and complaint statements can be used for business intelligence.

Aspect	Aspect Level Praise Word Cloud	Aspect Level Complaint Word Cloud
Staff		
Room		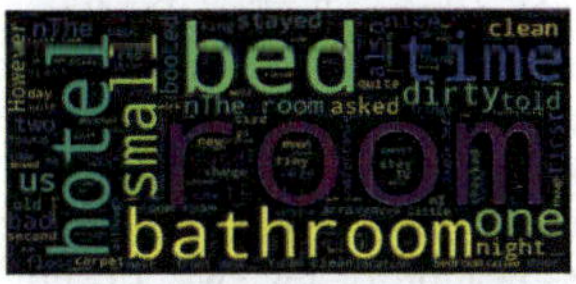
Location		
Facilities		
Value		
Service		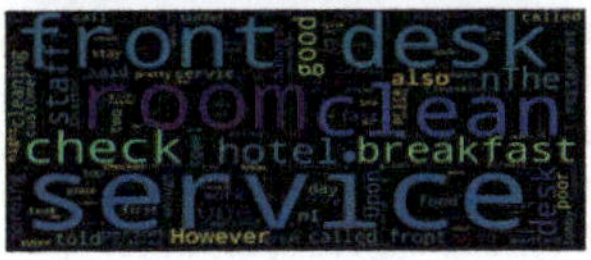

Fig. 9 Aspect level praise and complaint word cloud

In the future, we will evaluate other ensemble classifiers like decision trees, XGB classifier, AdaBoost classifier etc. We will evaluate the use of trained word embeddings and applicability of deep learning models in identifying extreme opinions from review corpus. We will also develop a methodology to analyse praises and complaints at more granular level for getting meaningful insights from business intelligence point of view.

References

1. Moschitti, A., Basili, R.: Complex linguistic features for text classification: a comprehensive study. Adv. Inf. Retr. 181–196 (2004)
2. Liu, Y., Jiang, C., Zhao, H.: Using contextual features and multi-view ensemble learning in product defect identification from online discussion forums. Decis. Support Syst. **105**, 1–12 (2018)
3. Alaei, A.R., Becken, S., Stantic, B.: Sentiment analysis in tourism: capitalizing on big data. J. Travel Res. 004728751774775 (2017)
4. Hu, N., Zhang, T., Gao, B., Bose, I.: What do hotel customers complain about? Text analysis using structural topic model. J. Tour. Manag. **72**, 417–426 (2019)
5. Liu, B.: Sentiment analysis and opinion mining. Synth. Lect. Hum. Lang. Technol. **5**, 1–167 (2012)
6. Turney, P.D.: Thumbs up or thumbs down? Semantic orientation applied to un-supervised classification of reviews. In: Proceedings of the 40th Annual Meeting on Association For Computational Linguistics, Philadelphia, PA, USA, pp. 417–424 (2002)
7. Pang, B., Lee, L.: Seeing stars: exploiting class relationships for sentiment categorization with respect to rating scales. In: Proceedings of the 43rd Annual Meeting on Association for Computational Linguistics, pp. 115–124 (2005)
8. Ganesan, K., Zhou, G.: Linguistic understanding of complaints and praises in user reviews. In: Proceedings of NAACL-HLT (2016)
9. Pang, B., Lee, L.: Opinion mining and sentiment analysis. Found. Trends Inf. Retr. **2**, 1–135 (2018)
10. Tripathy, A., Agrawal, A., Rath, S.K.: Classification of sentimental reviews using machine learning techniques. Proc. Comput. Sci. **57**, 821–829 (2015)
11. Lacy, E., Kowald, D., Lex, E.: High Enough?: Explaining and predicting traveler satisfaction using airline reviews. In: Proceedings of the 27th ACM Conference on Hypertext and Social Media, pp. 249–254 (2016)
12. Abrahams, A.S., Fan, W., Wang, G.A., Zhang, Z.J., Jiao, J.: An integrated text analytic framework for product defect discovery. Prod. Oper. Manag. 975–990 (2015)
13. Zhao, Y., Xu, X., Wang, M.: c. Predicting overall customer satisfaction: big data evidence from hotel online textual reviews. Int. J. Hosp. Manag. (2018)
14. Saumya, S., Singh, J.P., Baabdullah, A.M., Rana, N.P., Dwivedi, Y.K.: Ranking online consumer reviews. J. Electron. Commer. Res. Appl. (2018)
15. Vishal, A., Kharde, Prof, Sonawane, Sheetal: Sentiment analysis of twitter data: a survey of techniques. Int J. Comput. Appl. **139**(11), 5–15 (2016)
16. Krishnamoorthy, S.: Linguistic features for review helpfulness prediction. Expert Syst. Appl. **42**(7), 3751–3759 (2015)
17. Reddy, C.S.C., Uday Kumar, K., Dheeraj Keshav, J., Prasad, B.R., Agarwal, S.: Prediction of star ratings from online reviews. In: 2017 IEEE Region 10 Conference TENCON, pp. 1857–1861 (2017)
18. Almatarneh, S., Gamallo, P.: Linguistic features to identify extreme opinions: an empirical study. In: Lecture Notes in Computer Science, pp. 215–223 (2018)
19. Almatarneh, S., Gamallo, P.: A lexicon-based method to search for extreme opinions. PLoS ONE **13**(5), e0197816 (2018)

Big Data & Disruptive Computing Platforms Braced Internet of Things: Facets & Trends

Naveenkumar Jayakumar and Dhanashri P. Joshi

Abstract The Internet of Things (IoT) has been evolving in tandem with centralized cloud computing to fog computing, edge computing, Semantic computing, etc. Pervasive IoT applications like Healthcare applications generate a huge amount of sensor data and imaging data that needs to be handled rightly for further processing. In traditional IoT ecosystem, Cloud computing ensures solution for efficient management of huge data with its ability to access shared resources and provide common infrastructure in a ubiquitous manner. Majority of the IoT applications are highly sensitive to time and necessities latency bounded execution. There is time delay introduced when the data transmission occurs between the cloud and the application which is unacceptable. This chapter will disclose the various facets and trends in the integration/coupling of evolving computing platforms and disruptive technologies like Fog/Edge Computing, Big Data, blockchain with IoT that overcomes challenges existing in the traditional deployment of IoT environment. Further, there are several issues with IoT and Cloud Computing framework like each individual part of the IoT architecture could act as a point of failure that can interrupt the whole network and secondly, the centralized cloud model is vulnerable to loss of integrity of data. This chapter will review and reveal the trends in bracing computing approaches, managing the huge data using datacenters in Internet of Things ecosystem. The review is advocated with a case study on waste management system applied with the fog/edge computing and cloud computing. Further, various trends in applicability of block chain in the IoT ecosystem is also reviewed. This chapter discusses the fundamental facets of various computing paradigms and approaches that may help to address issues of big data through creating IoT ecosystems.

N. Jayakumar (✉)
Department of Computer Engineering, Bharati Vidyapeeth (Deemed to be University), College of Engineering, Pune, India
e-mail: naveenkumar@bvucoep.edu.in

D. P. Joshi
Research Scholar, Department of Computer Engineering, Bharati Vidyapeeth (Deemed to be University), College of Engineering, Pune, India
e-mail: dhanashrijoshi28@gmail.com

N. Dey et al. (eds.), *Internet of Things, Smart Computing and Technology: A Roadmap Ahead*, Studies in Systems, Decision and Control 266,
https://doi.org/10.1007/978-3-030-39047-1_6

Keywords Internet of Things (IoT) · Cloud computing · Block chain · Big data · Machine learning · Coupling

1 Introduction

IoT is viewed as a global infrastructure foundation for the information society, enabling benefits through services by interconnecting things dependent on existing and developing interoperable IT resources and technological advances. The IoT infrastructure is incessantly growing with the scaling and spanning of number of things (devices) used by end-users [1].

In present scenario Internet of Things facilitates in solving many real-world problems. IoT applications range from consumer IoT and enterprise IoT to manufacturing and industrial IoT (IIoT). Nowadays many traditional application areas are transforming from their conventional ways of implementation to smarter ways.

In developed and developing countries many IoT solutions are designed and deployed that provides smart solutions in the field of agriculture like field irrigation system, field monitoring, pesticides sprinkling (By using Drones) etc. In many countries' cities are transforming from Metro Cities to Smart Cities like Pune, Bhopal, Kochi in India etc. Under smart city initiatives taken by Government of India cities are becoming smart with initiatives like Smart Waste management Sys-tem, Smart Transportation System and so on. IoT has evolved in big way through E-Governance [2] and citizen's services, Smart waste management, Smart Water Management, Smart Energy Management, Smart Parking Management and Smart Parking System, Tele Medicine System, Incubation Centres etc. Initiatives are getting shape under various Smart City Schemes [3]. In Ahmedabad, Ahmedabad Municipal Transport Services has built up a transportation model using dockless bikes for Bus Rapid Transit System (BRTS) [4].

There are many IoT solutions that are solving real world problems as cited above. As the technology gets exploited and trends change the technology building, new issues and challenges emerge in existing solution leading to a real-world problem that needs to be addressed. Some of the challenges are [5]:

1. Tininess of Devices
2. Facilitating Longer Battery Life for IoT Devices
3. Support for Infrastructural Standards
4. Getting Robust Privacy and Security Within the system
5. Creating Lightweight Protocols Without negotiating Reliability and efficiency
6. Dealing with the issues Introduced by Big Data generated from sensors
7. Support for all through Service etc.

To overcome these challenges many researchers are undertaking research in the field of IoT and integrating it with other computing paradigms. Some of the recent trends in research of IoT are discussed in next section.

2 Big Data and IoT

IoT Infrastructure and data

Focus in this area of research is to design components of the IoT system, scalable networks infrastructure for the IoT, to achieve seamless and reliable data communication, routing, security, and quality of service (QoS) within the IoT. For efficient data management, processing and to reduce infrastructural requirement of large IoT applications, IoT infrastructure is integrated with Cloud computing and Fog computing, the elaboration on the same can be seen in Sects. 3 and 4 respectively. Datacenters provide facilities of huge data processing as well as cloud-based services in terms of software recourses by using virtualization. To deal with large amount of data which is generated in IoT system Datacenters are most important component of IoT ecosystem. Use of datacentres is further explained in Sect. 5.

Security

Kaspersky Lab ICS CERT identified 63 vulnerabilities in industrial and IIoT/IoT systems in 2017 like shared responsibilities, SSL Vulnerabilities etc. [6]. The main area of focus is acquiring real world data in a secure and reliable way. To overcome these security threats, research on integration of IoT and blockchain technologies are explored. Blockchain is a kind of distributed ledger technology or a data structure that makes it possible to create a tamper-proof digital ledger of transactions and share them. Section 6 of this chapter will explain the applicability of blockchain in IoT ecosystem.

Machine Learning and Big Data:

For the application areas where sensing and generation of data, collection of data, data storage and management are requirement, big data analytics is majorly used. Machine learning could be integrated with IoT where intelligent applications are required. Active research is done in this area as well [7]. Section 5 of this chapter will explore how Big data is facilitating IoT to handle huge amount of proliferating data in the digital ecosystem.

3 IoT Braced with Cloud Computing

It is visible from various researches that IoT is evolving in tandem with centralized cloud computing to fog computing, edge computing etc. The data is abruptly proliferating and requires an efficient way of processing and managing the data. In traditional approach of IoT, Cloud computing ensure solution for efficient management of such data with its ability to access shared resources and common infrastructure in a ubiquitous manner.

The sensation of the IoT ecosystem requires ubiquitous, reliable, efficient and scalable service provision. In order to achieve this, the approach to be followed is to integrate the IoT and Cloud computing, i.e., facilitates an "Everything as a Service". Innovative IoT centric Cloud infrastructure is evolved to address challenges of IoT and Cloud computing individually or as an integrated system.

Cloud computing is dependent on sharing of resources, which is a pivotal demand for IoT setup. The Cloud Computing system not only shares the resources but also increases the availability of resources. It is location independent; the users can access the services of cloud from any locality and with any device or infrastructure through the internet.

Cloud computing complement IoT in many aspects as follows:

- Cloud infrastructure in integration with IoT provides security to the data generated or collected by IoT devices by using effective authentication protocols.
- Many small initiatives taken to develop IoT system and remote monitoring are now not dependent on large infrastructure like servers and networking because of use of cloud infrastructure.

The virtualization of devices is important characteristic of cloud computing, virtualization facilitate end users to easily share the physical devices. Because of the virtue of design, cloud computing is also more standardized. In cloud computing multiple instances of same application or service can be used for different users. This enables sharing of resources to multiple users over location and for the allocated definite period. Cloud infrastructure offers flexibility, scalability of resources and services. The service and resources are easily accessible and available. Hence the integration of IoT and Cloud can offer opening for both technologies to overcome their challenges [5, 8].

Most of the data generated by IoT devices or sensors is unstructured and of massive amount. IoT-Cloud Computing integration needs to facilitate real-time data processing and service provision in view of such growing data. Another challenge is to provide dynamic resources management and coordination technique for dynamically unloading of data from device to cloud and vice versa. Providing distributed storage and processing of such huge amount of data is another issue with this system. In addition to the above mentioned, there are other issues which are (a) Virtualization of IoT devices- Access to specialized resources or services, including actuators, sensors, etc. (b) Portability of the services (c) Real-time and reliable correspondence from devices to applications.

The Internet of Things (IoT) and cloud integration has created many opportunities but also has its constraint due to centralized compute, storage, and networking infrastructure in a very few datacentres, and due to the relatively more distance between remotely placed datacentres and the devices. Conventional cloud based IoT infrastructure can cause more latency than many real time applications can endure. Fog or edge computing will help in tackling this issue.

4 Challenges and Trends in IoT Braced with Fog/Edge Computing

The IoT braced with cloud computing is moving rapidly towards creating a holistic digital ecosystem which will have potential to connect billions of edge devices which could generate very large quantity of data at a high speed and real time applications may experience low latency. As mentioned in the previous section the conventional cloud based IoT infrastructure can cause more latency issue than many real time applications can tolerate. Fog or edge computing can tackle this challenge by taking network and compute nearer to the network edge or users, thus diminishing the need of data to get transferred to cloud which results in decreased latency and increases communication and computational efficiency. Such system can be empowered by advanced technologies such as Network Function Virtualization (NFV) and Software Defined Networking (SDN) [9, 10].

Generally, in Fog Computing service providers like Internet Service Providers (ISP), infrastructure providers, cloud-service providers etc. provide fog services via customer premises equipment (CPE) designed specifically for the provider. This differentiates Fog or Edge Infrastructure from Public cloud infrastructure where Service provider may provide generic platform that can be used with many applications [9].

The IoT is obliged by devices such as sensors, wearable devices and smartphones connected to the Internet. A few peculiar highlights of the IoT system make it appropriate for applications dependent on edge computing platforms. The characteristics facilitate by edge or fog computing are as follows [11, 12]:

- Low-latency communication: It is critical for Real time applications where real time data should get processed in very small amount of time and user requires result immediately. E.g. Health monitoring applications, military applications etc.
- Geographical distribution: It is characteristic of Edge Computing, where Edge devices could be able to process data locally.
- Bandwidth Intensive: In the application where data uploading needs more bandwidth can process some data locally. E.g. in real time videos used in sensitive situations can be processed locally and can provide immediate results to police nearby.
- Device Mobility: This gives additional requirement on low latency processing of data. Edge computing facilitates relocation of virtual resources based on the mobility of nodes and processing of data generated by such mobile devices locally with good quality.

With open edge fog infrastructure, characteristic of low latency can be addressed by facilitating local computing, storage, and networking resources to aid the often asset poor IoT devices. In latency sensitive applications, the delay occurred due to transferring data to the cloud and back to the application is not tolerable because of time consumption. It is not efficient to upload large data to the cloud for storage and processing, as it would use network bandwidth and it is slow scalable. To address these challenges, Fog computing was proposed to use computing resources for local storage and data pre-processing. This would decrease network congestion, as well as speed up analysis and decision making. However, because of the limited resources edge devices can't handle multiple IoT applications at a time, which results in resource conflict and increases processing latency. Fog computing which without interrupt integrates edge devices and cloud resources to overcome the above discussed limitations.

The Internet of Things (IoT) connects billions of smart things to the Internet, which can deliver a bright future to initiatives like smart cities or smart homes. These things or devices are creating very huge amounts of data and send that data to the cloud for processing, and to discover knowledge from it. This leads to unnecessary engagement of resources (e.g. network etc.). The Fog or Edge computing scenario is developed to counterpart this limitation by using data analytics to the edge devices. Yet edge devices have limited computational capabilities. Due to these advantages and limitations, neither Cloud computing nor do Edge compute paradigm can overcome these limitations individually. Therefore, both need to get integrated to create a sustainable IoT infrastructure [13]. With that sustainable IoT infrastructure there are mainly two issues, one is security of data and second is handling large amount of data flowing through network and system. In security aspect, a main limitation of IoT platforms is IoT infrastructure relies on a centralized cloud.

4.1 Trends in Architectural Designs

Two architectural designs are proposed in [9, 14] as FocusStack and IndieFog respectively. FocusStack can use various groups of IoT edge devices to deploy diverse applications. These devices (mostly mobile) have limited computational capacity, energy requirement, and connectivity. FocusStack provides opportunity to developer to give attention on application rather than deployment of devices. There are two architectural components of FocusStack namely The Location-based Situational Awareness (LSA) subsystem and OpenStack Extension (OSE) subsystem. LSA provides needed information to OSE when system focuses on geographical region. OSE allows management and deployment of small edge computing devices with limited networking capabilities. Fog Vehicular Computing uses similar concept with the use of Openstack in implementation [15].

Unlike FocusStack, IndieFog is used for deployment of stationary devices or sensors for application that requires data processing in short period of time and mostly in urban area. Vehicular IndieFog is IndieFog architecture that is not restricted to stationary devices. In urban areas, vehicles could also be used as nodes. This could help the Inter-net of Vehicles. Vehicular nodes could collect and process data while in moving state and enabling continual data streams on environment changes. In Smartphone IndieFog devices could not be able to create a stable cluster computing environment as smartphones are moving with user and every time connected to different base station.

In previous section of this chapter, it was discussed that Cloud platforms are not applicable for applications involving real-time operation or low latency, and high quality of service (QoS). Recently, mobile-edge computing (MEC) is evolved to address the mentioned constraints. MEC is a unique infrastructure that extends cloud-computing services to the edge cellular network. Due to dense geographical distribution, closeness of network to consumers, high mobility support, and open platform, MEC can be used in applications and services with low latency and high QoS. MEC is network infrastructure designed for implementing at the cellular base stations or other edge devices. It enables flexible deployment of services and applications [16].

4.2 *IoT Reference Layered Model*

The above discussed architectural designs adapt to the Fig. 1 shows IoT reference layered model that has seven layers. These seven layers encompass all the elements into it as per their working and needs. The data collected or generated by the IoT devices at high speed can be stored and processed by the local edge infrastructure and only a processed data which is now in small chunks are sent back to data-centers. Characteristic of Geographical distribution can be achieved by offloading this processing responsibility of edge device to the edge servers if edge devices are overloaded with data processing [10].

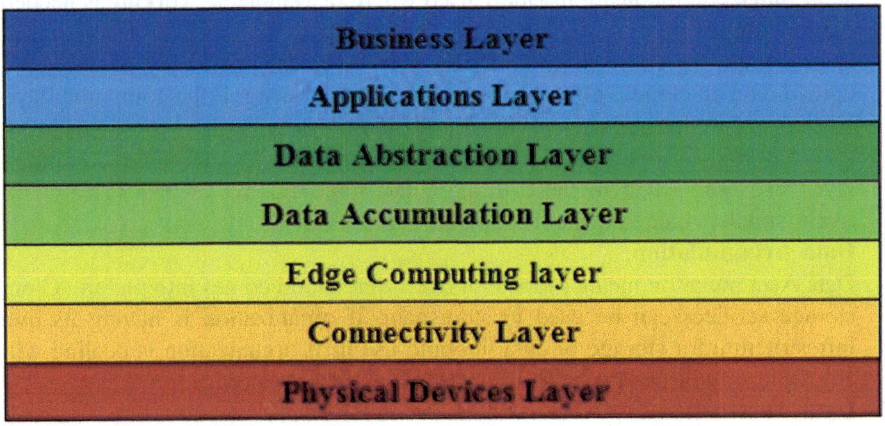

Fig. 1 IoT reference layer model [17, 18]

It additionally encourages the management of computing, networking, and storage services between datacenters and devices. In addition, it supports user mobility, distributed data analytics to address the prerequisite of different applications that requires low latency.

Sensors send data to IoT server, applications on fog devices access data and process it, and the output of data processing are converted into signals sent to actuators. Devices on the layer of resource-management use information of each resource from the resource monitoring device or service to record the state of available cloud computing and fog computing resources and identifies the best resource to process the task. Fog Computing and Cloud Computing resources communicate to each other through machine-to-machine (M2M) standards such as MQTT (formerly MQ Telemetry Transport) Protocol and the Constrained Application Protocol (CoAP) Protocol. Software Defined Networking (SDN) helps the system with the management of fog networks by centralized control and separate data and control flow [19, 20].

The layers are shown below in the Fig. 1.

They are as follows:

1 **Physical Devices**:
Physical devices consist of basically sensors and actuators. Sensors can be able to sense or collect data and actuators gives response.
e.g. Humidity sensor, Temperature Sensor etc. There are different protocols working to connect physical devices like LoRaWAN

2 **Connectivity**:
Data collected through sensors are sending to edge devices on third layer by using various network protocol and devices. RFID, Bluetooth, Zigbee are some of the examples of protocol used for communication amongst IoT system [21]. ZigBee smart energy is used for large IoT applications like Industrial IoT, Smart Waste Management, and Smart Home. Bluetooth low energy protocol used for short range communication widely used where in-vehicle networking is needed.

3 **Edge Computing**:
Edge computing devices are generally used for local processing of data or to upload data on cloud platform for storage and processing. Edge computing layer is used in low latency or real time applications where system could not afford time required for communication between cloud and devices. Edge computing layer also used when we need to aggregate huge amount of data to comparatively smaller scale and then to upload it to server.

4 **Data Accumulation**:
Data Accumulation means storage of data. Here cloud comes into picture. Cloud storage services can be used to store data. If organization is having its own infrastructure for storage of data, it could use it. If organization is dealing with Big Data, it will use distributed data centers for data storage.

5 **Data Abstraction**:
Data abstraction means analysis and analytics of data. Here data could get analyze by using different methods such as machine learning or artificial

Intelligence. Some decisions could be taken depend on this analysis of data. Processed data is represented in the form of Graphs, charts etc. Machine Learning or Big Data Analytics frameworks and algorithms could assist decision support system on business layer.

6 **Application**:
Here application working on user side or server side is present. Application represents processed data. Application can be used to interact with this IoT system.

7 **Business/Collaboration**:
On this layer business for which application is developed is rely. Here the business authorities can take decisions based on collected and analyzed data for future actions.

Let's see example of Smart agriculture system to see how data communication is done in all layers of IoT layered model. Physical Layer contains devices such as sensors, actuators or anything that need to connect to internet. Physical devices collect raw data depending on external factors in the real world. In case of Smart Agriculture system physical device layer would contain environmental sensors like humidity, water level sensor etc. These sensors would collect information about environment around crop and some information about soil [22]. Connectivity layer is used for communication of physical devices to network. Bluetooth or Zigbee can be used at connectivity layer to send raw data from underlying layer to edge devices. Edge computing layer consist of edge devices which would be used for processing data locally. In agriculture-based system if we required that watering system should start automatically based on soil moisture level then our application doesn't require all values collected by sensors; it will create data traffic on network. Edge device could process data collected and if edge device finds out soil moisture below threshold value then edge device would send that data to above layer. Data accumulation layer consist of technology used for storage of data. This could be Data centers or cloud storage as a service. In smart agriculture system data collected from various places could be used afterword for analytics and decision making. So, it might be able to store by using cloud storage server. Data abstraction layer is used for analytics of data which can further be useful for Decision Support System (DSS). In agriculture-based application, system could contain module which could be able to solve queries about diseases of crops submitted by farmers. These queries could be solved by all information available on server about crops and environmental conditions as well as data collected from sensors (Environmental conditions, disease symptoms, images of contaminated crops etc.). Data abstraction layer can work on all this data. Next layer is application layer. Application layer contain actual application that would be interacting with end user as well as having actual processing capacity. In agriculture-based system application is able to take some decisions like to start watering system or to give alert to farmer regarding intrusion in farm. Application could be useful for query solving of farmers. On application layer protocol like MQTT or COAP is used to communicate. Business layer is applicable for making business decision. System could suggest farmers regarding how to overcome disease of his/her crop. We can see how all layers are working together for Smart systems.

IoT connects millions of devices or things to the internet. These devices gather or create massive data and transmit that data for further processing to cloud via network. This process results into redundant usage of network resources. Edge computing is a infrastructure where this limitation of cloud is complemented. In edge computing, edge devices first process the data locally and then processed data is transmit via network. But edge devices have limited computing capacity. To overcome these limitations in Fog and Cloud computing and to create sustainable IoT framework both cloud and edge computing need to assimilate [13].

The coupling of computing platforms with IoT devices are made possible through the above discussed layers of IoT ecosystem. In next section, as a case study waste management system is considered to analyze how the design and deployment of system has adapted to the above discussed seven layers and have enabled IoT coupled with Fog/edge and cloud computing.

4.3 Case Study: Waste Management System

Handling Garbage has always been one of the main problems anywhere around the world. As garbage tends to accumulate more and more, the management of the garbage becomes more tedious tasks. Garbage sources are spanned across the city and to collect that garbage across the city, dust bins are kept at every starting and end of the streets. These dust bins needed to be emptied in a week's time or as per the scheduled period. The garbage trucks need to collect the garbage from these dust bins kept at multiple corners of various streets and dump the total wastes into the dump yard. This manual process costs huge. The fuel is wasted in just travelling between the corners of streets and dump yards unnecessarily even when the dustbins are empty or half empty. What if the above said process is automated and the cost is reduced to handle the overall management of wastes. Waste management process includes set of tasks like collection, transportation, processing, disposal, managing and monitoring of wastes. If all these processes are automated and optimized the money and time could be saved that can be utilized to overcome other issues. Automation of the process can be done using the IoT platform. Earlier the waste management system was deployed using the IoT and cloud [23]. The sensors were installed on the dustbins; the dustbins sensors were connected to small powered mobile kits which get connected to cloud through the data card/Sim card on the mobile kit. Similarly, these sensors enabled kits with the power of connecting to internet were installed on garbage trucks, recycling devices, dumping area. The data were collected on the cloud, and the decisions were taken on how to collect, when to collect (scheduling of trucks), where to dispose the trash etc. This whole process was automated using IoT and Cloud creating the smart garbage/ waste management system but still there are some demerits to its deployment. There is still scope of improving the efficiency of end to end system. Since IoT is the cluster of billion devices and smart entities connected to network of networks, it is expected that the data generated would be of huge volume creating a big data scenario. In case of waste management, huge quality of data might be generated from each smart sensor units installed on bins, trucks, dump area. These massive data would be transmitted to the cloud for processing. Decision making is done from the information gained as a result of processing. This process reduces the potential usage of the smart devices since, all the raw data is sent to cloud out of which some never gets processed by the cloud and those data that are not useful is regarded as noise data. These noise data are less useful or not at all useful. Secondly, the data transfer from these devices to cloud will consume time thereby increasing the latency and impacting the performance of end to end system. Fog and edge computing are proposed to overcome the demerits of the earlier scenario. Fog and Edge computing are not a replacement of cloud, but they will be deployed in tandem with each other. The fog and edge computing are deployed to reduce the time taken to make the decisions and to reduce the noise data being transferred, thereby reducing the size of data to be transferred to cloud for the processing. Some data from edge devices will also be used to execute small tasks by edge devices itself. The following section will discuss the modelling of waste management systems using edge, Fog and cloud computing that can be deployed in smart cities.

4.3.1 Waste Management System Model

Figure 2 illustrates various components of the smart waste management system that can be deployed. In smart cities, there are several groups or entities interested to work in waste management domain. For example, a city municipal corporation may use sensor data to develop future garbage collection and management plan, so they can optimize path of garbage truck. In addition, recycling companies can use data collected from sensors to predict the amount of garbage coming into their plants for recycling or processing to optimize processes. These processes are executed by the edge devices rather than sending the data to cloud.

Figure 2 illustrates various components of the smart waste management system that can be deployed. Sensor units on dustbin or in Garbage collection Vehicles or Trucks are used to collect information by using sensors. Processors connected to these units can be worked as Fog devices which will be able to process some data locally and can communicate to cloud via gateways. These gateways could be network devices and connectivity. MQTT Broker which is working on cloud platform will publish data processed or generated at publisher (sensor units and Fog devices) and subscriber (Server which wants to process it) would be subscribing it via broker.

In order to develop smart waste management system, different sensors and small processing kits are required to be set at different places such as garbage bins and garbage collection vehicle. These sensors used to collect data like amount of garbage, types of garbage, and level of garbage collection bin etc. At Garbage bins or at Garbage collection Vehicle this data gets processed locally to avoid overhead of

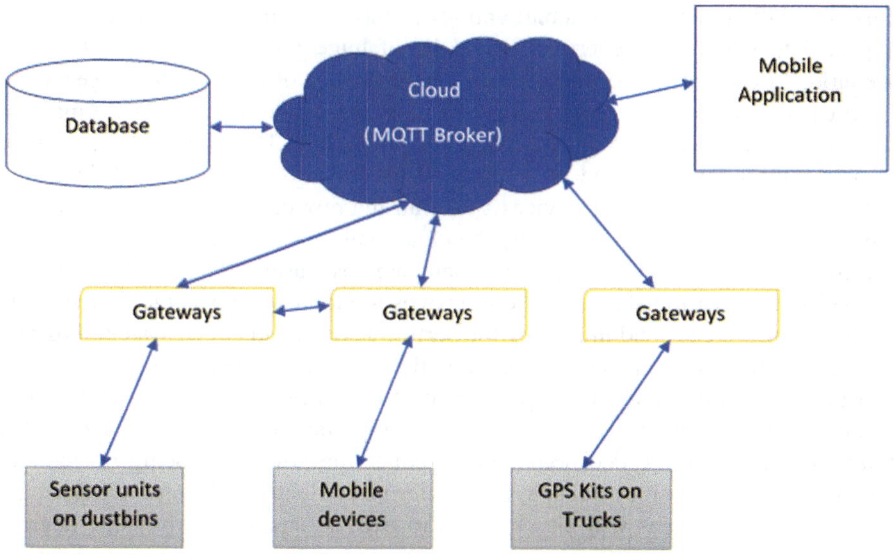

Fig. 2 Waste management system model

data on cloud. As sensor will take readings time to time there will be large amount of data generated and which is not useful till Garbage bin or vehicle is full. This will minimize latency.

Physical and Data Link Layer Protocols like LoRaWAN or Zigbee can be used for communication between sensors and gateway. Gateways are fundamentally fog devices. This can be used to collect and send sensor data to the cloud through gateway for processing and storage. Energy efficient and low powered sensors can be used, and data can be offloaded to the cloud with the help of wireless network infrastructure. In addition, when communication technology needs to be long range and it is unavailable, data can be uploaded to the cloud with the help of devices attached to the vehicles, like garbage collection vehicles, intra city buses that operate in the various areas of city i.e. all edge devices that are available.

Here Cloud can be used as MQTT Broker for mobile applications of vehicle driver, Municipal Corporation Web application and analytical server. In this system, garbage bins are fitted with sensors and garbage collection vehicle is acting as a gateway. In addition to this we can use big data Analytics to analyze data collected from sensor so that municipal corporation can predict garbage generation in near future and can plan to act on it for processing and recycling. The detailed architecture of the waste management is shown below in the Fig. 3. It shows the working model of the above discussed elements in a layered framework.

As discussed in above sections the computing spectrum can be integrated for evolving an innovative and efficient IoT based waste management system. Different protocols can be used on different layers of IoT and Fog infrastructure plays a vital role in waste management system. To decide which protocol are suitable for the

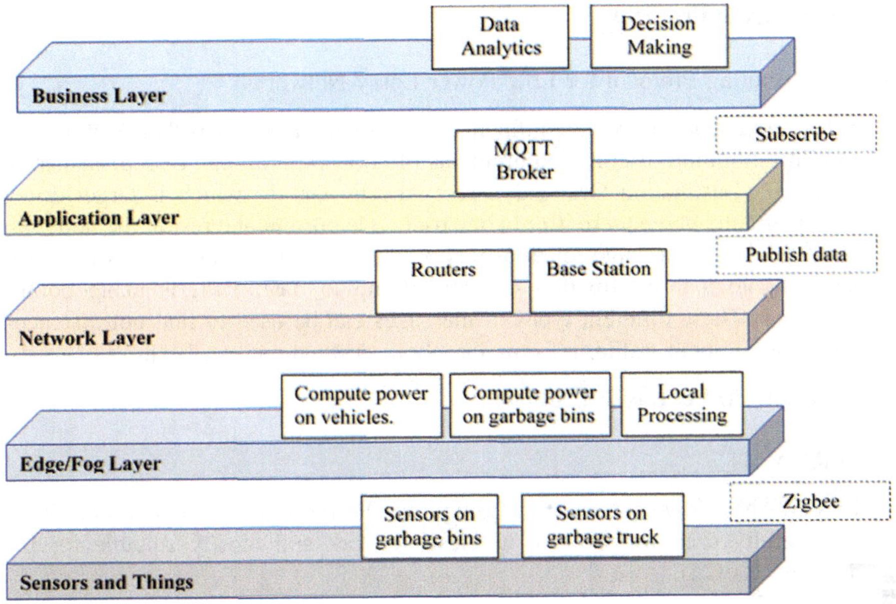

Fig. 3 Waste management system layered model

system, it is required to consider and analyze constraints like devices used, net-work bandwidth, where the devices are deployed, application deployment platform etc.

Importance of Protocols

Application layer protocols:

- MQTT (Message Queue Telemetry Transport)

  MQTT [24] is a TCP server-client architecture-based publisher-subscriber transport protocol on used for the communication between embedded applications and middleware. Protocol providing flexibility and easy deployment required for IoT application by using different routing mechanisms. MQTT has a fixed header of length 2-byte suitable for devices used in IoT system having low bandwidth and low battery power. In MQTT 2 devices can communicate with each other as publisher and subscriber with the help of broker as a middle agent.

- COAP (Constrained Application Protocol) [24]

  REST (Representational State Transfer) is transport layer protocol generally suitable for IoT applications where devices are with low bandwidth, low power and mobile. REST can be used over functionalities of HTTP. Instead of using REST if we linked COAP with the UDP, COAP is becoming lighter protocol and suitable for IoT applications. COAP can adapt some functionalities of HTTP.

Network Layer Protocol:

- RPL (Routing Protocol for Low Power Lossy Networks)

  When in network links have influence of noise and network is then RPL is used to fulfill minimum routing requirements of network. RPL can be represented by DODAG (Destination Oriented Direct Acyclic Graph) which is target driven acyclic graph, where there is only one root node present and rest of the nodes are child. Children can have relationships of parent child to each other. In order to keep topological information of network up to date, RPL is using control messages. These different types of messages can be used to find out distances, information about traffic, information about destination, etc. [24].

Physical and MAC Layer Protocol:

- LoRAWAN

  LoRaWAN is low cost, secured Low power WAN. It also supports mobility. It is generally used for scalable wireless network and mostly suitable for IoT applications [25].

- Zigbee Smart Energy

  ZigBee [21] smart energy is used for large IoT applications like Industrial IoT, Smart Waste Management, and Smart Home. It uses star, peer-to-peer, or cluster-tree network topologies. A coordinator in the network controls the routing in the network. It can be central node in a star topology, the root in a tree or cluster topology or may be any node in peer-to-peer communication.

- Bluetooth

  Bluetooth low energy protocol used for short range communication widely used where in-vehicle networking is needed. Its energy requirement is 10 times lower and latency is 15 times lower than normal Bluetooth. It has master-slave architecture as conventional Bluetooth. Nodes are awake only when they are communicating with another node otherwise, they will be on sleep mode to optimized usage of power.

Similarly, other applications are also deployed realizing the above layered reference model. The applications are discussed below in the following section.

4.3.2 Proposed Framework Model for Waste Management Model

The above discussed waste management model is based on the reference layered architecture of IoT which can be further improved and deployed in an efficient way using the trending technology in this field. The issues that exists using the layered reference model is of handling data generated from various sensors and moving that data from source to cloud for processing through fog entities. As discussed in the earlier section, sensors are installed on the dustbins which will send data to the fog entities, these dustbins will be in many numbers and the data generated would be sent from all these locations in stream form to the edge entities or to cloud for processing. Similarly, the same data streams can be used for garbage truck movements between the dump yard and various garbage bin locations. The proposed framework model uses vehicular fog computing along with volunteer edge computing and path computing.

The vehicular fog computing encompasses layers that supports waste management applications and its services followed by policy management layer and perception layer as shown in Fig. 4. The first layer i.e. application and service layer facilitate the real time execution of applications and host various services. The policy management layer is responsible for managing the resources and using the various resources in the fog/edge network and execution of the tasks right from allocating it. The perception layer manages hardware layer i.e. managing the devices, interfacing the devices and securing the devices. When all three layers are made to work together by installing them on a garbage truck, the truck is used for processing the data as well for communication between the various edge devices like smart garbage/waste bins. This vehicular fog computing techniques will help in creating a cluster of vehicles along with cluster of smart waste bins using the geo location of the waste bins. These clusters will enable to plan the movement of the garbage trucks on the specific route for set of smart garbage bins in each locality. This will also help in covering a major locality as well pre-plan the trip of the trucks to move and collect the garbage from specific garbage bins which are in the cluster. These algorithms can be created and deployed in the policy management layer of vehicular fog computing. Further, there can be a logic embedded with each truck for receiving the data about the emptiness and fullness status of the garbage bins based on which the decisions can be taken to re-route the path of travelling to collect the garbage from the specific garbage bins. If a truck is busy in other location and if the cluster has a spare truck, the task of the collection of garbage can be offload to the spare truck whereby achieving the load balancing in real time.

Since, the vehicles have battery power, devices can be embedded with storage and small computation unit which can execute data intensive applications. Since the trucks are moving, they stay close to the sources of garbage data that originates from garbage bins. The distance of the source and processing are reduced here, and the decision is taken on the edge device itself. The cluster of vehicles (garbage trucks) and garbage bins management shows the data sources and compute sources along with their location which leads to location aware processing and storing of data. This is volunteering edge computing whereby using an application offloading framework, the tasks can be shared or offloaded to the data source device for processing of the data on the source (edge device) itself.

Path computing is a way where the datacenters or cloud computing paradigm can be connected to the edge/fog network spanned throughout the breadth and length of the network. For instances, these cloud computing infrastructure or mini data centers can be installed on local garbage management municipal offices, dump yards etc. The offloading of data and applications can be done to these set of

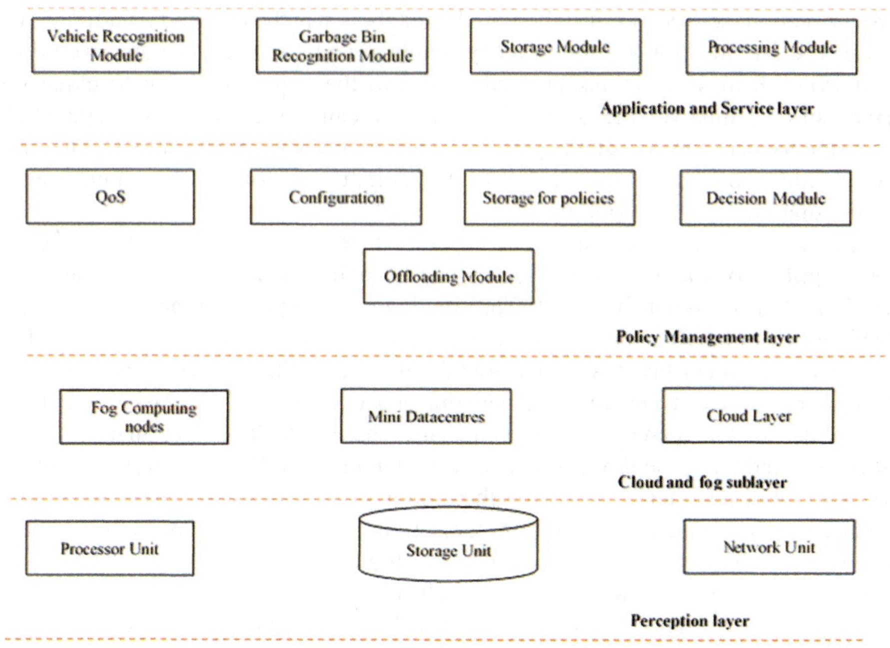

Fig. 4 Waste management system layered model using vehicular fog computing

scattered data centers and cloud infrastructure though the edge device for processing and decision making on moving the trucks and emptying the garbage bins.

All the above three framework model can be deployed for garbage management system and which further improve the quality and efficiency of the overall system.

4.4 Case Study: Smart Healthcare System

From last few years, conventional healthcare system is transformed to technologically advance healthcare system very rapidly. In healthcare domain historical data of patients and experiences of doctor plays important role during diagnostic and therapy as well. It is very difficult task for healthcare professionals to observe patients continuously by using manual processes. Smart healthcare management system can help doctors and professionals in all these aspects of healthcare system.

IoT supported healthcare systems can connect things anywhere and at any time. Sensors attached to patient's body can gather huge amount of data or useful information about that person's health. Such collected information from multiple sensors from multiple persons can affect healthcare domain positively. With such huge data availability with auto automation and system intelligence, healthcare system can (i) reduce costing of healthcare management and treatment (ii) provide assistance to professional in the field of medicines (iii) personalize treatment. This all would results into improved performance of smart system [26].

Personalized healthcare is major part of smart healthcare management system. In personalized healthcare sensors could be connected to person's body and they would sense and collect data related to his/her health like heartbeat rate, pulse rate, oxygen level etc. There are two types of applications present in personalized healthcare according to its use.

Remote monitoring is one type in which person is not present at clinic or hospital, but health professionals can access data collected by sensors and processed by system and can give recommendations to user. Remote monitoring application or system generally uses wireless network connectivity to communicate to system reside at server. This system is useful for early detection of symptoms of disease. Second type is Clinical care. This type of system takes care of patients hospitalized. System collects physiological data of patient and process it using cloud-based application. It helps professionals and hospital to take intense care of their patients with minimal cost [27].

Medical body area networks (MBANs) [27] or a Wireless body area network (WBAN) [28, 29] is network of devices which are connected to patient body. These devices can measure body parameters which are needed to be observed. Generally, these devices are connected to processing unit via wireless media. But there are few restrictions on usage of MBAN such as MBAN devices must be used with proper prescription of professionals and MBAN devices should only be used for diagnostic assistance purpose.

Following activity diagram Fig. 5 explains Smart healthcare system for hospitals and patients.

As shown in Fig. 5 Body sensors sense data from patient's body like body temperature, Heartbeat rate, Pulse rate, Blood Pressure etc. Sensors will push that data to Edge devices associated with sensors using PushPatientData() method.

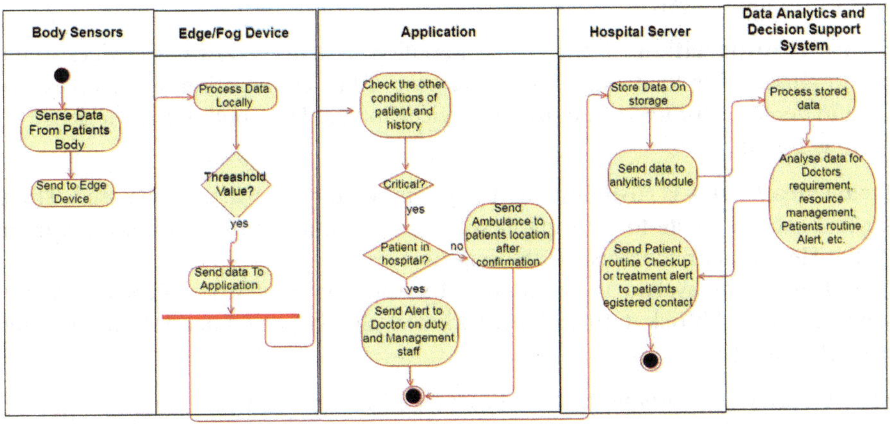

Fig. 5 Activity diagram for smart healthcare system

Edge device will process that data. Edge devices lies on Fog or Edge Layer as shown in layered architecture (Figs. 1 and 4) Each parameter of patient's health will be set with some threshold value and that threshold value will be checked with method CheckThres(). If patient crossed that threshold value that data will be send to hospital data storage unit as well as to application server. Application server will check other activities and body conditions of patient; using Machine Learning Technique application will send alert to doctor if patient is critical using SendAlert () method if patient is admitted to hospital. If patient is not admitted then it will check fact and location of patient using FindLoc() and will send alert to ambulance with same SendAlert(). Hospital server will used to store all data of all patients and would used to send data to analytics module of hospital. Data analytics module will be used to take some policy decisions like Resource management, doctor's duty and requirement etc. and also used to send alert to patients for their health depending on data coming from sensors and patient history. This Data Analytics Module is working on Application and service layer (As shown in Fig. 4).

4.5 Other Applications

The IoT is moving rapidly towards creating the future digital eco-system which will have potential to connect billions of edge devices which could generate very large quantity of data at a high speed. Further these generated or collected data can be processed automatically to provide services in many different areas from smart healthcare, to intelligent irrigation systems and many others. The applications [14, 30] that possibly can adapt to the above-mentioned framework model is discussed below:

Smart Agriculture: Fog devices can be used to sense and collect in-formation about Plant growth, environmental factors and other data that can be useful for smart agriculture system.

Smart Transportation: In Smart transportation system we can use devices in the bus to sense and upload data about movement of bus and location. Bus stop can be used to deploy Wi-Fi connectivity for devices in bus. Here both the devices can be used as Fog devices.

Smart Health and Well-Being: In this system wearable sensors can be used. These sensors can send data to Smartphone of user and Smartphone can upload data on cloud. Here Smartphone can be the Fog devices.

PERSONA is system proposed by Motorola for Healthcare. PERSONA will sense if any health risk during rehabilitation of elderly person and inform the same to doctors [31].

Smart Water Management: Here sensors used for ground water monitoring, water harvesting can be a Fog device.

All these applications create a huge amount of data with high velocity and variety of data. The bottleneck now shifts to the storing and processing of the data in an IoT embraced with Fog/Edge computing ecosystem. The following section explains the trends in handling the methods of bigdata in IoT environments.

5 Inclinations in Handling Big Data in IoT Ecosystem

The IoT expand data volume, variety as well as velocity. The issue here is to solve the problem of data gathering, integration, storage, and data analytics, which are caused by the IoT. A layer in Fig. 1 IoT Reference Layer Model facilitates the huge data processing and supporting in reducing the big data to processing of smaller chunks using fog/edge computing devices.

Large-scale and real time IoT applications require managing and processing huge data stored or generated across geographically distributed datacenters. Despite the prerequisites presented by IoT big Data applications, the capacity of existing IoT technologies and computing infrastructure is limited. For instance, such system could only be able to process data on computing and storage resources within a centralized and local area network. In addition to this, system doesn't facilitate to process different types of data uninterruptedly stored across network and various sources. Hadoop framework allows applications to run on large clusters and provides transparent reliability and data transfer [32, 33]. IoT big data is also called as smart IoT. It is a novel system that enables smart applications that provide higher-quality and timely analytics and actions depending on it. Existing datacenters have following capacities:

Datacenter Cloud Computing Infrastructure Service Stack

Public datacenters and Commercial datacenters such as Amazon Web Services (AWS) provide facility of computation, storage facility, and software resources as cloud-based services based on virtualization. Examples of such datacenter services includes virtual machine instance systems such as Amazon Elastic Compute Cloud (EC2), data storage and processing systems such as the Hadoop distributed file system (HDFS), and Amazon Simple Storage Service (S3).

As Private Datacenters are not exactly like public datacenters, private datacenters generally create basic infrastructure facilities by using already available services and tools. This includes cluster management systems and parallel file system or storage systems such as storage area network (SAN) [32].

Massive Data Processing system or Frameworks

Because of its low price and fault-tolerant process the MapReduce model is widely used for huge data computing inside datacenters. Hadoop framework permits applications to run on massive clusters and facilitates knowledge transfer. Implementations of such system are Compute Unified Device design (CUDA), field programmable gate array (FPGA), virtual machines. MyHadoop provides on-demand service of Hadoop instances on High Performance computing (HPC) resources [32].

Data Management Service across Datacenters

The storage service abstractions supported by cloud computing service providers differ in how they index, store and execute queries: First is Binary Large Object (Blob) used for unstructured data such as Amazon S3. Second is key-value storage for semi-structured data such as-MongoDB, and BigTable. Third are message queuing systems such as SQS. Forth is relational database management systems where structured data come into picture such as Oracle and MySQL. All the four support ACID properties i.e. atomicity, consistency, isolation, durability.

Because of some issue related to service stack for datacenters and huge data processing system, big IoT data processing across numerous distributed datacenters is difficult [32]. Those issues are explained in Sect. 7.4.

6 Trends in Applicability of Blockchain in IoT Ecosystem

With the sustainable IoT infrastructure, there are two issues, one is security of data [34] and second is handling large amount of data flowing through network and system. In security aspect, a main limitation of IoT platforms is IoT infrastructure relies on a centralized cloud. It is challenging to find out whether information about users and infrastructure is stored securely or not.

Blockchain is technology or a data structure used to create an inviolable digital ledger of transactions and to share them. Blockchain uses public-key cryptography [35] for transactions among users. The transactions are then stored on a distributed digital ledger or data structure. It consists of cryptographically linked blocks of data, which form a blockchain. It is almost impossible to modify or remove blocks of data stored on the blockchain digital ledger.

As we can see in Fig. 6 Blockchain is based on primarily four building blocks Ledger, Consensus, Cryptography and Smart Contract. Ledger provides list and detailed information about transaction inside networks. Consensus is the proof of work (PoW), it verifies every action within the networks. Cryptography makes sure that data in the ledger should be encrypted and user can only decrypt the information after authentication and authorization. Smart contract is used for validation and verification of the participants in the network [36].

Blockchain could be only vulnerable due to clock needed for time stamping in blockchain. Blockchain's integrated into IoT can be useful to strengthen security of IoT system or application.

As blockchain is distributed framework it requires more processing capacity than the available processing capacity of IoT nodes.

Blockchain nodes are connected through network, contributing in maintaining the transactions in the blockchain network. Validation of transaction can be done through mining process. Nodes can be computers, stand alone and cloud-based

Fig. 6 Fundamental entities of blockchain

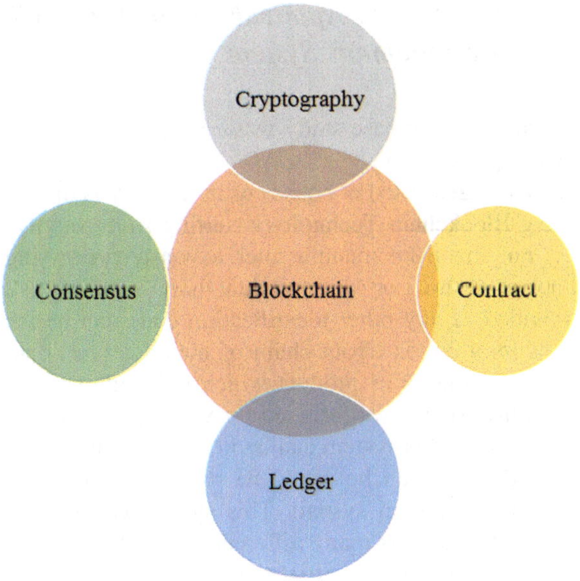

servers or IoT devices. Transaction data is stored on distributed ledger and ledger stored on nodes of Blockchain except client nodes (here client nodes are mostly IoT sensors or devices) Client nodes communicate with each other through APIs. IoT devices generate transactions and these transactions are placed on nodes for further processing and storage of data into the distributed ledger that rely on nodes. Generally, HTTP protocol is used for communication between blockchain nodes [36].

Blockchain-based identity and access management systems can be useful to reinforce IoT security. Such systems can be used to store data about user's identity, credentials, and digital access rights securely. It is challenging to ensure that properties of devices, user's credentials, resource utilization like energy and bandwidth usage through smart devices, and other relevant events are stored securely. This limitation can be overcome for most IoT devices. For example, a blockchain can be used to store cryptographic hashes of each device firmware. This system generates a persistent record of configuration of devices. This can be used for verification about genuineness of device and for the verification of its software and settings that have not been changed or modified. After verification only that device can communicate to other services or devices. From this it is believed that blockchain can play a vital role in finding the sources for security threats in sensitive IoT applications. Blockchain-based identity and access management systems can overcome IoT security issues such as IP spoofing [37].

6.1 *Use Cases: Applicability of Block Chain in Waste Management System*

Considering the case study waste management system, here are the feasible trends of its applicability.

Waste generated at houses or in industry is difficult to differentiate and track. By using Blockchain Technology starting from Supply chain Management's first step i.e. ranging from manufacturer to waste processing industry, it will be easier to track everything or every product that can be in waste generated. A QR code can be provided or any other identification code can be issued to each entity which contains information about chain of manufacturer, distributor and vendor. So that to maintain quality of product which will indirectly affect waste processing. Unique identification tag can be used for waste collection bag to identify location of waste generation. This system can be integrated with waste segregation system. Dumping ground waste can be segregate by using smart garbage categorization system or Waste segregation system. This system can be consisting of multiple sensors to segregate waste as many different types of waste could be there [24]. There can be following advantages of system:

1. Fertilizer is good byproduct from waste.
2. Members could feel that they are contributing to good social cause.
3. System can contribute towards odorless society.

As shown in Fig. 7, sensors can take timely readings of waste bins level. This collected data is sent to edge devices by using protocols like Zigbee with function PushData(). At edge device data pushed by sensors will be processed locally to find out whether bin is full or not with the help of function BinFull(). BinFull function will return Boolean type of value. If function gets true value, then message with data associated to fully filled bin is send to MQTT broker by using underlying network Zigbee. MQTT broker will publish data to subscriber using method Publish(). MQTT subscriber subscribe data published by broker and Process that data to find out suitable vehicle to be assigned by using function AssignVehicle() and will find out optimized route according to vehicle location and assigned tasks by PathFinding() method. Here MQTT Broker will be application running on application layer and MQTT Subscriber will be application or data analytics application running on server. Assigned task and optimized route can be seen from vehicle driver's Android application. Android application will have functionalities such as Getting alert from server, Display path to driver (Using Maps), Notify Garbage weight and collection time to server etc. Here Mobile application for vehicle driver is also working on application layer. Mobile application will send said data to server using PushGarbageData() method. When Garbage is collected at dumping ground application running there will send segregation data along with associated house or organization name to server using function SendSegregationData(). As discussed above if Blockchain technology is applied to Supply Chain Management then data collected by server would be used to credit

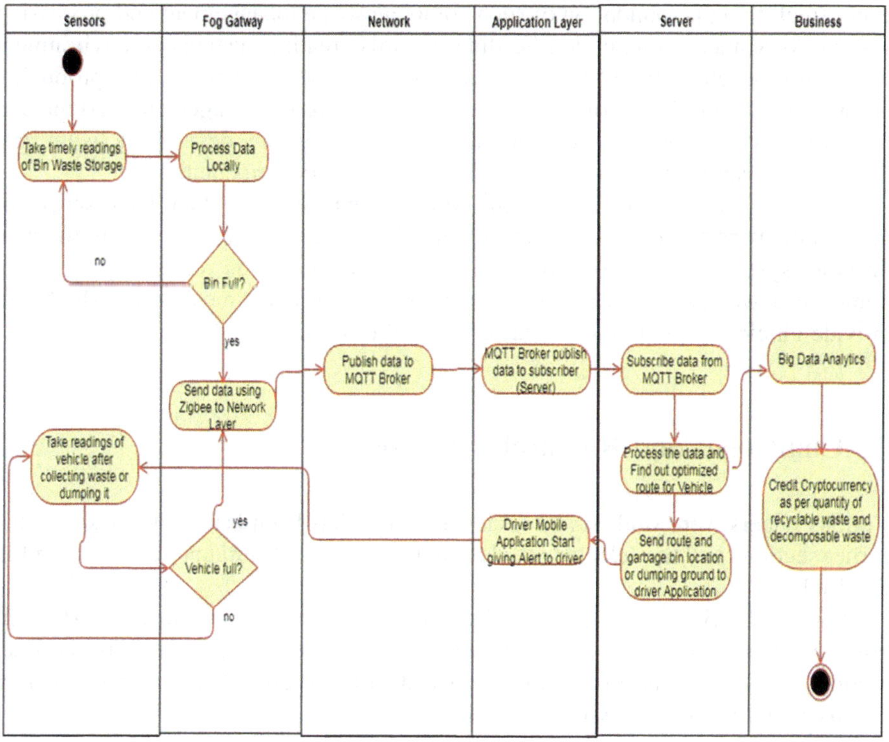

Fig. 7 Activity diagram for smart waste management and segregation system

cryptocurrency to citizen or organization according to their type and amount of garbage as discussed above using method CalcCurrency() and CreditCurrency(). Some cryptocurrency example is discussed below.

One of the crypto-currency examples is Waste Token. It is based on Ethereum Blockchain as an ERC20 token. The Earth token is type of waste token that is credited to wallets of earthtoken holders and they can generate revenue through it. These revenues could be in the form of bitcoin/ethers and the other supported cryptocurrencies. Vendors can provide environmentally sustainable services to earthtoken holders in the form of waste recycling and production of byproduct. Many similar initiatives are taken in different countries. Some of the names are Recereum, Waste token, Swachhcoin, Agro Tech Lab etc.

Swachhcoin is a blockchain based system generated by Swachhcoin Foundation which is non-profitable organization. Swachhcoin is generated by managing wastes from industry as well as households and recycling or converting it into useful products and by-products. Swachh Ecosystem is governed by foundation based on predefined rules and instructions which are present in the form of smart contract. Initially, foundation is aimed to explore into the do-main of green-energy, power generation through gases and fertilizers. Decentralized system and management of every small thing is foundation of such complete waste management and processing system. Swachhcoin foundation heading towards creating sustainable environment by using blockchain based full proof system for processing waste and producing product and byproduct from it [38]. Managing waste management system and processing system is very critical task in many of the developing countries. Swachhcoin can give pathway for such systems for implementation.

Still there are some areas in Waste Management System which have scope for improvement and some issues are still unaddressed. In future, scope is there to develop system which will provide unique identification efficiently to each domestic as well as industrial waste bin. System needs some mechanism which will provide efficient segregation of waste as per the need.

7 Challenges and Research Roadmap

Various papers were studied and reviewed which has been briefly discussed in the earlier sections and the selected articles application based summary can be found in the Table 1.

While reviewing and modeling the waste management system and IoT technologies, challenges were observed which open further research areas in the integration of IoT with various computing and data paradigms. These research scopes are discussed in the following sections.

7.1 Resource Management Schemes

IoT applications generate huge amount of data which increases need of resources to store and process data. In IoT applications many real-world things are connected to internet. But in this most of the things relate to embedded devices or systems which are insufficient with respect to the need of processing the data as per the application requirements. Due to constraint on resources available and complex deployment there are many issues on each layer of IoT architecture. In IoT system Applications are deployed on application layer and analysis modules are deployed on business layer. But due to resource constraints Edge or Fog computing come into picture where we can deploy partial application on edge devices for local data processing. Also, we can migrate resources to use for other applications when not in use. There are various policies like Binding policy, Neighborhood Latency Monitoring Policy etc. are proposed by various authors [42] who could be used to deploy applications on edge devices based on characteristics of edge devices and for efficient resource management. Still there are many issues in this resource management system such as to utilize free resources which are used only for specific time, to avoid looping in migration of resources etc. So, there is a scope for improvement in resource management schemes.

Table 1 Study of IoT systems

Sr. No.	Title of the paper	Computing platform	Application	Merits	Limitations
1	IndieFog: an efficient fog-computing infrastructure for the Internet of Things [9]	Fog computing	Service oriented architecture	1. Interoperability of nodes using SDN 2. Scalable system 3. Secured system using sandbox-based software deployment	1. Quality of service 2. Standardization of IndieFog deployment environment
2	Edge computing for the internet of things: a case study [11]	Edge computing	Gaming anywhere cloud gaming platform Network: WiFi and LTE	Scalability using NFV and SDN	Delays in the system
3	FocusStack: orchestrating edge clouds using focus of attention [14]	Edge computing	Location-based situational awareness (LSA) subsystem implemented based upon AT&T labs geocast system, VMs	Scalable model	1. Security of edge nodes is not addressed in FocusStack 2. Multiple stakeholders' control on access is assumed
4	Fog computing: helping the Internet of Things realize its potential [19]	Fog computing	iFogSim simulator (Cisco IOx, Cisco data in motion, LocalGrid's fog-computing platform, Cisco ParStream's fog-computing platform)		1. Real time analysis of data 2. Power consumption 3. Security
5	Adaptive transmission optimization in SDN-based industrial Internet of Things with edge computing [20]	Edge computing	MATLAB to simulate system	1. Smallest average time delay 2. Goof performance in terms of download time, Throughput and Goodput 3. Solution to urgent situations for data (Latency)	–
6	Blockchain: a game changer for securing IoT data [36]	Cloud computing	Blockchain technology	Gives model useful for secured transmission of data in nodes connected to network in IoT system	Only provide security model for internodes transmission but proposed model doesn't enlighten security issue with massive data at data centers

(continued)

Table 1 (continued)

Sr. No.	Title of the paper	Computing platform	Application	Merits	Limitations
7	A blockchain-based storage system for data analytics in the Internet of Things [39]	IoT, data analytics	Sapphire (proposed system): blockchain based system Location and type sensitive (LTS) hashing mechanism for better data analytics	Better interoperability, secured	–
8	IoT enabled smart waste management using machine learning analytics [40]	IoT and machine learning	Hardware: Raspberry Pi software: python and Google Maps API	Fuel efficiency increased up to 46%	Scalability should be increased
9	Bringing IoT and cloud computing towards pervasive healthcare [41]	Cloud computing	Simple real time telemonitoring system: Lilypad (hardware) and Java EE application (cloud computing)	Scalability, interoperability, light weight access	1. Security and privacy of data 2. Energy efficiency
10	Waste management system using IoT [23]	Cloud computing	Google Maps API: for location actual setup using Wi-Fi ESP 8266 module, arduino UNO, ultrasonic sensor	Simple to implement and cost effective	1. Vehicle tracking is not implemented 2. Big data management is not given

7.2 Enumerate the Resource Management Policies

There are many resources in an IoT ecosystem which needs to be managed. These resources are categorized into two parts: Physical resources and virtual resources. Physical resources are actual characteristics of network devices such as memory, storage capacity, processing capacity etc. Whereas Virtual resources are those resources which are used to manage network scenario like protocols, algorithms etc. [43]. Depending on need of application performance metrics of policies of resource management consist of Energy, throughput, latency.

7.3 Load Balancing Using SDN

In any network architecture there are two parts running on every network device. i.e. Network controller and Data Controller. Network controller is for deciding path in network or next hop to which current device is going to send data and data controller is only for which data packet is to be sent. But if system is implemented with SDN (Software Defined Network), SDN is working as central controller which will have view of full network stored on SDN controller. SDN can balance load on Fog devices for efficient resource utilization [44]. In this system there could be a problem in scalability and mobility of nodes. This issue can be overcome by applying NFV in same system. NVF can virtualize network nodes and resources [45]. Still there are some issues unresolved fully in SDN-IoT system such as security. Researchers are working on security of such network. Another issue is to transform current IoT system to SDN enabled IoT system.

7.4 Big Data Handling in IoT Ecosystem

Huge amount of data generated from IoT infrastructure or applications is processing at multiple and geographically distributed datacenters. Public cloud services to make compatible with existing Private Datacenters infrastructure is challenge for integrating large IoT applications with cloud and big data processing [32].

Service Stacks in a Multi-datacenter Computing Infrastructure

Regardless of significant advances, public cloud services are challenging technically to serve Industrial or large IoT applications across datacenters. To overcome this issue public cloud services should be combined with the resource management and file systems of private datacenter infrastructures to provide cloud services for large IoT applications [32].

Massive Data Processing Models for Datacenters

There are several constraints in the use of MapReduce or Hadoop framework for distributed large-scale applications where huge data processing is required. First constraint is, framework can only compute within infrastructures within a local area network (LAN) or single datacenter and can't be directly used for geographically distributed data-center's huge IoT applications. Second limitation is, Because of need of a high-performance parallel and distributed file system that can flawlessly process across different datacenters, MapReduce faces performance degradation problem. Third limitation is, MapReduce uses a process "fork" mechanism which is not directly compatible with private datacenters with local task managers with lack of security majors. Fourth limitation is, MapReduce and Hadoop both are not supported by large data- demanding systems [32].

Optimized Data Management across Datacenters

Research work undertaken by researchers is facilitating combination of heterogeneous cloud services with transparent interfaces. Though, these services and interfaces could not be able to guarantee da-ta security both in storage and while transferring through network. To store and process huge data across geographically distributed datacenters while providing required quality of service (QoS) might raise some issues regarding the same. First issue is current scenarios of cloud storage services are not secured in aspects of data exposure, modification, and denial of data access. Ensuring data confidentiality, integrity, and availability is a great concern. Second problem is about the intelligence used to automate the choice of the best storage services and network routes for optimal application QoS [32].

Benchmark and Application Kernels

Currently, available performance for executing large-scale Industrial IoT applications in distributed datacenters has no performance measure. Even the benchmarking for understanding the performance of large scale geographically distributed Industrial IoT application kernels are also not existing [32].

These all issues could be future research work for many researchers.

7.5 Other Computing Paradigms and Its Scope

Semantic Computing

Semantic computing overcomes the issue of heterogeneity in IoT and enhances interpretation of data generated and processed in IoT application. Semantics is associated to ‘meaning with data’. Semantic Computing deals with the data to interpret data in context, despite representation or structure and syntax differences [46]. Semantic computing encloses the technologies used for representing concepts and their interrelations that is like the interrelation of concepts in the human brain. Illustrate data and interpret knowledge from data could be achieved by using this conceptual knowledge, represented formally as part of a structured knowledge base (also called as knowledge graph). In addition to this semantic computing plays a vital role in achieving interoperability at various system architecture levels, which contains integrating multisensory and multimodal data from heterogeneous sources.

Cognitive Computing

Cognitive Computing reason, use represented knowledge, learn from experience, accumulate knowledge, explain itself, accept direction, as well as respond in a robust manner to surprises. Cognitive systems be aware of its own capabilities and behaviour, use represented knowledge, learn by from experiences and then perform better while repeating tasks [46]. Cognitive computing is rapidly moving towards ability to answer complex questions by using data mining, Prediction, pattern recognition as well as natural language processing. Cognitive computing is useful in projects where system can answer queries of farmers about their crops or can be able to detect diseases based on photographs.

Perceptual Computing

Perceptual computing provides complementary ability to cognitive computing and semantic computing for developing intelligence from collected data by providing a technology to complement human capacity to ask contextually relevant questions

[46]. Perceptual computing can be useful where machine needs to interact with human intellectually in some context.

Machine Learning

Whenever the IoT sensors collect data, there must be some application at the backend for classification of data, processing it and ensure information is sent back to the device for decision making. If the data set is massive Machine Learning comes to play with it [28]. Machine Learning is useful when we have analyzed massive data and based on that data want to make some decisions or predictions.

IoT and Artificial Intelligence (semantic, cognitive, and perceptual computing) together develop a smart IoT system that could be act as a human [46].

8 Conclusions

The IoT ecosystem requires ubiquitous, reliable, efficient and scalable service provision. In order to achieve this attribution, the solution is to integrate the IoT and Cloud Computing concepts, i.e., facilitate an "Everything as a Service" representation. Innovative IoT centric Cloud infrastructure is evolved to address challenges of IoT and Cloud computing individually or of their integrated system. As the number of connected devices is increasing, IoT network is growing and the data generated is exponentially increasing leading to big data. Fog/Edge Computing and its various framework model is one of the best possible solution to address this exponential data growth. Fog Computing is used for low latency and analytics of huge amount of data generated and collected by IoT system. There are several securities issues in these infrastructures such as Identity management, Data Security, Data Integrity and they could be overcome by using Blockchain technology in integration with these technologies.

This chapter covers a detailed review of various facets and trends in the IoT with respect to cloud infrastructure, architectural design, applicability of fog/edge computing, handling of big data using data centers and applicability of the blockchain in the IoT ecosystem. Further, the review includes a case study on waste management system explaining the applicability of various framework models in fog computing and addressing the handling of data growth. These reviews are analyzed and concluded with an outline of research directions and trends in fog computing to address the issue of big data.

Acknowledgements Our sincere gratitude to all the people who have supported us directly or indirectly in completing and documenting this review and study research. We sincerely thank all the editors of this Book, Reviewers and the publication house.

References

1. International Telecommunication Union: Internet of Things global standards initiative. https://www.itu.int/en/ITU-T/gsi/iot/Pages/default.aspx
2. Almeida, V.A.F., Doneda, D., et al.: Governance challenges for the Internet of Things. IEEE Internet Comput. 56–59 (2015)
3. Ministry of Urban Development Government of India: Mission statement and guidelines. https://punesmartcity.in/wp-content/uploads/2018/01/SmartCityGuidelines.pdf. Accessed June 2015
4. Smart Cities Council India: Smart cities council India. https://india.smartcitiescouncil.com/
5. Chernyshev, M., Baig, Z., et al.: Internet of Things (IoT): research, simulators, and testbeds. IEEE Internet Things J. **5**(3), 1637–1647 (2018)
6. Tikhonov, A.: IoT security for critical information infrastructures (2018). https://www.itu.int/en/ITU-T/Workshops-and-Seminars/20180604/Documents/S3_8_Tihonov%20Kaspersky.pdf
7. Bellagamba, S.: Trust by design: the Internet of Things. https://www.itu.int/en/ITU-T/Workshops-and-Seminars/20180530/Documents/Presentations-Forum/Session5/Pres1-SebastianBellagamba.pdf. Accessed May 2018
8. Sarowar, M.G., Kamal, M.S., Dey, N.: Internet of Things and its impacts in computing intelligence: a comprehensive review–IoT application for big data. In: Big Data Analytics for Smart and Connected Cities. IGI Global, pp. 103–136 (2019)
9. Chang, C., Srirama, S.N., et al.: Indie Fog: an efficient fog-computing infrastructure for the Internet of Things. Computer **50**(9), 92–98 (2017)
10. Pan, J., McElhannon, J.: Future edge cloud and edge computing for Internet of Things applications. IEEE Internet Things J.IEEE Internet Things J. **5**(1), 439–449 (2017)
11. Premsankar, G., Di Francesco, M., et al.: Edge computing for the Internet of Things: a case study. IEEE Internet Things J. **5**(2), 1275–1284 (2018)
12. Cicirelli, F., Guerrieri, A., et al.: Edge computing and social Internet of Things for large-scale smart environments development. IEEE Internet Things J. 5(4), 2557–2571 (2018)
13. Perera, C., Qin, Y., et al.: Fog computing for sustainable smart cities: a survey. ACM Comput. Surv. (2017)
14. Amento, B., Hall, R.J., et al.: FocusStack: orchestrating edge clouds using focus of attention. IEEE Internet Comput. 56–61 (2017)
15. Sookhak, M., Yu, F.R., et al.: Fog vehicular computing: augmentation of fog computing using vehicular cloud computing. IEEE Veh. Technol. Mag. **12**(3), 55–64 (2017)
16. Corcoran, P., Datta, S.K.: Mobile-edge computing and the Internet of Things for consumers. IEEE Consum. Electron. Mag. 73–74 (2016)
17. Cisco: The Internet of Things reference model. White Paper-DRAFT—Controlled distribution, June 2014. http://cdn.iotwf.com/resources/71/IoT_Reference_Model_White_Paper_June_4_2014.pdf
18. Tabbane, S.: IoT standards, ITU Asia-Pacific centre of excellence training on 'Traffic engineering and advanced wireless network 1 planning'. https://www.itu.int/en/ITU-D/Regional-Presence/AsiaPacific/SiteAssets/Pages/ITU-ASP-CoE-Training-on-/Session3_IoT%20Standards.pdf. Accessed 17–19 Oct 2018
19. Dastjerdi, A.V., Buyya, R.: Fog computing: helping the Internet of Things realize its potential. Computer **49**(8), 112–116 (2016). IEEE Computer Society
20. Li, X., Li, D., et al.: Adaptive transmission optimization in SDN-based industrial Internet of Things with edge computing. IEEE Internet Things J. **5**(3), 1351–1360 (2018)
21. Riaz, A.: IoTs—Overview of the ecosystem. Training on planning Internet of Things (IoTs) networks. https://www.itu.int/en/ITU-D/Regional-Presence/AsiaPacific/Documents/Events/2018/IoT-BDG/Day%201%20-%20IoT%20-%20Aamir.pdf
22. Varghese, R., Sharma, S.: Affordable smart farming using IoT and machine learning. In: Second International Conference on Intelligent Computing and Control Systems (ICICCS 2018), pp. 645–650 (2018)

23. Adam, M., Okasha, M.E., et al.: Waste management system using IoT. In: 2018 International Conference on Computer, Control, Electrical, and Electronics Engineering (ICCCEEE) (2018)
24. Pardini, K., Rodrigues, J.J.P.C., et al.: IoT-based solid waste management solutions: a survey. J. Sens. Actuator Netw. 1–25 (2019)
25. Salman, T., Jain, R.: A survey of protocols and standards for Internet of Things. Adv. Comput. Commun. **1**(1), pp. 1–20 (2017)
26. Dey, N., Ashour, A.S., Bhatt, C.: Internet of Things driven connected healthcare. In: Internet of Things and big data technologies for next generation healthcare, pp. 3–12. Springer, Cham (2017)
27. Bhatt, C., Dey, N., Ashour, A.S. (eds.): Internet of Things and big data technologies for next generation healthcare (2017)
28. Dey, N., Hassanien, A.E., Bhatt, C., Ashour, A.S., Satapathy, S.C. (eds.): Internet of Things and big data analytics toward next-generation intelligence. Springer, Berlin (2018)
29. Elhayatmy, G., Dey, N., Ashour, A.S.: Internet of Things based wireless body area network in healthcare. In: Internet of Things and big data analytics toward next-generation intelligence, pp. 3–20. Springer, Cham (2018)
30. Sethi, P., Sarangi, S.R.: Internet of Things: architectures, protocols, and applications. Hindawi J. Electr. Comput. Eng. 1–26 (2017)
31. Schaller, A., Mueller, K.: Motorola's experiences in designing the Internet of Things. Int. J. Ambient. Comput. Intell. (IJACI) **1**(1), 75–85 (2009)
32. Wang, L., Ranjan, R.: Processing distributed Internet of Things data in clouds. IEEE Cloud Comput. **2**(1), 76–80 (2015)
33. Dey, N., Wagh, S., Mahalle, P., Pathan, M. (ed.): Applied Machine Learning for Smart Data Analysis. CRC Press, Boca Raton (2019)
34. Yadav, P., Yadav, H., et al.: IoT: challenges and issues in indian perspective. In: 2018 3rd International Conference on Internet of Things: Smart Innovation and Usages (IoT-SIU), pp. 1–5 (2018)
35. Kimbahune, V.V., Deshpande, A.V., Mahalle, P.N.: Lightweight key management for adaptive addressing in next generation internet. Int. J. Ambient. Comput. Intell. (IJACI) **8**(1), pp. 50–69 (2017)
36. Singh, M., Singh, A., Kim, S.: Blockchain: a game changer for securing IoT data. In: 2018 IEEE 4th world forum on Internet of Things (WF-IoT), Singapore, 5–8 February 2018
37. Kshetr, N.: Can blockchain strengthen the Internet of Things? IT Prof. **19**(4), 68–72 (2017)
38. Swachhcoin Foundation. https://swachhcoin.com/
39. Xu, Q., Aung, K.M.M., et al.: A blockchain-based storage system for data analytics in the Internet of Things. In: Yager R.R., Espada J.P. (eds.) New Advances in the Internet of Things, Studies in Computational Intelligence, pp. 119–138. Springer International Publishing AG (2018)
40. Bakhshi, T., Ahemad, M.: IoT enabled smart waste management using machine learning analytics. In: 2018 International Conference on Energy Conservation and Efficiency, pp. 66–71 (2018)
41. Doukas, C., Maglogiannis, I.: Bringing IoT and cloud computing towards pervasive healthcare. In: 2012 Sixth International Conference on Innovative Mobile and Internet Services in Ubiquitous Computing, pp. 922–926 (2012)
42. Avasalcai, C., Dustdar, S.: Latency-aware decentralized resource management for IoT applications. In: ACM IOT '18, Santa Barbara, CA, USA, 15–18 October 2018
43. Zahoor, S., Mir, R.N.: Resource management in pervasive Internet of Things: a survey. J. King Saud Univ.–Comput. Inform. Sci. 1–15 (2018)
44. Tomovic, S., Yoshigoe, K., et al.: Software-defined fog network architecture for IoT. Wirel. Pers. Commun. **92**(1), 181–196 (2017)

45. Ojo, M., Adami, D., et al.: A SDN-IoT architecture with NFV implementation. In: 2016 IEEE Globecom Workshops (GC Wkshps), Washington, DC, USA, 4–8 December 2016
46. Sheth, A.: Internet of Things to smart IoT through semantic, cognitive, and perceptual computing. IEEE Intell. Syst. **31**(2), 108–112 (2016)

Healthcare Information Technology for Rural Healthcare Development: Insight into Bioinformatics Techniques

Satya Narayan Sahu, Jagannath Panda, Rojalin Sahu, Tejaswini Sahoo, Shanta Chakrabarty and Subrat Kumar Pattanayak

Abstract Continuous investments made by government of India in healthcare information technology as well as various health care programmers allow a revolution in healthcare and technology sector. This revolution helps in improvement of quality and reduction of cost in this field. However in the present scenario the healthcare systems are still in pitiable condition in many rural areas. Most of the time the lack of effective communication between patient and doctor creates ample of health hazards in rural areas. To avoid the communication gap and early prediction of diseases in rural community a pipeline is modeled in the present study, which is based on human whole genome sequencing to protein sequence and the diagnosis of disease by using different biomarkers and computational techniques. Advances in computational methodologies have ushered of innovations in calculation, visualization and prediction of factors relating to different diseases with novel strategies like protein interactive networks comparative genomics, bioinformatics and different online/offline software and tools. By the help of this studied modeled strategy doctor can easily predict the diseases and refer to the patient for early treatment. The health hazards in rural areas can be minimized with constructive application of this method in health care system.

Keywords Healthcare information technology · Bioinformatics · Human whole genome sequence

S. N. Sahu · J. Panda · R. Sahu · T. Sahoo
School of Applied Sciences, Kalinga Institute of Industrial Technology, Deemed to Be University, Bhubaneswar 751024, India

S. Chakrabarty
School of Mechanical Engineering, Kalinga Institute of Industrial Technology, Deemed to Be University, Bhubaneswar 751024, India
e-mail: chakrabarty.shanta@gmail.com

S. K. Pattanayak (✉)
Department of Chemistry, National Institute of Technology, Raipur 492010, India
e-mail: skpiitbbs@gmail.com; skp10@iitbbs.ac.in

N. Dey et al. (eds.), *Internet of Things, Smart Computing and Technology: A Roadmap Ahead*, Studies in Systems, Decision and Control 266,
https://doi.org/10.1007/978-3-030-39047-1_7

1 Introduction

Health care sector is a very eminent area in terms of revenue generation and considered as one of the profit making sector for the Government. This sector needs a strong labor force along with health experts and more innovative technologies [1]. From the literature survey it was found that, in India the health care sector is predicted to a growth of 220 billion dollar by 2020 and a workforce of 7.4 million dollar by 2022 [2]. It is also observed that in various developed as well as developing countries; people are habituated with using wearable personal electronic health record systems [3]. In the current scenario, the market for wearable personal electronic health record systems is emerging in India which is immensely popular not only with younger generation, but also paying more purposeful attention to their health [4]. Electronic mode of health care system is quite challenging to implement everywhere, but it is another great opportunity for world to lift up in health care sector [5, 6]. The increased demand of patient recovery along with their safety, reduction of medical error and the emerging technologies in medical field needs an improved Healthcare information technology (HIT) [7]. The enhanced HIT helps in maintaining the quality and effectiveness of health care services along with safety of patient. To make a transformation in improving health care system, the technology transfer could be the better option where the availability of experts/doctors is less. In comparison to urban area the quality of education based on health and medical field is very limited in rural areas [8–10]. Especially in rural areas people have less inclination towards education. The major reason behind is poverty, no permanent source of income and most importantly ignorance of the concerned authorities who have orphaned these areas. The collective impact of these above mentioned factors is on the healthcare of the rural people. As, mentioned earlier the literacy rate being very poor they are least aware of even general health hygiene [11]. This lack of awareness pushes these unfortunate souls in the trap of diseases like dysentery, common cold, typhoid, malaria, tuberculosis etc. [12]. Further due to unavailability of proper medical facility and improperly managed health care center and irresponsible doctors, these people fall prey to such diseases which can be avoided to far extent, if properly provided with training regarding precautions against these diseases [13]. Though with the constrained surroundings, the 19% of students from rural areas come for doctoral degree, but after completion of degree they prefers to move to urban areas for betterment of the career, which leads to shortage of health experts in rural areas and creates an imbalance between urban and rural areas [14]. This imbalance affects the rural health care infrastructure significantly. Health care informatics can act as blessing in this pretext. Local women, child and health awareness center in rural areas like "Angan wadi" can be installed with well set up of health care information technology systems. These workers should be given training to make the people aware of various do's and don'ts of general health care hygiene and other information for their benefit [15]. Creating awareness among the villagers visually through presentation will be more productive than doing it verbally [16]. Here is the list of few targets which can be achieved by digitalizing the rural health care bodies and related centers:

- Government is making many schemes for public benefit specially for developing rural health but due to involvement of number of middle person, the deserving people remain unaware of these facilities and get very less benefit. With the advent of health care informatics we can omit the middle man system creating a straight line between the provider and benefiter i.e. government and the rural mass. Like "Adhaar Card" government can make" digital health card". For this health card there would be proper set up like ATM machines in rural government health centers. By accessing these cards people can get medicines in very nominal rate and all these could be done without the involvement of any third party [17].
- Moreover, a separate department can be formed and its authority can be given to senior, responsible, experienced and specialist doctors from reputed health institute like AIIMS. Their job will be to keep an eye on the activities of the rural health care center doctors and practitioners. This can be achieved by these health cards. Through these cards all the medical data of the concerned patient will be in record. Rules should be made that the medical bills should be paid through these cards. These data will be in record and can be accessed by higher authorities and hence unnecessary harassment caused by rural health care centers doctor for money can be checked. Also, digital copy of prescription can be checked by the concerned authorities whether diagnosis and treatment is correctly done or not. By this initiative medical negligence can also be reduced to greater extent [17, 18].
- Brain drain is one of biggest hurdle in the development of a nation especially in India. In search of better job opportunity and financial stability many talented doctors deny serving in rural areas. Health care information technology can help this problem. These cards will carry points for the doctors attending the patients. Government can make rules that these points will be counted and accordingly increment in salary will be given for their service to the patients. This effort will attract migrated doctors to give their service in rural areas. Various effects of health care informatics is shown in Fig. 1.

Health care informatics will be definitely proved as blessing to the mankind. Its contribution will mend all the defective areas in the health care system in rural areas. It will add up perfection in all stages of treatment of the patients [19]. No doubt medical facility has been developed to greater extent but it has yet not reached to many interior parts of the nation.

During medical emergency rural people face lots of problem to reach the hospitals due to absence proper constructed roads and lack of transport accessibility to those areas. It causes delay in required medication to the patient, ultimately causing death [20]. Health care informatics can ease this problem by telecommunication facility so that until the patient reaches hospital initial treatment can be provided in local health care center by guidance of experienced doctors remotely [21].

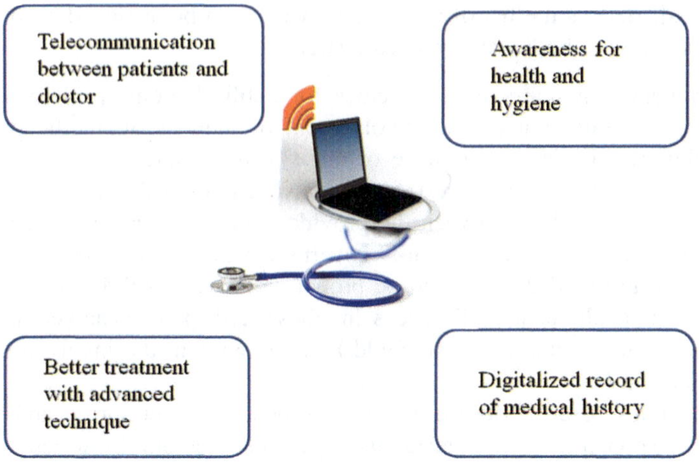

Fig. 1 Health care information technology in rural areas

There are some health conditions which require regular check-up and keeping of record of health parameter data in order to compare the results and keep a track of the health [22]. Health condition like Blood pressure, blood sugar etc. needs a regular check-up and tracking of records for smooth functioning of the body. If the blood pressure or blood sugar level exceeds or gets lower from the normal value then fatal consequences can occur for the patients. Due to the shortage of doctors in rural areas some digital tools can be provided to the people so that they can keep a track of their health condition and can also take proper precaution by the help of the information provided from these tools [23].

There are number of barriers in establishing all these. Some are:

- Financial Restriction
- Income Inequality
- Excessive cost of usage fees and telephone time
- Politics i.e. problem caused by concerned authorities in the distribution of fund.
- And other socio-economic causes.

There are many such factors which can be a hurdle in setting of e-health in rural areas.

But, with proper planning, awareness and support of the Government, doctors and people a massive change can be brought in the field of health care. Developing country like India has a major sector of village and if the nation needs to be developed, then its core should be first reformed and this can be only done by improving health condition and medical facility of these areas. For, this goal health care informatics will definitely act as a boon. It will create a different recognition of India globally in the field of rural health thus adding colors to its fame and national identity [24–26].

Health informatics has improved to give quality, safety, effectiveness and proper delivery of health care services in rural area. This can make a bridge between rural patients and the health services provider specialist in the urban areas. Undoubtedly some existing system may be from urban areas are superior compared to this one. However, the health risk factors are more in rural areas. It is often related to increasing rate of chronic diseases. For the treatment of chronic diseases the patients should be preferred for regular consultant with physicians. Due to some major factors like minimum physicians available in rural residents and increasing transportation barriers are affecting the patients visit to a physician. The limited resources and expertise available in the rural areas can be eradicated by implementing, updating, maintaining and optimizing the health informatics [27]. The technology has been used to store, retrieve, secure and transfer the important health information electronically within health care systems in a protective way. The key systems for the health informatics are discussed as follows;

I. Electronic health records instead of hard copy records.
II. Digitalization for sharing the up to date records to the clinic whenever or wherever required to the patients.
III. Tele-health application to increase the more accessibility to the people.
IV. Confidential health report portal to access all the patient to their health reports
V. Communication between patients and health care provider, electronically.
VI. In order to ignore and minimize medical errors the prescription and ordering can also be performed electronically.

Technology transfer could be the better way to make a transformation in improving health care services, where the availability of doctors is very less. A trouble making imbalance exists between in urban and rural area. Quality education based on health and medical education is not available in the rural areas. Though 19% of doctors are from the rural area but after completion of the degree they are migrating to the urban areas. Hence the rural health infrastructure is significantly affected. The health care sector has huge opportunity to improve and provide a quality health care. The change of the technology can improve the critical process in this area. The collaboration and conversation between the doctor and health workers provide better and more accurate diagnosis and managing operations [28–30].

The manpower shortage of health care sector has improved by the convergence of problems related to technology and their solutions with using of cloud computing, data analysis and wireless technology [31]. The main benefit of these technologies is easy accessibility irrespective of geographical locations, error free data, and quick response to emergency and improves the experience of the patients. The cost of treatment for the rural people is rising gradually now days. As the use of HIT technological revolution better assimilates with health care delivery, it will permit scale and lower costs to rural people, lashing up adoption. And adoption will be further obsessed by the automation of critical situations at hospitals in administration, finance, billing, patient records and pharmacies [31]. Telemedicine have a greater interest in the rural health care sector which is a proven technology [32].

From last three decades it has been used by the different countries. According to the report of NIH in the year 2008, NASA played important role for the development of this technology [33]. This technology has provided good health care to the space scientists. This technology is very crucial for unreachable areas through the satellite technology. This technology is a powerful tool which will allow the medical peoples and experience doctors to give the quality healthcare beyond the boundaries of medical facility. It could save the money for rural patients. The healthcare systems across the globe have an opportunity to create a valuable impact by this technology. The spread of this technology is only possible by the sustainable financial support with initiative of public private partnerships (PPP) and the social workers. For example, Apollo Tele-Health Centre has taken a good initiative for the healthcare sector by PPP mode.

Among the different profit making sectors healthcare sector is also a prominent in terms of its revenue generation and its need for a strong labor force and innovation in the sector. From the previous literature survey it was found that, the healthcare sector is expected to witness a growth of $220 billion by 2020 and a workforce of $7.4 million by 2022 in India. The people belong to developing countries, are using a personal wearable and electronic health record systems. In India, the wearable market is immensely popular with not only younger audience, but also those paying more deliberate attention to their health. For healthcare providers, this means lower error rates in diagnosis and other medical errors, fewer visits to the doctor saving time and money, and better overall care for the patient population [34]. Though more challenging to implement, electronic medical record systems is another big opportunity for world to leverage. This makes health care more efficient, safer, and cost effective over time. Having everyone's medical records electronically saved has far-reaching impact that outweighs the heavy cost of this transition. Today, it is not far behind to adopt electronic record keeping in medical facilities and is further along the pathway of digitizing health care records. We need an advance database for the improvement of the rural health care [35].

2 Applications of the Bioinformatics

Over the past decades, there is a new branch of interface molecular bioscience with the information technology introduced with the other branches like information technology, statistics and computer science. It gives the data from new high-throughput technologies like gene expression, whole genome sequencing, bimolecular structure analysis and image processing [36]. The biological invention has been predicted with the help of bioinformatics techniques on new mechanics for data management, computing and communication, data capture and integration and data analysis [37]. When we compare the system biology and bioinformatics, the bioinformatics techniques are said to focus on the development of software to manage, analyze and visualize biological data, whereas the system biology is the application of all the methods and software to theoretical or applied biological questions [38]. The bioinformatics field was first started in the year 1970, by Paulien Hogeweg and Ben Hesper to accelerate the research. The large number of data which is related to produce extensive information to gene/protein sequence, structure and their annotation information data, based on these information of scientists are esteemed number of database like European molecular biology laboratory (EMBL), GenBank, DNA data bank of Japan (DDBJ), Universal protein resource (Uniprot) and Protein data bank (PDB) [39]. Now a day the hot topic of current research to accelerate bioinformatics is targeted to whole genome sequencing to protein sequence analysis, functional annotation and disease diagnosis in clinical sector. Molecular biology technology gives the big-data which will be the main aim of bioinformatics [38]. From the big-data we will produce position of protein coding and areas and array in large scale. The whole genome sequencing and multiple gene expression annotation concepts are used to uncover the thousands of nucleotide fragments within DNA [40]. The single gene sequence, which is used for recognizing the coding of DNA from the sample and locate the subset of gene sequence. The genomics and proteomics are more useful for big-data management and data mining methods in the field of bioinformatics. Discovery of effective noble drugs has a successful role in identification of genes and various disorders related to it [41]. The electronic health records for patients helps to store the genetic information which has been taken care by many charities, autonomous, national, international health care bodies [42]. These groups are encouraged to make the database for the improvement of the health care for rural patients. The various interpretation and computation problems associated with molecular level are solved by the use of mathematics and statistics. The information technology and computer science are used to make the suitable computational tool which will be very helpful for the bioinformatics [38]. The various applications of the bioinformatics are discussed below.

A. **Sequence Analysis**

Buerman and Den Dunnen [43] reviewed about next generation sequencing technology. Owing to sequencing reaction and template generation their review threw light on the availability of commercial NGS platform and their technical background. One of the superiority of sequencing data is the quality as well as low noise. They also described the various implementations of NGS towards biomedical research field. Deurenberg et al. [44] focused on the application of next generation sequencing. They also described outbreak management, molecular finding, and characterization of pathogen. Heather and Chain [45] reviewed on the sequence of sequencers. Sequencing on DNA and RNA has been paid attention towards impact on biology as well as medicine. NGS has reduced the price of producing sequence data. It has been used for diagnosis and treatment in health care sectors [46]. The recent technology advances and details methods in NGS and current large scale sequencing towards the evolution of genomics space were studied by previously [47]. In genomics and proteomics the sequencing is the process to determine the order of nucleotide base in DNA and amino acid residues in protein. Sequence analysis is the process of subjecting to nucleotide sequence (DNA or RNA) and peptide sequence to any of the wide range of analytical method to understand the structure, function and phylogeny behavior of a living organism. Several computational tools and software are available to analyze such types of evolutionary distance and sequencing outputs in the field of bioinformatics. Genetic mutation of an organism can be identified by using bioinformatics tools. We can also learn how to interpret sequence alignments in terms of disease mutation in human and related sequence and structural information. We might even use available structural data to make better sequence alignment. The sequence analysis of numerous fragments can be done by using whole genome sequencing technology [48].

B. **Protein structure prediction**

Proteins are the most catalytic biomolecules of the living organism. To know its activity in different function and involvement in different catalytic activity, we should go through the functional and structural annotation of the protein. Proteins are structurally derived in four types like primary, secondary, tertiary and quaternary. We can use either bioinformatics tools or crystallography for the prediction of these structures [49]. Applications of the bioinformatics are discussed below.

C. **Genome annotation**

Genome annotation can be used for determining regulatory sequence of gene and coding of corresponding protein. Genome annotation is a process to identify the gene locos, chromosomal position stop and start coding region of the gene sequence [41, 49].

D. **Comparative genomics**

The comparative genomics is the process by which one can find the complete information of whole genome sequence of the organisms. In order to determine the structure of genome and its function related with various biological sources, comparison of gene prediction, sequence alignments of DNAs and RNAs can be done. The researchers use the genomic maps in order to trace the whole path of evolution occurring in different organisms. The details of the large chromosomal segments and mutations can be determined by these maps [41, 50].

E. **Health and drug discovery**

Drug discovery and appropriate target identification for drug delivery based on molecular disease is still a challenging process for biologist. The entire process of drug discovery starting from lead identification to clinical trial it teaks many years. To minimize the time and experimental cost the computational biology tools are used in drug designing process. Based on the molecular basics of diseases and diagnosis management of patients the researcher will develop the medicines which will benefit more than 500 genes [41]. The computational drug discovery or drug design process depends on the protein to be target and various bioinformatics tools used. After the development of digital computers the scientific society assists the digitalization for calculation and information retrieval. After some decade the health professionals and physicians began to know about the dramatic effects of information technology. Encouraged by the health informatics technology vendors of most of the clinics and medicals have developed computational information management environment. Health information technology is the mixture of information technology (IT) and medical data of patients and health care administration. Recently wire less technology [51], clustering of biomedical signal processing [52], ambient intelligence based multi agent system [53] and communication technology are used in heath care systems for rapidly diagnosis [54]. The progress in health information technology gives the freedom to the scientists and researchers to upgrade the health care system, public health care services and biomedical field. The main objective of the health information technology is to maintain the complete information of the medical history of the patients to physicians. This information makes the efforts of doctors simple in decision making and proper treatment at right time. Furthermore by using health informatics the rural people can take consultation from the specialist doctors remotely and can get assistance during emergency. The innovation in the computational expertise of electronic health records, allow the development of high reach data warehouse. Different computational tools are used to analyze the big-data produced by biological professionals to understand the diseases. In current scenario researchers are targeted in health informatics procedures and algorithms for analyzing proteomics and genomic data used in various fields like drug design, medicine and health care services [55].

Productive implementation of health care service in rural community can be upgraded through understanding of each patient's genetic composition and better diagnosis of his/her symptoms and diseases. This revolution helps in improvement of quality and reduction of cost in the field of health care diagnosis system. To avoid the communication gap and early prediction of diseases in rural community, this study is done. It is based on human whole genome sequencing to protein sequence as well as the diagnosis of disease by using different biomarkers and computational techniques. Sequencing is a technique to maintain the exact order of occurrence of nucleotides in DNA and Manoacid in proteins. Understanding the gene sequencing is important for all branches of biological science and its uses have grown popularly in the past decades. The first gene sequencing method was established by Edward sanger, called Sanger sequencing. It is a powerful method for DNA sequencing that utilizes single stranded DNA as template. Life science research has changed quickly since the approach of sequencing technologies and all these are broadly known as Next generation sequencing (NGS). Next generation DNA sequencing is the most dynamic and rapidly growing area of modern life science technology. NGS technology is tremendous sequencing method, which can be describe as DNA sequencing technology accelerate genomic research. This new technologies have given permission to produce millions of short sequencing fragments, so it is also called massively parallel sequencing [38]. Using NGS technology the whole genome of human can be sequenced within a single day. In contrast to the previous sequencing technology, used to decode the human whole genome, required over a decade to deliver the final draft [56]. NGS technology used for DNA sequencing has the prospective to accelerate biomedical and bioinformatics research by allowing the complete analysis of genomes to become cheap and global rather than demanding significant production scale efforts. The NGS sequencing technology gives output as high-throughput sequencing data at a manageable cost in a minimum time period and advances in biological research in the field of genomics, transcriptomics, RNA sequencing technology, protein binding site identification and SNP detection. Next-generation sequencing technology was started with the initiation of second-generation sequencing. There are many companies, those who deal with second generation of sequencer machines including Hiseq, Miseq from Illumina, 454 sequencer from Roche, Heliscope from Biosciences, solid and Ion torrent from Life technology and the third generation of sequencing machines are provided by oxford's nanopore, oxford's gridion and oxford's minion from oxford. These all the sequencing machines have the common application like parallel sequencing method, which increase the amount of data produced in a single run and generate short readings. All the above mentioned sequencing platforms generate short-read sequences and quality score value of the output sequence and the several NGS platforms generate paired-end reads [57].

3 Emerging Next Generation Sequencing Technology

The third generation sequencing technology was mostly acceptable by many researchers and in health care as well as clinical sector due to time consumption, less expenditure and user friendly. Nanopore sequencing is the third generation sequencing technology, which has a different sequencing method, currently in advancement phase. In the nanopore technology the scanning tunneling electron microscope measures the conductivity of the machine. Amount of the current flow through the nanopore at a given moment, is depended on size, shape and length of the nucleotide blocking the flow of ion through the pore. The DNA molecules passes through the nanopore represents the direct reading of the DNA sequence detect by the changes in current flow through the nanopore [58]. An outer most nuclease enzyme is used to cleave or cut the individual nucleotide base from the DNA and the appropriate observation of the system and nucleotide bases could be in a correct order. Furthermore, the nanopore technology has some drawbacks as compared to other platforms through the elimination of optical observation and synthesis of DNA and even amplification of target DNA. The implementation of NGS technology has formed several assembler. The first whole genome sequencing assembly was proved in bacterial genome by employing de Bruijn graphs [59]. After that there are some developed tools to improve assembly process in terms of speed, accuracy, contigs and scaffold length to avoid the error rate. The data analysis of whole genome sequence was totally dependent on bioinformatics tools and software, like Minia, Velvet, SOAPdenovo2 and for visualization the tools like sequence alignment/map (SAM) tools, tablet and for quality check tools like FastQC and NGS QC are used in data analysis of whole genome sequence. In public health sector we do not have the bioinformatics experts to analyze the data generated from the sequencers. That's why they are unable to take full advantages of the whole genome sequencing technology [60]. The objective of this chapter is to provide an effective method to understand the bioinformatics analyses, which are needed to use in whole genome sequence data and computational analysis and can be implemented in public health sector. The implementation of computational dependent analysis is the requirements for large scale data analysis, complex molecular analysis and dedicated bioinformaticians to perform this analysis. All that are needed is a high performance computing system with a browser and a well-connected Internet [61].

4 Health Informatics Pipeline

The proposed pipeline is designed for the early diagnosis of the human diseases and symptoms by using bioinformatics software. This pipeline consists of six step: like (a) read quality check or control, (b) read quality filter or trimming, (c) denovo assemble, (d) evaluate the quality of assemble file, (e) annotation of assembled genome, (f) find the diseased mutation position. The whole pipeline was shown in Table 1. The data, which are generated from the machine, are in the form of FASTQ file format. The computational pipeline processes data from the hard devise deliver by sequencer.

Read quality check

The raw read generated from the sequencer is processed for the quality check of the sequence by using FastQC tool, which gives the simple way to perform quality control check on sequenced data generated from high-throughput sequencing pipeline. The quality check process provides the introductory set of analysis which can be used to give a first impression of the data whether that output has any problems or not, which should be aware before doing any further analysis. The

Table 1 Health informatics pipeline showing various steps and tools

Methods	Software/Tools/Hard devise
Whole genome sequencing data generate from sequencer in fastq file format	Oxford nanopore or Illumina
Read quality check or control	Fast QC tool
Read quality check and filter	NGS QC toolkit
De novo assemble	SOAPdenovo2
Quality evaluation	N50 matrix using N50Stat.pl (script)
Genome annotation	AUGUSTS → BLASTp
Findings of Mutation positions	LALIGN

quality check process can be performed by downloading the FastQC tool from the web link http://www.bioinformatics.babraham.ac.uk/projects/fastqc/.

Read quality filter or trimming

The raw read sequence generated from the sequencer was processed for filter or trimming by using quality control (QC) tool of NGS QC toolkit. It is a toolkit for the control of quality of the sequencing data. The NGS QC toolkit can accept the sequence data in different format as input and perform quality check by using default parameters set by user. At the end quality check reports for unfiltered data as input file and filtered as output data are produced in different formats. In contrast the Illu QC quality control tool can only detect the fastq format as input file and process both paired-end and single-end sequencing data for quality check [62]. All the above discussed tools are set for the quality evaluation system as set for fastq format and perform quality check using default parameters set by the user. Quality filter process can perform by the QC tool of NGS QC toolkit downloaded from the web link http://www.nipgr.res.in/ngsqctoolkit.html.

4.1 De Novo Assemble

De novo assemble sequencing is the genome of a particular organism which is without reference genome sequence. It assists in prediction of coding of protein regions and the different pathways. Raw read genome sequence of human whole genome sequence is assembling using denovo assembler SOAPdenovo2 software. SOAPdenovo2 is an algorithm based software package, which is designed to perform the de novo genome assembly [63] for and sequence alignment for short read sequencing data. This process is perform through the direction of de Bruijin graphs for whole genome sequence assembly by the error removal and simplify the repeated regions by using different K-mers values. This software handles reads separately depending up on their length, pairing and library. The fragments of sequence reads are denoted as K-mers. As notable that the smaller K-mers will give greater connectivity and larger K-mers will gives better specificity. The hash lengths are referred to as word length or K-mer length, corresponding to the length of the words which are stored and compared in the construction of the de bruijin graph. The hash length must be an odd number, which is shorter than most reads in the dataset [60].

4.2 Evaluate the Quality of Assemble

The performance of every assembly was measured N50 length matrix. Higher the value of N50 length better will be the performance of the assembly [38]. The statistic value for set of contig or scaffold lengths is N50. The N50 can be

Table 2 Example of SOAPdenovo2 statistic results of assembly [38]

Sl. no.	Hash length	N50 length	Performance	Total no. of sequence	Maximum sequence length
1	79	44,880	Better	821	129,560
2	81	45,080	Better	804	129,564
3	83	47,198	Better	806	129,568
4	85	50,416	Better	799	129,572
5	87	50,420	Better	780	124,761

statistically defined similar to a mean or median of lengths, but it has greater weightage given to the longer contig file. This statistical value is used for the whole genome sequencing assembly, especially with reference to contig lengths with in a draft assembly. N50 statistic value explains quality of assembled set of contigs with its own length. The selected de novo assembly tool is applied to assemble the read sequence files convert to contigs. In paired-end assembly, tools are support paired-end reads to perform an additional step of scaffold construction to get the final output contigs. Here we show an example of result coming from N50 matrix value in Table 2. In the example here we show hash length (K-mer) 79–89 and hash length 89 showing best N50 length comparison to other hash length. As per the statistic result the N50 length 50,424 showing better performance and the K-mer 89 is the best contig file [38].

4.3 Annotation of Assembled Genome

The assembly file generated by the SOAPdenovo2 assembly tool, the final refined contig file should be used for further genome annotation or gene prediction process.

Gene prediction annotation should be performed by AUGUSTS online server (http://bioinf.unigreifswald.de/augustus/submission.php) which can analyze the human whole genome de novo assemble data. AUGUSTS web server is a family of genome annotation programs and to the sets of per-computed species specific algorithm parameters (model parameters). This process makes use of hidden Markov models (hMM) of protein-coding region. The Gibbs sampling multiple alignment programs is also used as a part of an iterative routine that creates a two-component statistical model conserved region of upstream sequence. AUGUSTS program finds the maximum likelihood method of a given unknown sequence into protein-coding region. Newly predicted sequence is utilized to update the models used for the next analysis [64].

Then the protein sequence, which is resulting from the AUGUSTS web server was proceeded for protein sequence alignment against non-redundant protein sequence database. The protein sequence alignment should be done by using online basic local alignment search tool for protein (BLASTp). From the BLASTp result we can find the corresponding healthy proteins of human. Most of the annotated healthy protein sequences of human are already present in national center for biotechnology information (NCBI) and uniprot database. Uniprot is a universal Protein resource database, which is an all-inclusive resource for protein sequence and annotation data. The uniprot database is maintained with collaboration between the Swiss institute of bioinformatics (SIB), European bioinformatics institute (EMBL-EBI) and Protein Information resource (PIR) [65].

4.4 Finding of Diseased Mutation Position

After collecting healthy proteins of human with respect to diseased protein should be taken for pair wise sequence alignment. Pair wise sequence alignment performed in a symbolic form of each base or residue in each sequence is represented by a letter. The arrangement is to find the single letter codes for the component monomers in order to a fixed font. This process is based on the supposition that all the monomers uniformly spaced along the single dimension of the molecules primary structure. By the help of this process one can find the difference between healthy protein sequence and diseased protein sequence based on mutation. Mutation can be defined as sudden and discontinues change in specific character producing in body part and chromosomal level due to disturbance in nucleotide and amino acid sequence. Mutation may be removal of one or more nucleotide or amino acid from the sequence (deletion), addition of one or more residue in the sequence (insertion) and replacement of one or more residue in sequence (substitution). From pair wise alignment we can find the changes occurs in protein sequences and the mutation position, which can be applied as a bio-marker for specific disease. Previously Sahu et al. [66] perform a computational study and find some mutation positions to diagnosis the nephrotic syndrome in human, which is a type of kidney disease. They compare between healthy Podocin proteins with diseased podocin protein of human in structural level and find some mutations. But here we are showing the reported podocin protein sequence of human and compared between healthy and diseased protein in sequence level by using computational pair wise sequence alignment tool LALIGN (https://www.ebi.ac.uk/Tools/psa/lalign/) showing in Fig. 2. In Fig. 2 the healthy podocin protein sequence is assigned as HEA and diseased podocin protein assigned as DIS. Hear we shown some positions which can differ in healthy protein sequence and diseased protein sequence as denoted by red colored mark. The red colored mark are showing that the amino acid residues are changing in diseased sequence due to mutation in position R138Q, D160G, V180M and R291W. By using these bioinformatics tools and techniques a physician can be easily diagnose the diseases in human as described in proposed pipeline.

```
               10        20        30        40        50        60
sp|HEA MERRARSSSRESRGRGGRTPHKENKRAKAERSGGGRGRQEAGPEPSGSGRAGTPGEPRAP
       ::::::::::::::::::::::::::::::::::::::::::::::::::::::::::::
sp|DIS MERRARSSSRESRGRGGRTPHKENKRAKAERSGGGRGRQEAGPEPSGSGRAGTPGEPRAP
               10        20        30        40        50        60

               70        80        90       100       110       120
sp|HEA AATVVDVDEVRGSGEEGTEVVALLESERPEEGTKSSGLGACEWLLVLISLLFIIMTFPFS
       ::::::::::::::::::::::::::::::::::::::::::::::::::::::::::::
sp|DIS AATVVDVDEVRGSGEEGTEVVALLESERPEEGTKSSGLGACEWLLVLISLLFIIMTFPFS
               70        80        90       100       110       120

              130       140       150       160       170       180
sp|HEA IWFCVKVVQEYERVIIFRLGHLLPGRAKGPGLFFFLPCLDTYHKVDLRLQTLEIPFHEIV
       ::::::::::::::::: :::::::::::::::::::::: ::::::::::::::::::: 
sp|DIS IWFCVKVVQEYERVIIFQLGHLLPGRAKGPGLFFFLPCLGTYHKVDLRLQTLEIPFHEIM
              130       140       150       160       170       180

              190       200       210       220       230       240
sp|HEA TKDMFIMEIDAICYYRMENASLLLSSLAHVSKAVQFLVQTTMKRLLAHRSLTEILLERKS
       ::::::::::::::::::::::::::::::::::::::::::::::::::::::::::::
sp|DIS TKDMFIMEIDAICYYRMENASLLLSSLAHVSKAVQFLVQTTMKRLLAHRSLTEILLERKS
              190       200       210       220       230       240

              250       260       270       280       290       300
sp|HEA IAQDAKVALDSVTCIWGIKVERIEIKDVRLPAGLQHSLAVEAEAQRQAKVRMIAAEAEKA
       :::::::::::::::::::::::::::::::::::::::::::::::::: :::::::::
sp|DIS IAQDAKVALDSVTCIWGIKVERIEIKDVRLPAGLQHSLAVEAEAQRQAKVWMIAAEAEKA
              250       260       270       280       290       300

              310       320       330       340       350       360
sp|HEA ASESLRMAAEILSGTPAAVQLRYLHTLQSLSTEKPSTVVLPLPFDLLNCLSSPSNRTQGS
       ::::::::::::::::::::::::::::::::::::::::::::::::::::::::::::
sp|DIS ASESLRMAAEILSGTPAAVQLRYLHTLQSLSTEKPSTVVLPLPFDLLNCLSSPSNRTQGS
              310       320       330       340       350       360

              370       380
sp|HEA LPFPSPSKPVEPLNPKKKDSPML
       :::::::::::::::::::::::
sp|DIS LPFPSPSKPVEPLNPKKKDSPML
              370       380
```

Fig. 2 Protein sequence alignment of healthy protein with diseased protein

5 High-End Desktop Processor

For analyzing the disease diagnosis by using the proposed pipeline require high-throughput computing system with respect to high-end processing cores, more RAM and sufficient storage space than normal laptop or desktop. High-throughput computing systems are relatively expensive to purchase. We can used the normal desktop to single core of 3.2-GHz intel Core, minimum 32 GB of Random access memory (RAM) space, i5 processor and 2 TB of hard disk (storage space) for maintain the analysis. This system configuration may take 5–7 days to complete the one analysis for this proposed analysis.

6 Conclusion

Health care information technology has been used to store, retrieve, secure and transfer the collated, important and protected health information electronically within health care systems. The manpower shortage of health care sector improved by the convergence of problems related to technology and their solutions with the use of cloud computing, data analysis and wireless technology. Molecular biology technology gives the big data base which will be the main aim of bioinformatics. Biological research has changed rapidly since the advent of sequencing technologies and these are collectively known as Next-generation Sequencing (NGS). De novo assemble sequencing is the genome of a particular organism without reference genome sequence. It assists in prediction of coding of protein regions and the different pathways. The assembly file generated by the SOAPdenovo2 assembly tool, the final refined contig file should use for further genome annotation or gene prediction process. The arrangement is to find the single letter codes for the component monomers in order to a fixed font. This process is based on the supposition that all the monomers uniformly spaced along the single dimension of the molecules primary structure. By the help of this process one can find the difference between healthy protein sequence and diseased protein sequence based on mutation.

References

1. Stafinski, T., Menon, D., Philippon, D.J., McCabe, C.: Health technology funding decision-making processes around thed world. Pharmacoeconomics **29**(6), 475–495 (2011)
2. Cuckler, G.A., Sisko, A.M., Keehan, S.P., Smith, S.D., Madison, A.J., Poisal, J.A., Stone, D. A.: National health expenditure projections, 2012–22: slow growth until coverage expands and economy improves. Health Aff. **32**(10), 1820–1831 (2013)
3. Bhatt, C., Dey, N., Ashour, A.S. (eds.): Internet of Things and Big Data Technologies For Next Generation Healthcare (2017)
4. Dey, N., Wagh, S., Mahalle, P.N., Pathan, M.S. (eds.): Applied Machine Learning for Smart Data Analysis. CRC Press, Boca Raton (2019)
5. Elhayatmy, G., Dey, N., Ashour, A.S.: Internet of things based wireless body area network in healthcare. In: Internet of Things and Big Data Analytics Toward Next-Generation Intelligence, pp. 3–20. Springer, Cham (2018)
6. Dey, N., Ashour, A.S., Bhatt, C.: Internet of things driven connected healthcare. In: Internet of things and big data technologies for next generation healthcare, pp. 3–12. Springer, Cham (2017)

7. Chauhan, D.S., Singh, A.K., Kumar, B., Saini, J.P.: Quantization based multiple medical information watermarking for secure e-health. Multimed. Tools Appl. **78**(4), 3911–3923 (2019)
8. Hossain, N., Yokota, F., Sultana, N., Ahmed, A.: Factors influencing rural end-users' acceptance of e-health in developing countries: a study on portable health clinic in Bangladesh. Telemed. e-Health **25**(3), 221–229 (2019)
9. Adams, S.V., Mader, M.J., Bollinger, M.J., Wong, E.S., Hudson, T.J., Littman, A.J.: Utilization of interactive clinical video telemedicine by rural and urban veterans in the veterans health administration health care system. J. Rural Health (2019)
10. Wakerman, J., Humphreys, J.S.: "Better health in the bush": why we urgently need a national rural and remote health strategy. Med. J. Aust. **210**(5), 202–203 (2019)
11. Acharya, S., Robinson-Pant, A.: Women, literacy and health: comparing health and education sectoral approaches in Nepal. Comp.: J. Comp. Int. Educ. **49**(2), 211–229 (2019)
12. Gokul, M.S., Dahms, H.U., Henciya, S., Dhineshkumar, T., Muthukumar, K., James, R.A.: Socio-ecological studies on a tropical coastal area in southern India. Int. J. Environ. Sci. Technol. **16**(5), 2279–2294 (2019)
13. Caver, K.A., Shearer, E.M., Burks, D.J., Perry, K., De Paul, N.F., McGinn, M.M., Felker, B. L.: Telemental health training in the veterans administration puget sound health care system. J. Clin. Psychol. (2019)
14. Simpson, J.M.: Migrant doctors and the 'frontiers of medicine' in westernised healthcare systems. In: History, Historians and the Immigration Debate, pp. 173–189. Palgrave Macmillan, Cham (2019)
15. Scott, K., George, A.S., Ved, R.R.: Taking stock of 10 years of published research on the ASHA programme: examining India's national community health worker programme from a health systems perspective. Health Res. Policy Syst. **17**(1), 29 (2019)
16. Verdezoto, N., Bagalkot, N., Akbar, S.Z., Sharma, S., Griffiths, P., Mackintosh, N., Harrington, D.: Infrastructural artefacts in community health: a case study of pregnancy care infrastructures in South India. In: Infrahealth 2019-Proceedings of the 7th International Workshop on Infrastructure in Healthcare 2019. European Society for Socially Embedded Technologies (EUSSET) (2019)
17. Hansen, D.P., Dinger, M.E., Hofmann, O., Thorne, N., Boughtwood, T.F.: Preparing Australia for genomic medicine: data, computing and digital health. Med. J. Aust. **210**, S30–S32 (2019)
18. Chen, C.E., Harrington, R.A., Desai, S.A., Mahaffey, K.W., Turakhia, M.P.: Characteristics of digital health studies registered in ClinicalTrials. gov. JAMA Intern. Med. **179**(6), 838–840 (2019)
19. Hassan, M.K., El Desouky, A.I., Elghamrawy, S.M., Sarhan, A.M.: Big data challenges and opportunities in healthcare informatics and smart hospitals. In: Security in Smart Cities: Models, Applications, and Challenges, pp. 3–26. Springer, Cham (2019)
20. Hintenach, A., Raphael, O., Hung, W.W.: Training programs on geriatrics in rural areas: a review. Curr. Geriatr. Rep. **8**(2), 117–122 (2019)
21. Dandachi, D., Lee, C., Morgan, R.O., Tavakoli-Tabasi, S., Giordano, T.P., Rodriguez-Barradas, M.C.: Integration of telehealth services in the healthcare system: with emphasis on the experience of patients living with HIV. J. Investig. Med. **67**(5), 815–820 (2019)
22. McEwen, K. E., Roane, E. M., Pollock, C., & Cote, K. (2019). U.S. Patent Application No. 16/255,586
23. van Velthoven, M.H., Cordon, C., Challagalla, G.: Digitization of healthcare organizations: the digital health landscape and information theory. Int. J. Med. Inform. **124**, 49–57 (2019)
24. Sardi, A., Orozco-Urdaneta, M., Velez-Mejia, C., Perez-Bustos, A.H., Munoz-Zuluaga, C., El-Sharkawy, F., King, M.C.: Overcoming barriers in the implementation of programs for breast and cervical cancers in Cali, Colombia: a pilot model. J. Glob. Oncol. **5**, 1–9 (2019)

25. Al-Rawajfah, O., Tubaishat, A.: Barriers and facilitators to using electronic healthcare records in Jordanian hospitals from the nurses' perspective: a national survey. Inform. Health Soc. Care **44**(1), 1–11 (2019)
26. Stark, Z., Dolman, L., Manolio, T.A., Ozenberger, B., Hill, S.L., Caulfied, M.J., Boughtwood, T.: Integrating genomics into healthcare: a global responsibility. Am. J. Hum. Genet. **104**(1), 13–20 (2019)
27. Richesson, R.L., Andrews, J.E., Hollis, K.F.: Introduction to clinical research informatics. In: Clinical Research Informatics, pp. 3–15. Springer, Cham (2019)
28. Williams, F., Oke, A., Zachary, I.: Public health delivery in the information age: the role of informatics and technology. Perspect. Public Health 1757913918802308 (2019)
29. Durrani, M.I.A., Qureshi, N.S., Ahmad, N., Naz, T., Amelio, A.: A health informatics reporting system for technology illiterate workforce using mobile phone. Appl. Clin. Inform. **10**(02), 348–357 (2019)
30. Krakow, M., Hesse, B.W., Oh, A., Patel, V., Vanderpool, R.C., Jacobsen, P.B.: Addressing rural geographic disparities through health it: initial findings from the health information national trends survey. Med. Care **57**, S127–S132 (2019)
31. Khairat, S., Liu, S., Zaman, T., Edson, B., Gianforcaro, R.: Factors determining patients' choice between mobile health and telemedicine: predictive analytics assessment. JMIR mHealth uHealth **7**(6), e13772 (2019)
32. Myers, C.R.: Using telehealth to remediate rural mental health and healthcare disparities. Issues Ment. Health Nurs. **40**(3), 233–239 (2019)
33. Mervis, J.: Promising year ends badly after fiscal showdown squeezes science 18–19 (2008)
34. Patil, A.V., Somasundaram, K.V., Goyal, R.C.: Current health scenario in rural India. Aust. J. Rural. Health **10**(2), 129–135 (2002)
35. Balarajan, Y., Selvaraj, S., Subramanian, S.V.: Health care and equity in India. The Lancet **377**(9764), 505–515 (2011)
36. Eisen, M.B., Spellman, P.T., Brown, P.O., Botstein, D.: Cluster analysis and display of genome-wide expression patterns. Proc. Natl. Acad. Sci. **95**(25), 14863–14868 (1998)
37. Masys, D.R., Jarvik, G.P., Abernethy, N.F., Anderson, N.R., Papanicolaou, G.J., Paltoo, D. N., Levy, H.P.: Technical desiderata for the integration of genomic data into electronic health records. J. Biomed. Inform. **45**(3), 419–422 (2012)
38. Sahu, S.N., Moharana, M., Martha, S.R., Bissoyi, A., Maharana, P.K., Pattanayak, S.K.: Computational biology approach in management of big data of healthcare sector. In: Big Data Analytics for Intelligent Healthcare Management, pp. 247–267. Academic (2019)
39. Zou, D., Ma, L., Yu, J., Zhang, Z.: Biological databases for human research. Genomics Proteomics Bioinform. **13**(1), 55–63 (2015)
40. Buermans, H.P.J., Den Dunnen, J.T.: Next generation sequencing technology: advances and applications. Biochim. et Biophys. Acta (BBA)-Mol. Basis Dis. **1842**(10), 1932–1941 (2014)
41. Majhi, V., Paul, S., Jain, R.: Bioinformatics for healthcare applications. In: 2019 Amity International Conference on Artificial Intelligence (AICAI), pp. 204–207. IEEE (2019)
42. Rindfleisch, T.: Privacy, information technology, and health care. Commun. ACM **40**(8), 92–101 (1997)
43. Buermans, H.P.J., Den Dunnen, J.T.: Next generation sequencing technology: advances and applications. Biochim. et Biophys. Acta (BBA)-Mol. Basis Dis. **1842**(10), 1932–1941 (2014)
44. Deurenberg, R.H., Bathoorn, E., Chlebowicz, M.A., Couto, N., Ferdous, M., García-Cobos, S., Zhou, K.: Application of next generation sequencing in clinical microbiology and infection prevention. J. Biotechnol. **243**, 16–24 (2017)
45. Heather, J.M., Chain, B.: The sequence of sequencers: the history of sequencing DNA. Genomics **107**(1), 1–8 (2016)
46. McCombie, W.R., McPherson, J.D., Mardis, E.R.: Next-generation sequencing technologies. Cold Spring Harb. Perspect. Med. a036798 (2018)
47. Levy, S.E., Boone, B.E.: Next-generation sequencing strategies. Cold Spring Harb. Perspect. Med. **9**(7), a025791 (2019)

48. Alam, A., Ali, S., Ahamad, S., Malik, M.Z., Ishrat, R.: From ZikV genome to vaccine: in silico approach for the epitope-based peptide vaccine against Zika virus envelope glycoprotein. Immunology **149**(4), 386–399 (2016)
49. Baker, D., Sali, A.: Protein structure prediction and structural genomics. Science **294**(5540), 93–96 (2001)
50. Krzywinski, M., Schein, J., Birol, I., Connors, J., Gascoyne, R., Horsman, D., Marra, M.A.: Circos: an information aesthetic for comparative genomics. Genome Res. **19**(9), 1639–1645 (2009)
51. Dey, N., Ashour, A.S., Shi, F., Fong, S.J., Sherratt, R.S.: Developing residential wireless sensor networks for ECG healthcare monitoring. IEEE Trans. Consum. Electron. **63**(4), 442–449 (2017)
52. Dey, N., Ashour, A. (eds.): Classification and clustering in biomedical signal processing. IGI Glob., Hershey (2016)
53. Tapia, D.I., Corchado, J.M.: An ambient intelligence based multi-agent system for alzheimer health care. Int. J. Ambient. Comput. Intell. (IJACI) **1**(1), 15–26 (2009)
54. Baig, M.M., Gholamhosseini, H.: Smart health monitoring systems: an overview of design and modeling. J. Med. Syst. **37**(2), 9898 (2013)
55. Martin-Sanchez, F., Iakovidis, I., Nørager, S., Maojo, V., de Groen, P., Van der Lei, J., Baud, R.: Synergy between medical informatics and bioinformatics: facilitating genomic medicine for future health care. J. Biomed. Inform. **37**(1), 30–42 (2004)
56. Weber, J.L., Myers, E.W.: Human whole-genome shotgun sequencing. Genome Res. **7**(5), 401–409 (1997)
57. Yu, J.N., Won, C., Jun, J., Lim, Y., Kwak, M.: Fast and cost-effective mining of microsatellite markers using NGS technology: an example of a Korean water deer Hydropotes inermis argyropus. PLoS ONE **6**(11), e26933 (2011)
58. Mikheyev, A.S., Tin, M.M.: A first look at the Oxford Nanopore MinION sequencer. Mol. Ecol. Resour. **14**(6), 1097–1102 (2014)
59. Henson, J., Tischler, G., Ning, Z.: Next-generation sequencing and large genome assemblies. Pharmacogenomics **13**(8), 901–915 (2012)
60. Zerbino, D.R., Birney, E.: Velvet: algorithms for de novo short read assembly using de Bruijn graphs. Genome Res. **18**(5), 821–829 (2008)
61. Oakeson, K.F., Wagner, J.M., Mendenhall, M., Rohrwasser, A., Atkinson-Dunn, R.: Bioinformatic analyses of whole-genome sequence data in a public health laboratory. Emerg. Infect. Dis. **23**(9), 1441 (2017)
62. Patel, R.K., Jain, M.: NGS QC toolkit: a toolkit for quality control of next generation sequencing data. PLoS ONE **7**(2), e30619 (2012)
63. Zerbino, D.R.: Using the velvet de novo assembler for short-read sequencing technologies. Curr. Protoc. Bioinform. **31**(1), 11–15 (2010)
64. Stanke, M., Steinkamp, R., Waack, S., Morgenstern, B.: AUGUSTUS: a web server for gene finding in eukaryotes. Nucleic Acids Res. **32**(suppl_2) W309–W312 (2004)
65. UniProt Consortium: Activities at the universal protein resource (UniProt). Nucleic Acids Res. **42**(D1), D191–D198 (2013)
66. Sahu, S.N., Moharana, M., Sahu, R., Pattanayak, S.K.: Impact of mutation on podocin protein involved in type 2 nephrotic syndrome: insights into docking and molecular dynamics simulation study. J. Mol. Liq. **281**, 549–562 (2019)

Internet of Things

Green Internet of Things Schemes and Techniques for Adaptive Energy Saving in Emergency Services

Rachana Borawake-Satao and Rajesh Prasad

Abstract In future, the objects such as vehicles, smartphones, watches and electronic devices will be able to automate the tasks along with ubiquitous communication capabilities. Saving and reducing energy is challenging research area. Various energy saving approaches and techniques motivates researchers towards Green-Internet of Things (G-IoT). A smart word ubiquitously gathers and disseminates the information using various existing protocols. Modern applications demands energy efficient devices and protocols for G-IoT. Most of the time the emergency service applications are at remote places and their lifetime is a crucial part of the IoT enabled devices and applications. In modern applications, the multimedia information is a vital part in communication. Sensing and communication of multimedia information needs to be controlled in efficient way to achieve better performance and energy saving. In this chapter various techniques and approaches used for energy saving in Green-Internet of Things (G-IoT) are discussed in detail. This chapter discusses the various vital applications of IoT (Internet of Things) such as border security and battlefield systems. This paper also elaborates the energy saving approaches and techniques proposed by various researchers for such applications for ubiquitous environments. The broad classification of the applications of Internet of Things and the approaches used for energy saving are elaborated.

Keywords Internet of Things (IoT) · Green Internet of Things (G-IoT) · Energy saving · Sleep scheduling · Duty cycling

R. Borawake-Satao (✉)
Department of Computer Engineering, Smt. Kashibai Navale College of Engineering (Savitribai Phule Pune University), Pune, India
e-mail: rachana.borawake1@gmail.com

R. Prasad
Sinhgad Institute of Technology and Science (Savitribai Phule Pune University), Narhe, Pune, India

N. Dey et al. (eds.), *Internet of Things, Smart Computing and Technology: A Roadmap Ahead*, Studies in Systems, Decision and Control 266,
https://doi.org/10.1007/978-3-030-39047-1_8

1 Introduction

Green Internet of Things (G-IoT) is a demand of future applications. As shown in Fig. 1 the applications in future internet are ubiquitous. Most of the applications demands energy saving mechanism for longer lifetime.

The type of data involved in communication is not fixed. Every application needs different type of data transmitted through network. Various applications have text, image and video data for communication.

Multimedia data transmission and its applications are the leading factors for the future internet. Wireless Multimedia Sensor Network (WMSN) is efficient technology which supports multimedia data gathering and disseminations. In WMSN, the network consists of the combination of multimedia nodes and the scalar sensor nodes. As multimedia information is critical part of future applications, it is very essential to manage large amount of data effectively. In sensor network nodes are placed at remote areas for monitoring purpose. Multimedia nodes are assumed to be high capacity and high processing capability nodes. In variety of applications it is not possible to provide continuous power supply for all multimedia nodes. Hence high capacity battery operated nodes at remote areas required to work in two modes i.e. sleep mode and awake mode. In sleep mode the energy of the multimedia node can be saved by keeping it in sleep mode when its functioning is not required. These nodes are controlled by the scalar controller nodes for saving the energy when functioning of the multimedia nodes is not needed.

Energy efficiency is a fundamental requirement in Wireless Sensor Network (WSN). Duty cycling is used to maintain the sleep and awake schedule as per requirement of the network. Duty cycling for WSN cannot be extended for WMSN as multimedia node schedules need to be taken care separately. The multimedia nodes can be switched to the sleep mode whenever its functioning is not needed. Most of the time the sleep time for the multimedia node is dynamic and triggered by the scalar nodes or regular schedules for regular monitoring are applied. Information fusion, data compression and sleep scheduling are the common strategies applied to save energy and improve lifetime. Enhancing the G-IoT techniques and approaches is a demand of current research and a challenging task.

Fig. 1 Applications of future internet

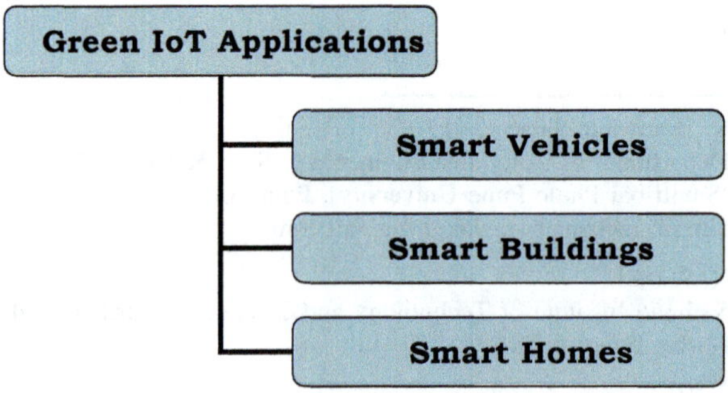

The event based transmission takes place only when the event takes place in a sensor field. The multimedia node switch from sleep mode to active (awake) mode when instruction is given by controller node and starts transmitting multimedia data to the respective nodes. It is assumed that in emergency situations it is necessary to collect multimodal information for accurate analysis of the event instead of saving energy. Saving energy and covering the event both are critical things for the network performance.

The event based sleep scheduling helps in critical event based emergency services. In various applications such as forest fire detection, disaster management and battlefield monitoring systems the event based sleep scheduling can be applied to save energy along with full coverage of the event.

2 Background and Related Work

In order to increase network lifetime for various applications, different techniques are used. The duty cycling approach varies as per the applications. In [1] authors proposed technique for Underwater Wireless Sensor Networks (UWSN). The battery powered underwater acoustic nodes are kept in sleep mode to maximize the lifetime of the network. In [1] authors proposed two types of receiver initiated duty cycling. In first protocol, duty cycle is based on Residual Energy (RidE), each sender node to adjust its duty-cycle based on its residual energy. In the second protocol, Duty Cycle scheduling based on Next Wake-up Time (NeWT), based on last awake node schedule the schedule of current node is decided. This reduces collisions and utilizes the sleep periods efficiently.

The Fig. 2 describes the example of Internet of Things enabled emergency services as a battlefield monitoring scenario. As shown in Fig. 1 the battlefield consists of multiple nodes with variety of functionalities for communication and control of the emergency situation. The nodes in the example are static and mobile. When the event is detected in the battlefield the monitoring and the tracking can be done for specific objects using mobile and static nodes. In such situations the accurate information about the event is communicated to the destination using scalar and multimedia nodes. It is also necessary to save energy whenever possible or whenever emergency situation are not present. As multimedia nodes in the network, consumes more energy the saving energy is an important issue in design of various protocols.

Tracking objects in one of the dynamic application of sensor network. Due to mobility there can be rise in the delay for data transmission. Mobility aware duty cycling will definitely reduce delay in communication and also increase lifetime of the network [2]. The Mobility Aware duty cycling mechanism [2] is tested for mobile object tracking at a tunnel excavation site. The duty cycle ratio is adjusted as per the (Received Signal Strength Indicator) RSSI value.

In (Wireless Multimedia Sensor Network) WMSN the audio and visual information is collected. To solve the energy problem in WMSN is it essential to

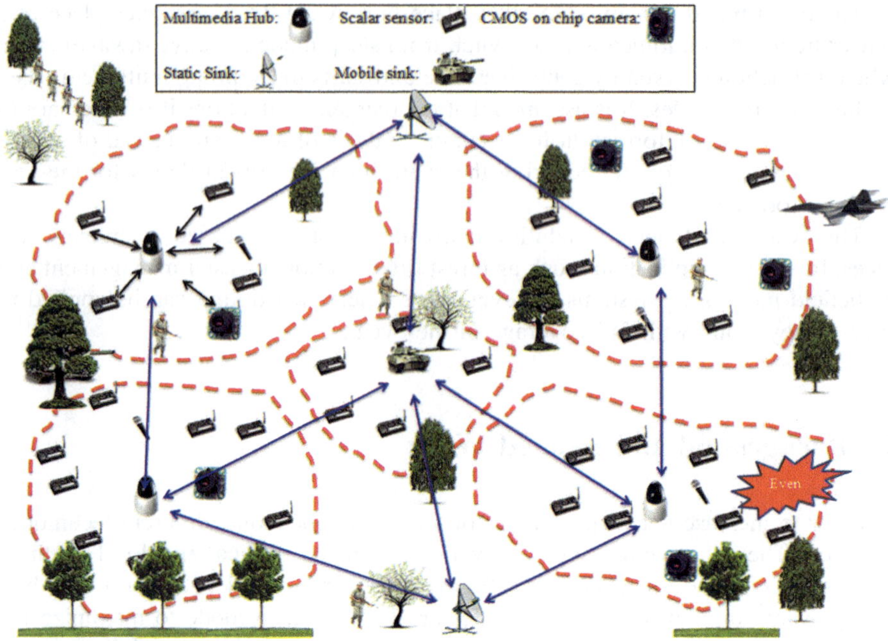

Fig. 2 Battlefield monitoring scenario

schedule energy oriented duty cycling for the network. In [3] the target surveillance is based on particle swarm optimization (PSO). Target motion is extracted by forecasting algorithm based on hidden Markov model (HMM). The forecasting results are used for awaken the nodes in vicinity of future target position. The target detection is done in distributed manner.

In [4] authors proposed an adaptive technique for sleep/awake periode of the nodes. Based on reinforced learning technique every node decides its own sleep/ awake periode independently. Every node in the network decides the sleep/awake time based on the current condition and approximation of its neighbour. Communication among neighbours is avoided for decision making. This new approach is completely based on game theory and reinforcement technique. This technique reduces communication overhead for sleep/awake scheduling and outperforms than other similar scheduling schemes.

In opportunistic networks mobile devices establish communication with each other using contact probing. In [5] authors propose wakeup scheduling for mobile nodes. When communication between nodes is likely to happen the mode is changed to the wakeup otherwise node remains in sleep mode to save energy. Energy saving is achieved by making node sleep during inter-contact times when contact probing is unnecessary.

Instead of implementing a fixed schedule for the complete network it is always better to deploy adaptive sleep schedule for the network based on the traffic at the particular node [6]. In [6] authors proposed sleep/awake schedule based on traffic load. Three important factors are observed

1. Sink node distance
2. Connectivity aspect of node
3. Event proximity

Using these heuristics the scheme reduces delay and increases throughput by minimizing the load at a particular node.

The peer to peer communication of the devices such as smartphones is communicating through opportunistic networks [7]. The past recorded communication contact history is used to decide the adaptive working schedule of the nodes. This information is used to decide future communication contact periods. As per collected information the adaptive schedule is applied to the opportunistic network [7]. Results show that adaptive schedule is better than random schedule in terms of delay and packet loss.

To save energy in various applications is a major concern in future. The sleep scheduling is the key method to achieve energy saving. In [8] authors propose a novel scheme for the sleep scheduling, which is based on Decentralized Partially Observable Markov Decision Process (Dec-POMDP). Author targets coverage connectivity and tracking rewards. This technique saves energy by switching redundant node to sleep node.

As per the application and the emergence of the environment the duty cycle scheduling must be changed adaptively. Table 1 shows various applications where dynamic duty cycling is implemented. It also shows the parameters considered for dynamic sleep scheduling schemes.

In [9, 10] we proposed mobility aware adaptive routing for WMSN. The routing strategy includes three phases:

1. Path discovery phase
2. Priority based traffic classification
3. Event based sleep scheduling.

Table 1 Dynamic sleep scheduling implementations

Paper	Parameters considered for implementation	Applications
[1]	Residual Energy (RidE), wakeup time	Underwater Wireless Sensor Networks (UWSN)
[2]	Received Signal Strength Indicator (RSSI)	Tracking objects during tunnel excavation
[3]	Signal energy feature and residual energy	Target tracking
[4, 5]	Reduces communication overhead	Smart environment
[6]	Traffic load	Event based
[7]	Energy	Detecting social interactions, location tracking, crowdsourcing

The addition of adaptive sleep scheduling in routing methodology is due to the need of emergency services required by future internet applications. The energy saving can be achieved by adding sleep scheduling to the mobility aware routing. In WMSN, the multimedia nodes need not be switched on all the time. As per event occurrence and demand of application turning the multimedia nodes will save lot of energy. A synchronous sleep scheduling can be applied in regular situations and adaptive schedule can be implemented in emergencies [9, 10].

3 Event Based Duty-Cycle Scheduling Scheme

In Wireless Multimedia Sensor Network (WMSN) camera sensor nodes consumes more energy as compare to the scalar sensor nodes [9, 10]. It is desirable to shift camera nodes in wakeup node only when it is required to gather the important data and should be in wakeup state for the specific duration of the recording.

Traditional duty cycle scheduling algorithms are not suitable for WMSN as shifting from sleep to awake mode and from awake to sleep mode is difficult to decide based on fixed pattern or schedule. It is necessary to observe the environment and take dynamic decision accordingly.

There are basically three types of applications in WMSN (1) Event based (2) query based and (3) Regular Monitoring applications. We are targeting the applications which are event based, query based and regular monitoring. Events can occur in any type of applications. In case of regular monitoring the camera nodes will be always on nodes for observing the environment. In case of query based schedule based duty cycle scheduling is possible. But in case of event based scenario assigning efficient duty cycle for camera node is a critical task. All the types of the application needs adaptive sleep scheduling for handling emergencies.

Duty cycle operation is a promising approach for energy saving, in which nodes operates in sleep and wake state alternatively. Duty cycle scheduling is basically classified into two broad categories: synchronous and asynchronous.

In synchronous duty cycle scheduling nodes in the networks adapts the schedule in synchronous with other nodes in the network. This technique has drawbacks of synchronization overhead in the network.

In asynchronous duty cycle scheduling node selects its independent schedule and work with it. This technique can degrade the performance of the network as communication overhead will increase in this case.

As shown in Fig. 3 the duty cycle scheduling works in three phases

1. Active Mode.
2. Transmission and Receiving mode.
3. Sleep mode.

In active mode the sensor node either senses the physical characteristics or process the data sensed by the node. The processing of data involves data extraction

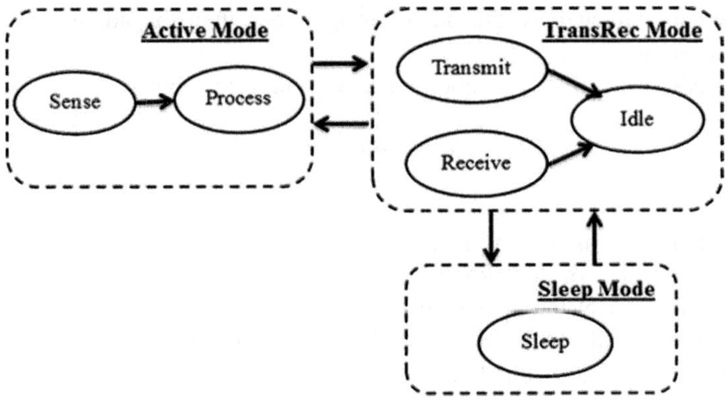

Fig. 3 State transitions for duty-cycling

as per requirement of the sink node in case of query based information retrieval. In event based systems fast transmission of event related important information is done [9, 10].

In TransRec mode the nodes communicates with the neighboring node for data transmission. If there is no data to receive and transmit nodes switched to idle state.

Sleep mode is activated as per the synchronous sleep schedule of the network.

Figure 4 shows the scenario for event occurrence in WMSN for proposed scheme. When event is detected by immediate neighbors, these nodes report the event to the sink node using dynamic routing scheme [9, 10].

In proposed scheme all the nodes in the network follows the synchronous schedule as shown in Fig. 5 during normal conditions for all type of networks.

Fig. 4 Event detection in WMSN

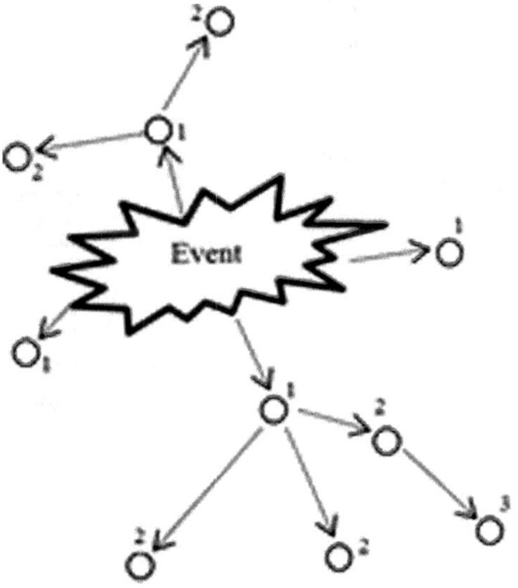

When event is detected by the neighboring nodes then the adaptive schedule is followed as shown in Eq. 1.

Every node has a working schedule represented by tuple

$$Si = (Wi, Ti, \Delta t)$$

where,

Wi—bitmap of wakeup and sleep period.

Ti—the total time of the bitmap.

Δt—the duration of the single bit in the bitmap.

We assume that all nodes in the network has same Ti, $\forall$ i $\in$ N. Figure 5 shows the example of the sleep schedule for the nodes with the tuple Si.

$$Si = (110001, 180\,s, 30\,s).$$

Marked slots represent the sleep state and other represents awake state.

$$f_\tau(\alpha 1, \alpha 2) = \begin{cases} \tau/\alpha 1 & \text{if } E = 1 \\ \tau * \alpha 2 & \text{otherwise} \end{cases} \quad (1)$$

Duty cycle is represented by the Eq. 2, which is a ratio of wakeup time and total time. The duty cycle describe the working schedule of the node.

$$D = T/(T + \tau) \quad (2)$$

For proposed system we consider the dynamic environment with n nodes at any given time,

where,

N = {1, 2, 3, …, n}.

Each node i $\in$ N.

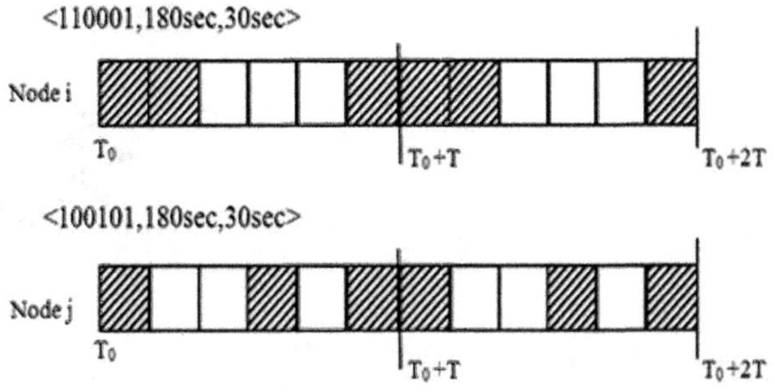

Fig. 5 Duty cycle representation

3.1 Energy Consumption Calculation

Expected energy consumption of periodic duty cycle with fixed sleep time (τ) can be calculated as:

$$E_{periodic} = E_w * D$$

$$E_{periodic} = E_w * (T/(T + \tau)) \quad (3)$$

As given in Eq. 3, when there is a fixed scheduled duty cycle for the deployed WMSN the energy consumption is calculated with fixed sleep time (τ). The energy consumption is directly proportional to the energy consumption in wakeup state and the duty cycle.

Expected energy consumption of event based duty cycle scheduling with variable sleep time (τ) can be calculated as:

$$E_{adaptive} = E_w * T/((T + \tau/\alpha 1)) * \rho + E_w * T/((T + \tau * \alpha 2)) * (1 - \rho) \quad (4)$$

As per Eq. 4 the energy consumption in adaptive mode can be calculated as addition of the energy consumption in event occurrence plus energy consumption in absence of event occurrence. ρ is the probability of event occurrence.

Table 2 illustrates the parameter description for all equations. Delay calculation is done using Eq. 5 for k nodes of the network.

Table 2 Parameter description

Symbol	Meaning
D	Duty cycle
T	Wakeup time
τ	Sleep time
$\propto 1, \propto 2$	Adjustment values for sleep scheduling
E	Event value
Δt	Duration of each bit
$f_\tau (\propto 1, \propto 2)$	Sleep time adjustment function
Si	Duty cycle of the node i
Wi	Bitmap of sleep and awake period
Ti	Total time duration of bitmap
$E_{periodic}$	Energy consumption when periodic sleep scheduling
E_w	Energy consumption when adaptive sleep scheduling
P	Probability of event occurrence
d_{proc}	Processing delay
d_{queue}	Queuing delay
d_{trans}	Transmission delay
d_{prop}	Propagation delay

$$D_{total} = \sum_{i=1}^{k} delay_i \tag{5}$$

4 Results and Discussions

The experimentation is done in NS-2 for implementation of mobility aware routing [9, 10]. After achieving the results for routing, the sleep scheduling is added to the current implementation for energy saving.

The proposed algorithm in [9, 10] is implemented for three types of applications:

(1) Regular monitoring applications
(2) Query based networks applications and
(3) Event based applications.

For supporting event based applications efficiently and saving energy whenever necessary the event based duty cycling is proposed.

The performance of mobility aware routing without event based sleep scheduling is compared with mobility aware routing with event based sleep scheduling.

The Table 3 describes the various parameters used for implementation in NS-2. The traffic generated in the network includes text, audio, image and video information for routing.

The parameters evaluated as Average energy consumption, packet delivery ratio, packet drop and residual energy. The synchronous sleep scheduling is carried out using regular sleep schedules as per WSN synchronous sleep schedules. The event based sleep scheduling is applied as per proposed methodology.

(1) **Average Energy Consumption**:

The average energy consumption is calculated at the end of the simulation periode on the basis of the average value of consumed energy of all the nodes in the network.

The Table 4 describes the implementation results for nodes 100 to 150 for synchronous sleep scheduling and event based sleep scheduling. The analysis

Table 3 Parameter description—implementation

Name of parameter	Value
Type of data traffic	CBR/Evalvid
Packet generation rate	5 packets per sec
Size of simulation area	500 m × 500 m
MAC protocol	802.11
Initial energy	100 J
Data rate	10 Mbps
Density of nodes/no of nodes	100, 120, 130, 140, 150

Table 4 Experimentation values for average energy consumption

Number of nodes	Sleep scheduling (J)	Event based sleep scheduling (J)
100	0.414048	0.295645
110	0.521851	0.372502
120	0.631871	0.449291
130	0.632311	0.444768
140	0.636026	0.434175
150	0.686909	0.477872

shows reduction in energy consumption when event based sleep scheduling is applied. The multimedia nodes are switched to the active mode only when some event occurs in the vicinity. Hence the energy consumption of the nodes increases only in presence of event.

AS numbers of nodes are increasing the proposed event based scheduling is giving efficient result as compare to the synchronous sleep scheduling.

As shown in Fig. 6 the mobility aware routing implemented without sleep scheduling consumes more energy. If we are adding adaptive sleep scheduling to the mobility aware routing it consumes less energy. Hence addition of adaptive sleep scheduling leads to the network lifetime improvement.

(2) **Packet Delivery Ratio**:

The packet delivery ratio is calculated as the ratio of number of packet sent to the number of packets received. The Table 5 describes the analysis of packet delivery ratio for both the scenario. The packet delivery ration is increased when event based sleep scheduling is applied. We are increasing the awake time for the multimedia nodes and scalar nodes when event occurs, hence the data transmission is faster and efficient. Hence the packet Delivery ratio is improved.

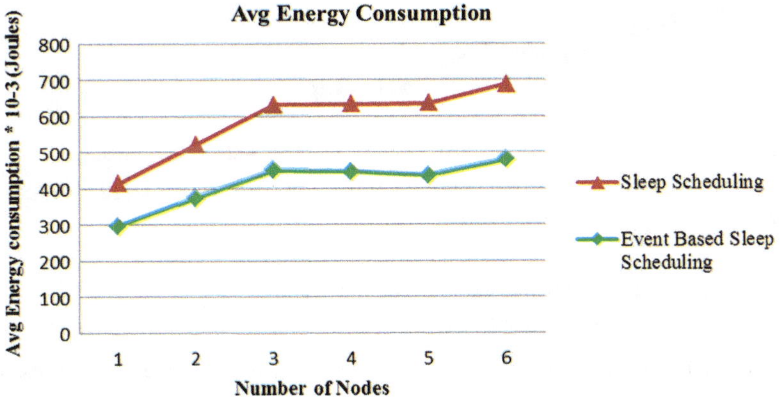

Fig. 6 Comparison of energy consumption for synchronous sleep schedule and event based sleep schedule

Table 5 Experimentation values for packet delivery ratio

Number of nodes	Sleep scheduling (%)	Event based sleep scheduling (%)
100	93.8895	94.9325
110	86.9301	89.0866
120	77.5855	80.6141
130	80.7308	81.562
140	66.4513	67.056
150	64.9377	65.6823

Figure 7 shows improvement in packet loss ratio in event based sleep scheduling. Addition of event based sleep scheduling also improves the data accuracy in case of multimedia data.

(3) **Packet Drop**:

The packet drops are calculated based on the number of packets dropped during the data transmission. As shown in Table 6 and Fig. 8 the packets drops are less when we apply event based sleep scheduling.

(4) **Residual Energy**:

Average value of remaining energy of every node is calculated for the very case. As shown in Table 7 and Fig. 9, the residual energy is improved.

(5) **Throughput**:

As shown in Fig. 10 and Table 8 the throughput is improved in event based sleep scheduling. As number of nodes are increasing the throughput also increases.

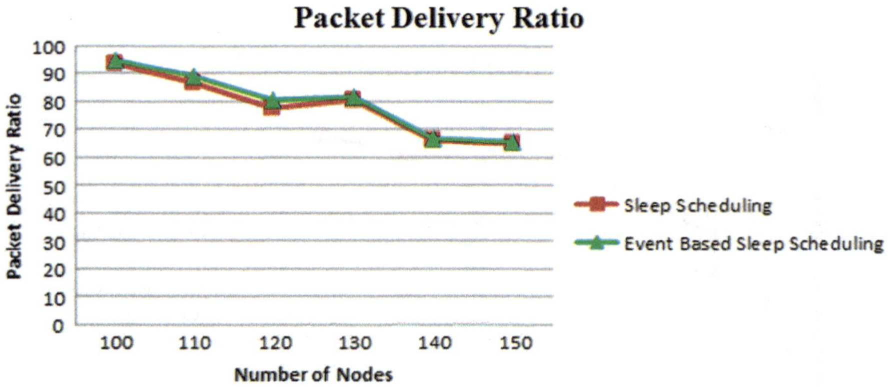

Fig. 7 Comparison of packet delivery ratio for synchronous sleep schedule and event based sleep schedule

Table 6 Experimentation values for packet drop

Number of nodes	Sleep scheduling	Event based Sleep Scheduling
100	6.1105	5.06747
110	13.0699	10.9134
120	22.4145	19.3859
130	19.2692	18.438
140	33.5487	32.944
150	35.0623	34.3177

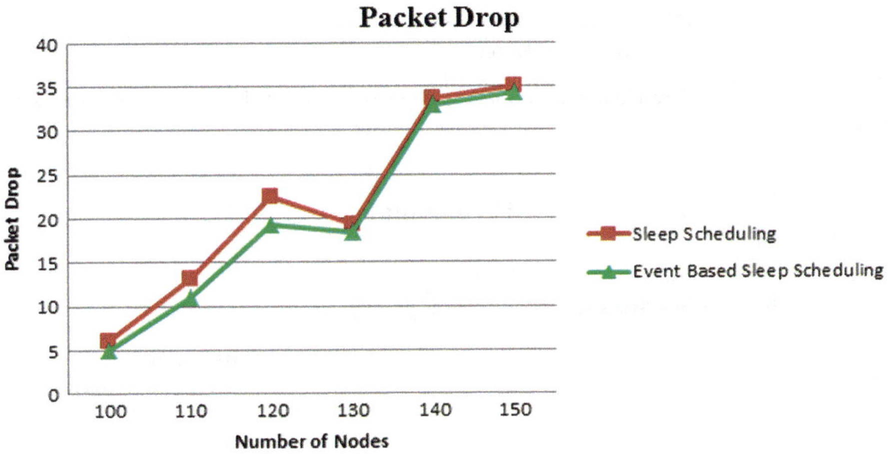

Fig. 8 Comparison of packet drops for synchronous sleep schedule and event based sleep schedule

Table 7 Experimentation values for residual energy

Number of nodes	Sleep scheduling (J)	Event based sleep scheduling (J)
100	50.091	50.2094
110	49.9369	50.0862
120	49.7883	49.9709
130	49.7553	49.9428
140	49.7237	49.9255
150	49.6487	49.8577

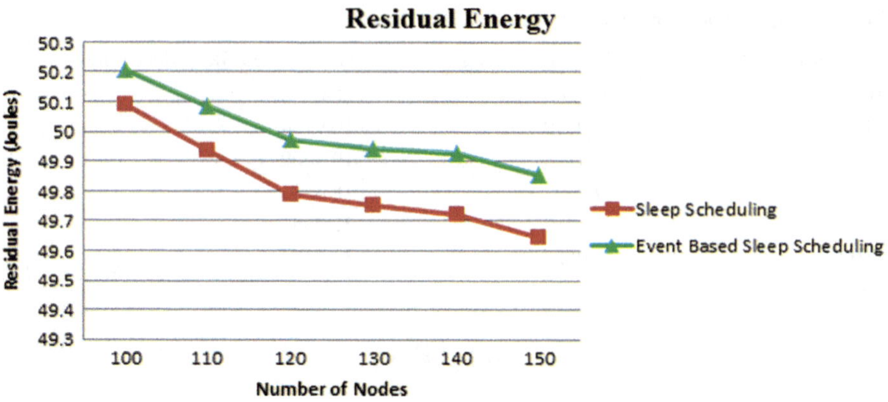

Fig. 9 Comparison of residual energy for synchronous sleep schedule and event based sleep schedule

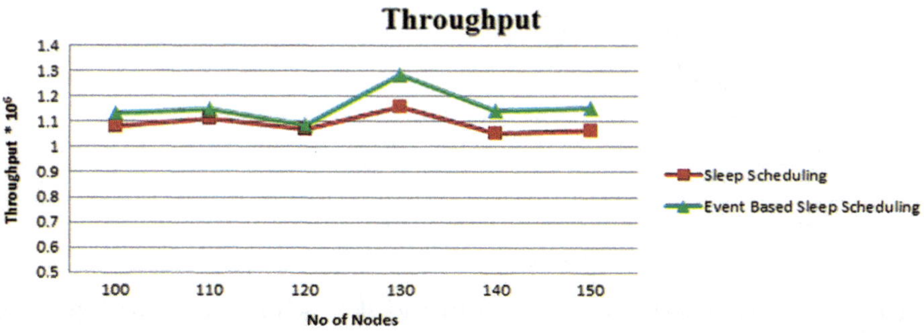

Fig. 10 Comparison of residual energy for synchronous sleep schedule and event based sleep schedule

Table 8 Experimentation values for throughput

Number of nodes	Sleep scheduling	Event based sleep scheduling
100	1.07911	1.13352
110	1.11072	1.15017
120	1.07229	1.08724
130	1.15935	1.28358
140	1.05486	1.14225
150	1.06526	1.15364

5 Conclusions

In this paper, we have proposed event based duty cycle scheduling for WMSN. The proposed sleep scheduling is working for three categories of applications query based, regular monitoring and event based applications. The event based duty cycle scheduling is implemented along with mobility aware routing. Results show that energy consumption is reduced in case of adaptive duty cycle scheduling and other parameters are not affected due to this addition. Improvement in lifetime can be achieved in WMSN through the addition of event based duty cycle scheduling.

Addition of event based sleep schedule in routing increases lifetime of the network and makes this algorithm suitable for wide range of applications for future internet.

References

1. Yaqub, M.A., Khan, M.T.R., Ahmed, S.H., Kim, D.: Receiver-initiated dynamic duty cycle scheduling schemes for underwater wireless sensor networks. In: 2018 15th IEEE annual consumer communications & networking conference (CCNC), pp. 1–6. IEEE, Jan 2018
2. Kim, T., Min, H., Jung, J.: A mobility-aware adaptive duty cycling mechanism for tracking objects during tunnel excavation. Sensors **17**(3), 435 (2017)
3. Wang, X., Wang, S., Ma, J., Sun, X.: Energy-aware scheduling of surveillance in wireless multimedia sensor networks. Sensors **10**(4), 3100–3125 (2010)
4. Ye, D., Zhang, M.: A self-adaptive sleep/wake-up scheduling approach for wireless sensor networks. IEEE Trans. Cybern. **48**(3), 979–992 (2017)
5. Gao, W., Li, Q.: Wakeup scheduling for energy-efficient communication in opportunistic mobile networks. In: 2013 Proceedings IEEE INFOCOM, pp. 2058–2066. IEEE, Apr 2013
6. Nazir, B., Hasbullah, H., Madani, S.A.: Sleep/wake scheduling scheme for minimizing end-to-end delay in multi-hop wireless sensor networks. EURASIP J. Wirel. Commun. Netw. **2011**(1), 92 (2011)
7. Zhou, H., Zhao, H., Chen, J., Liu, C.H., Fan, J.: Adaptive working schedule for duty-cycle opportunistic mobile networks. IEEE Trans. Veh. Technol. **63**(9), 4694–4703 (2014)
8. Carrano, R.C., Passos, D., Magalhães, L.C., Albuquerque, C.V.: A comprehensive analysis on the use of schedule-based asynchronous duty cycling in wireless sensor networks. Ad Hoc Netw. **16**, 142–164 (2014)
9. Borawake-Satao, R., Prasad, R.: Comprehensive survey on effect of mobility over routing issues in wireless multimedia sensor networks. Int. J. Pervasive Comput. Commun. **12**(4), 447–465 (2016)
10. Borawake-Satao, R., Prasad, R.: Mobility aware path discovery for efficient routing in wireless multimedia sensor network. In: Proceedings of the International Conference on Data Engineering and Communication Technology, pp. 673–681. Springer, Singapore (2017)

Ms. Rachana Borawake-Satao Research scholar, Smt. Kashibai Navale College of Engineering research centre, Vadgaon, University of Pune. Graduated from University of Pune (Computer Engineering)—Dr. D.Y. Patil College of Engineering in year 2002 and M.E. from University of Pune (Computer Science and Engineering–Information Technology)—Sinhgad College of Engineering in year 2009. Working as Assistant Professor at Department of Computer Engineering, Smt. Kashibai Navale College of Engineering, Vadgaon, Pune. Total 12 years of experience in Technical Education and in the areas of Teaching, Area of Interest Wireless Multimedia Sensor Network. He is a Life member of ISTE.

Dr. Rajesh S. Prasad has received Masters (M.E. Computer Engineering) degree from College of Engineering, Pune in 2004 and his Ph.D. from SRTMU Nanded in 2012. He is working as Principal and Director in Sinhgad Institute of technology, Narhe, Pune, India. He is having 21 years of experience in academics. His area of interest is Soft computing, Internet of Things, Text Analytics and Information management. He has published over 70 papers in national and international journals. He is a Member of IEEE, life member of International Association of Engineers, Indian Society of Technical Education and Computer Society of India. He is also a fellow of Institution of Engineers, India.

Risk-Driven Analytics for Banking IoT Strategy

F. Khanboubi and A. Boulmakoul

Abstract New communication technologies have a significant impact on the management of banking activities and strongly influence their ecosystem. Key innovations include big data analytics, artificial intelligence, data science, digital currency, social media, blockchain and the Internet of Things (IoT). Many of these technologies are interdependent. IoT is the interconnection of uniquely identifiable integrated computing devices within the infrastructure of the surrounding computer network. In the field of banking services, the interconnection of those integrated equipment should allow the automation of several legacy processes. As the digital transformation driven by IoT begins to take root, new business models and products emerge. This opens new frontiers for innovation that can change customer behavior in the banking industry. The objective of this chapter is to highlight and illustrate the different uses of IoT in banking. We also analyze the impact of digital risks based on the Internet of Things on the traditional banking processes. It would be of interest to bring together the different types of digital risks that have a similar impact on bank's processes. We analyze those technologies to implement the digital transformation using new practices. We develop an approach based on a bipartite graph associating the processes and risks of the banking sector. From this, we deduce a strategy to safely lead the integration of connected objects into the banking industry.

Keywords Internet of Things · Bank · Digital risks · Digital strategy · Galois lattice · Fuzzy concept · Process modeling · Holistic analytics

---

F. Khanboubi (✉) · A. Boulmakoul
LIM Laboratory, IOS, Department of Computer Science, FSTM,
Hassan II University of Casablanca, B.P. 146, Mohammedia, Morocco
e-mail: khanboubi.fadoua@gmail.com

A. Boulmakoul
e-mail: Azedine.boulmakoul@gmail.com

N. Dey et al. (eds.), *Internet of Things, Smart Computing and Technology: A Roadmap Ahead*, Studies in Systems, Decision and Control 266,
https://doi.org/10.1007/978-3-030-39047-1_9

1 Introduction

More than 20 years later, the name of World Wide Web has never been so representative. Our daily lives, our activity, our objects, are caught in the web. Our whole life and interactions with the world will not be able to resist the sticky silk trap. The "Internet of Things" seems to be the proper illustrative of a future technological, economic and social revolution. We would be just at dawn. The ultra-connected and intelligent world which is controlled by dominant multinationals is growing exponentially. Now, most studies are targeting between 50 and 80 billion connected objects in the world in 2020. In the digital era, the financial actors have all interest to follow this new wave of digitalization.

The new manners and practices of the customers, as well as the mass of available data, imply inevitable digital transformation for banks and insurances. The IoT is above all a mean put at the service of a concrete use that must generate value. In the field of financial services, it represents a lever for growth and is based on a simple, personalized and more intuitive customer experience. IoT is able to fundamentally change the way banks interact with their customers through contact points that are agencies and vending machines. In the banking sector and perhaps more than in any other, the IoT offers a level of customization and a range of enriched services beyond banking products, and endless customer journey possibilities. Indeed, with connected objects, the banking service is more and more connected and intelligent, in other words accessible everywhere, at any time and from a very wide ecosystem of devices. Vehicles, refrigerators, scales, remote surveillance alarms and electronic toll tags are all likely to become banking service access interfaces. For banks, they can offer a continuous and instant means to know and anticipate the needs of customers based on data provided. The IoT is therefore one of the investment priorities of banks whose projects revolve around payment, biometrics and security [1].

The objective of this chapter is to present the digital transformation and to illustrate the different uses of IoT in banking. We also analyse the impact of digital risks based on the Internet of Things on the traditional banking processes. It is also interesting to bring together the different types of new inventions that have a similar impact on bank's activities. We develop then an approach based on a bipartite graph associating the processes and the digital transformation technologies of the financial sector. From this, we deduce a strategy to safely lead the integration of connected objects into the banking industry.

2 Digital Transformation in Banking

In the field of financial services, digitalization is often associated with the optimization of internal processes. Semantic analysis for example, makes it possible to speed up processing times and to offer more and more personalized solutions (credit, placement, etc.). Nevertheless, one point remains essential: technologies should not be conceived only as a lever for performance and productivity, but on the contrary, it contributes to increase customer experience that offers personalized advice and greater autonomy. The impact of digital risks on banking uses is gradually designing a new landscape in which institutions must adapt in a relevant way, despite economic, social and financial constraints.

2.1 The Banking Digitalization

2.1.1 The Different Trends of Digitalization

The digitalization of financial institutions is based on several trends that govern its deployment mode: mobility, immediately, simplicity. The first one, '**mobility**' makes people perform simple operations, such as consulting bank accounts or transfers from their smartphones. Mobility is one of the major dynamics, and we must be able to do everything, where and when we want it.

The second trend is about '**immediately**'. Among the most anticipated services, we find the mobile payment which will become the second common payment method by 2022 [2]. To distinguish themselves and respond to disruptive offers of new players, banks must engage in this field.

Finally, the third trend is the '**simplicity**' and acceleration of processes. Customers no longer want to wait when creating a file or responding to a credit application. They expect their bank to be as responsive as the e-commerce platforms.

2.1.2 The Digitalization in the Banking Sector

The digital has become essential in banking services to preserve current customers and conquer new ones. We create pioneering digital services to ameliorate customer knowledge and experience, for example:

- **Connected agencies**: provide advisors with tools to help them be more effective in the management of the customer relationship (loan simulator, configurator ...)
- **Customer relationship management and 360° customer view**: structure, consolidate and analyze customer data to offer them personalized services, like authentication using biometrics data [3] ...
- **Connected objects**: capitalize on the potential of IoT to imagine recent banking services and products
- **Digital performance**: examine and develop the performance of digital services
- **Data intelligence**: analyze clients' pathways and behaviors to predict their expectations, prevent and intercept frauds and offer the right service at the right time (higher education, marriage...).

2.1.3 The Digitalization Levers of Banking Processes

For a bank, a 100% digitized customer journey is paperless, instantaneous and automated from start to finish; that is to say without human intervention. This extreme simplicity and ease of use for the customer is based on advanced technologies: take the example of digitalization of the process of creating a bank account: it starts with the questioning in real-time of the different compliance and risk control files, then the digitalization of the documents via the application and finally the comparison of the selfie and the piece of identity by form recognition. Then, we create the bank account and generate in real-time the card number.

Also, there are many levers of digitalization use to satisfy the new customer expectations by removing the "irritants" complexities (lack of fluidity, slowness …). We can classified digitalization levers into eight broad categories (fig. 1):

- **Tracking and alerting**: in this new era, the customer wants to follow the progress of their requests and be warned proactively during a change …
- **Smart Automation**: Robotics Process Automation's tools enable you to quickly automate low value and repetitive tasks… While artificial Intelligence allows you to go further and to rethink the mode of customer interaction using virtual assistants.
- **Big data and Analytics**: new platforms and advanced analytics make it possible to personalize the customer's experience [4], for example: creating smarter interactions with customers: helping to find the best value for money, optimizing daily expenses or alerting you when a cheaper plane ticket is available!
- **Online visibility**: SEO (search engine optimization) can increase traffic and therefore the number of notoriety.
- **Open and interconnected IT**: we can improve our IT system with decoupling with legacy systems and creating APIs. For example, cloud and Devops bring flexibility and agility into development cycles.
- **Multichannel collaboration and mobility**: The multichannel capabilities (video, chat, co-browse, tablets) make it possible to rethink distance learning to better adapt to current customer requirements.
- **Customer and social experience**: The recognition (biometrics …) will allow to personalize the different banking tasks.
- **Dematerialization**: The paperless is the mainstay of a digital journey: for example, the Néo-banque N26 promises an account opening "guaranteed without paperwork".

2.2 The Internet of Things

Between business and personal wearables, the IoT impress. In the banking sector, connected objects are still too often perceived as gadgets. However, the potential is

Fig. 1 The digitalization levers of banking processes

immense and allows banks to improve the brand image and to develop the customer relationship … So, the digital and connected banks, what is the relationship?

2.2.1 The IoTs in Banking Sector

Connected objects are revolutionizing consumer life (activity trackers, connected watch, home automation, connected car …). They are part of the recent evolutions of the consumer's behaviour: being constantly connected and having access to a multitude of information from several supports. This ongoing revolution will have a significant economic impact. A McKinsey study estimates that the IoT will have a total potential economic impact of $ 3,900 billion to $ 11,100 billion by 2025 [5]. These forecasts can be explained in particular by favourable conditions:

- Decreasing IT costs
- Equipment in smartphone
- Universal connectivity
- Emergence of the cloud and Big data.

For the moment, the players in the banking sector resume this technological potential in the disappearance of the bank card. According to Juniper, more than

50% of worldwide POS transactions will be contactless (mobile or connected devices) within five years, compared to 15% in 2017 [6]. Africa is ahead of the rest markets by favouring mobile payment. For example, the percentage of individuals over the age of 15 with a payment account in sub-Saharan Africa has doubled in seven years, from 23% in 2011 to 47% in 2018 [7].

However, the potentials are much broader with three types of growth earnings for banks:

- A key enabler to secure a positioning on the major issues of tomorrow
- Incremental gains on current offers: the uses around the IoT make it possible to offer a range of enriched services beyond BtoC or BtoB banking or insurance products
- Increased operational performance [8].

Now, the question that arises is how IoT can be used in banking?

2.2.2 The Uses of IoTs in Banking

Consumers are increasingly using their mobile and personal devices to access their financial services. Beyond the practical aspect, this use generates additional data flows for financial institutions that must be managed and secured. Consisting of this set of connected devices, the Internet of Things (IoT) creates new opportunities and invents new modes of online communication and transaction. To use IoT in the financial field, several terminals are used: Laptops, mobiles, ATMs… (Fig. 2).

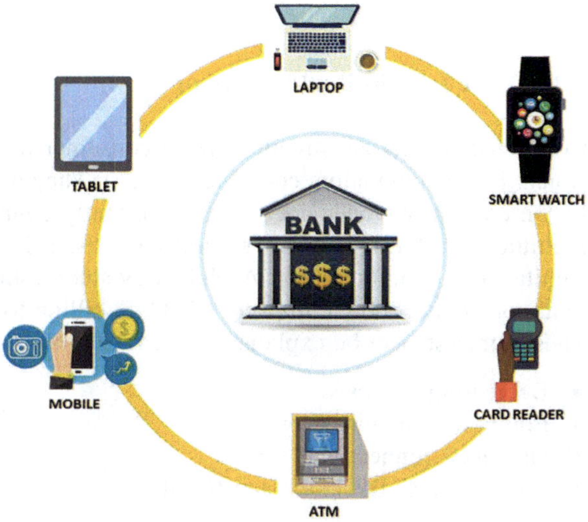

Fig. 2 How IoT can be used in financial field

2.2.3 The Profits of IoTs

Connected objects are special objects that benefit from internet connection to be more than a web access interfaces but rather a tool with additional value in terms of functionality, information and interaction with the environment. Everyday objects connect to the cloud to form the "Internet of Things". The use of IoTs counts several benefits like: having instant data, proposing customized services, using a real-time decision making, benefiting from communicate devices and using the customer IoT interaction (Fig. 3).

2.2.4 IoT and Fintech Industry

The internet of things offer the advantage of enabling the collection and the real-time processing of data flows. The sensor network provides important data on user behaviour. These measures are very useful for service providers and financial technology companies because they can improve customer service, risk management, process of decision-making… We cite some of the services offered by IoT in the field of fintechs:

- **The ability of making easy and invisible payments**: The mixture of sensors (wearable technologies, like voice recognition devices, RFID sensors, and smartwatches) and software facilitates the payment process to the point of

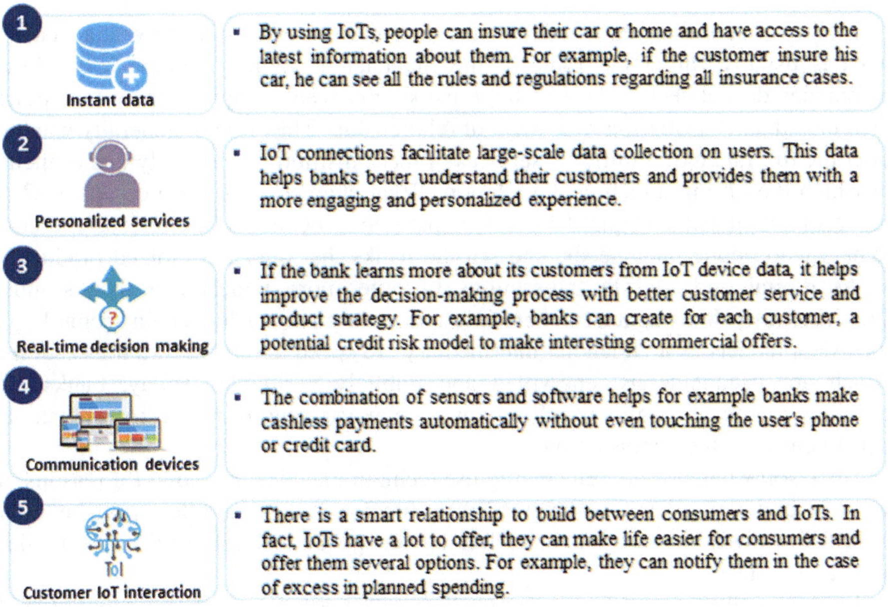

Fig. 3 5 profits of using IoT in banking services

making it invisible. Certainly, one of the best applications of IoT in the financial field remains the optimization of the payment process: you can pay anytime, wherever you are and from any intelligent device.

- **Wearable technology**: today, major financial institutions plan to make wearable devices their primary mean of payment. We can even say that we are preparing a new generation of portable devices: like virtual reality devices or high-tech clothing that one day will replace the current smartwatches…

2.3 The Influence of Digital Risks Using IoTs on Banking Activity

Banks continually monitor the latest technological inventions. They remain attentive to anything that can make them closer to consumers. Indeed, they began with mobile banking, which is used by the majority of banks. But not only, there are other digital risks using IoT that has an influence on banking processes.

Today, the smartphone has become indispensable, banks have relied on this tool to improve their customer services. They are investing in digital, which represents the future of the banking sector. Apps are now an integral part of the day-to-day relationship between a customer and their bank. In just a few steps, individuals can make transfers, transactions, view their accounts or get a RIB wherever they are. They can also make payments which represents also a major growth axis for sector players.

Moreover, the business model of the financial industry is deeply attacked by what is usually called the Fintech (the financial technology companies). These companies do not seek to be like large banks, but want to bypass them to capture their customers by offering new ways of relationship. They do not generally want to offer the full range of traditional services of these institutions, but only one of them, in which they bring a technical evolution. What characterizes these companies? A mode of relationship inspired by the collaborative economy where it is the customers who define their needs, and not the banks that offer customized products.

As digital and mobile transactions become more widespread, banks must simultaneously fight against cybercrime and data piracy. Blockchain technology, for example, acts as a "financial intermediary" to speed up and secure transactions, which are instantaneous, encrypted and publicly verifiable. Today, banks are watching this technology very closely and are experimenting with it to evaluate its effectiveness in their transactions.

Also, some banks were reluctant to use technologies like the cloud for reasons of security and compliance. However, they have to adapt and to optimize their information systems to try to take full advantage of technological opportunities such as Big Data.

In our analysis, we choose some digital technologies that we thought will undoubtedly affect the banking activities. First of all, we have the mobile technology which includes as well the mobile banking (m-banking) and the mobile payment. We have then, all the technological tendencies driven by fintechs such as: the crowdfunding, the cryptocurrencies and the high frequency trading firms. As a third group of digital risks, we select the cybersecurity and the prevention of frauds. Finally, we choose the data intelligence that is mainly concern by the big data and IT analytics and the last one which is the emergence of social media: facebook, twitter...Each digital risk chosen is explicitly explained below:

- **RD1: The mobile technology**
  RD11 Mobile payment: (also known as wallet or mobile wallet) are all transactions made from a mobile phone and debited either on a bank card or on an operator bill, or on an electronic wallet. Instead of paying by credit card or cash, a consumer can use a mobile phone or smartphone to purchase goods and services. In recent years, the technology to support these types of payment has grown in popularity and become widely available.
  RD12 M-banking: mobile banking includes all the techniques used to carry out banking operations from the support of a smartphone (consultation, management, payments, etc.).
- **RD2: The financial technology**
  RD21 crowdfunding: is an internet platform used to finance projects with collecting financial flows. Two parties are involved: the saver who wants to invest a certain amount of money in a project and the holder of the project who does not have the necessary funds to implement it. Both meet on the internet via a dedicated platform: The projects are presented by their holders and savers choose to finance those that interest them.
  Several modes of crowdfunding exist: the gift (with or without counterpart), the loan (with or without interest) and the investment in capital.
  RD22 Virtual money: the virtual currency is a unit of account that has no legal status, it is not regulated by a central bank and it is not issued by any financial institution. We distinguish three types of virtual money: the closed virtual currency used in video games and it has a limited existence in the game; the virtual currency with a unidirectional flow that can be purchased with a legal currency and the virtual currency with bidirectional flow, like Bitcoin.
  RD23 High frequency trading firm: is an ultra-fast mode of submission of orders of stock exchange (the order of the millisecond, the microsecond, or even the nanosecond). It is also fully automated and its positions, both for purchase and for sale, are made by sophisticated algorithms.
- **RD3: IT security**
- **RD31: Cybercriminality**: A criminal offense on a computer system generally connected to a network. It is therefore a new form of crime and delinquency that differs from traditional forms because it is located in a virtual space, the "cyberspace'. In recent years the democratization of access to computers and the globalization of networks have been factors in the development of cybercrime.

- **RD4: Data intelligence**
- **RD41: Big data and IT analytics**: are concepts that refer to the exploitation, processing and analysis of the data. Big Data refers to huge volumes of data (big volume and diversified content) that cannot be handled effectively with existing traditional applications. IT Analytics is the science of examining raw data, applying algorithmic or mechanical processes, in order to draw conclusions about this information.
- **RD5: Emergence of social media**
  RD51 Facebook, RD52 Twitter, RD53 Others…: A social network is a structure that brings together actors, companies or individuals. Social networks have an important place for marketing because they provide information on our activities and preferences [9]. They make it possible to perform analysis for marketers. This is, for sure, an opportunity to know more, to target and to approach customers.

3 Holistic Digital Transformation in Banking

Technology is just a part of the equation of digital transformation. Technological developments and innovations like cloud computing, Big Data, artificial intelligence, mobile applications, Internet of Things, or Blockchain technology are catalysts for digital transformation. They redefine the sector and are accelerators of this change. The holistic view of digital transformation will enable businesses to acquire the core skills they need to succeed in this new connected environment. The dynamism of change has led to a phenomenon that oscillates between technological acceleration and disruptive digital risk; this challenges the status quo of the traditional business model.

This lead us to the fact that only a holistic approach that combines the strategic vision of the digital business and all other areas of development in a unified roadmap, ensures the success of a digital transformation.

3.1 The Banking Processes

According to ISO [10], a process is defined as a sequence of interrelated activities that contribute to the creation of a good or service, which is intended for an end customer or other departments or processes of the enterprise. The standard AFNOR FD X 50-176 distinguishes three categories [11]:

- **Realization process**: they contribute directly to the realization of a product or service, from the detection of the customer's need to its satisfaction. They represent the core business of the organization.
  Examples include research and development processes, design, manufacturing, delivery, maintenance, training, etc.
- **Supporting process**: they contribute to the smooth running of other processes by providing them with the necessary resources, both material and immaterial. These processes include maintenance, provision of equipment or human resources, control of documentation and communication, etc.
- **Management process**: they design the strategy, objectives and drive the quality approach while ensuring its continuous improvement.
  These are, for example, the processes for planning actions, managing improvement or defining and monitoring objectives.

3.2 The Process Map and Digital Risks

The purpose of this chapter is to define a roadmap for digital transformation. To make this happen, we will determine the set of processes that will be transformed at the same time and which are influenced by the same set of digital risks. This practice is built using a matrix that links processes to digital technologies. Its elements represent the degree of impact which is calculated from the answers of a survey sent to the specialists in the financial field. The form of the matrix is displayed in Table 1.

Below the correspondence matrix of digital risks (previously defined) and banking processes (Table 2).

Table 1 The form of the correspondence matrix: process × digital risk

Process	Digital risk RDi			
	RD_{11}	RD_{21}	…	RD_{ik}
P_1				
P_2				
…			γ_{xy}	
P_L				

γ_{xy}: this coefficient denotes the intensity of the dependency between the process x and the digital risk y ($\gamma_{xy} \in [0, 1]$). It is calculated from the answers of specialists in the financial field. The matrix was sent as a survey and the coefficient represents the average of the responses received

Table 2 The correspondence matrix: process × digital risk

				Digital risks									
				RD1: Mobile technology		RD2: Financial technology			RD3: IT security	RD4: Data intelligence	RD5: Emergence of the social media		
				RD11	RD12	RD21	RD22	RD23	RD31	RD41	RD51	RD52	RD53
				Mobile payment	M-banking (Mobile banking)	Crowdfunding	Virtual money (bitcoin, ripple….)	High frequency trading firm	Cybercriminality	Big data and IT analytics	Facebook	Twitter	Others
Macro-processes	Realization processes	R1	Development and Improvement of Products and Services	0.7	0.8	0.5	0.4	0.4	0.8	0.8	0.8	0.8	0.8
		R2	Sell products and services	0.3	0.9	0.3	0.2	0.1	0.4	0.4	0.4	0.4	0.4
		R3	Manage the customer relationship	0.2	0.4	0.2	0.1	0.1	0.4	0.6	0.6	0.6	0.6
		R4	Perform banking operations	0.2	0.2	0.1	0.1	0.1	0.2	0.1	0.1	0.1	0.1
		R5	Realize the market operations	0.1	0.2	0.1	0.5	0.5	0.1	0.2	0.1	0.1	0.1
		R6	Perform security transaction	0.1	0.1	0.1	0.5	0.5	0.1	0.1	0.1	0.1	0.1
		R7	Realize international operations	0.1	0.2	0.1	0.1	0.1	0.2	0.1	0.1	0.1	0.1
		R8	Manage credit request	0.1	0.2	0.2	0.1	0.1	0.1	0.1	0.1	0.1	0.1
	Support processes	S1	Provide legal support and compliance	0.2	0.3	0.2	0.2	0.2	0.2	0.2	0.2	0.2	0.2
		S2	Develop agencies network	0.2	0.3	0.2	0.1	0.1	0.1	0.3	0.3	0.3	0.3
		S3	Develop human capital	0.1	0.1	0.1	0.1	0.1	0.1	0.1	0.1	0.1	0.1

(continued)

Table 2 (continued)

				Digital risks									
				RD1: Mobile technology		RD2: Financial technology			RD3: IT security	RD4: Data intelligence	RD5: Emergence of the social media		
				RD11	RD12	RD21	RD22	RD23	RD31	RD41	RD51	RD52	RD53
				Mobile payment	M-banking (Mobile banking)	Crowdfunding	Virtual money (bitcoin, ripple….)	High frequency trading firm	Cybercriminality	Big data and IT analytics	Facebook	Twitter	Others
		S4	Manage accounting and finance	0.2	0.2	0.2	0.2	0.2	0.2	0.2	0.2	0.2	0.2
		S5	Align and optimize the information system	0.5	0.5	0.5	0.5	0.5	0.7	0.5	0.5	0.5	0.5
		S6	Develop logistics and infrastructure	0.4	0.4	0.4	0.4	0.4	0.5	0.4	0.4	0.4	0.4
		S7	Manage communication	0.3	0.3	0.3	0.3	0.3	0.3	0.2	0.2	0.2	0.2
	Management processes	M1	Pilot and implement the strategy	0.8	0.8	0.8	0.8	0.8	0.6	0.8	0.8	0.8	0.8
		M2	Audit and ensure general and permanent control	0.1	0.1	0.1	0.1	0.1	0.1	0.1	0.1	0.1	0.1
		M3	Ensure the governance and management of the banking activity	0.4	0.4	0.4	0.4	0.4	0.4	0.4	0.4	0.4	0.4
		M4	Pilot and control risks	0.9	0.9	0.9	0.9	0.9	0.9	0.9	0.9	0.9	0.9
		M5	Develop growth drivers and share management	0.7	0.7	0.7	0.7	0.7	0.7	0.7	0.7	0.7	0.7

4 Formal Concept Analysis

The analysis of formal concepts is an analytical practice based on graph theory and strongly linked to Galois lattices [12–17]. It is a theory of mathematical algebra that makes it possible to formalize concepts as basic units of human thought and to analyze data in attribute-object form [18]. Created in the early 80s and over the last three decades, it has become a popular engineering tool for human-centered knowledge representation and data analysis with many applications in many fields.

In this section, we develop the Galois lattice foundations [14, 15] and we illustrate with a simple example the use and the interpretation of the elements of this algebraic structure. The main definitions regarding Galois correspondences and lattices are presented in [17]. Other results on Galois lattices redefined in the field of formal concept analysis appear in [13, 16]. We use here a broad presentation given in [17].

A context C is a triple (O, A, R) where O is a set of objects, A a set of attributes and R is a correspondence.

4.1 Galois Lattice

The set L of all concepts, provided with the order relation $\leq$, has the mathematical lattice structure and is called Galois lattice L(C) of context C. A Galois lattice is a formal concept derived from a relation R. It is a particular graph structure. A lattice being a directed graph, without a cycle and having a minimum node and a maximum node. The Galois lattice corresponds to a partial order induced by a binary relation R between two discrete sets, a set of objects O and a set of attributes A.

4.1.1 The Galois' Correspondence

Two functions Φ and Ψ make it possible to express the correspondences between the subsets of objects P(O) and the subsets of attributes P(A) induced by the relation R. The function Φ associates the set of attributes common to a set of objects, while Ψ, the dual function of Φ, associates the set of common objects with a set of attributes:

$$\Phi : \mathrm{P(O)} \rightarrow \mathrm{P(A)}, \Phi(\mathrm{X}) = \mathrm{X}' = \{\mathrm{a} \in \mathrm{A}/\forall \mathrm{o} \in \mathrm{X}, \mathrm{o\ R\ a}\}$$
$$\Psi : \mathrm{P(A)} \rightarrow \mathrm{P(O)}, \Phi(\mathrm{Y}) = \mathrm{Y}' = \{\mathrm{o} \in \mathrm{O}/\forall \mathrm{a} \in \mathrm{Y}, \mathrm{a\ R\ o}\}$$

The pair (Φ, Ψ) thus presented, defines the correspondence of Galois between the subsets of objects P(O) and attributes P(A) of the context.

4.1.2 Galois Closure

A closure on an ordered set (E, $\leq$) is an application R: E→E which for all (x, y) ∈E×E satisfies the following properties:

- x $\leq$ R(x) (R is extensive)
  if x $\leq$ y then R(x) $\leq$ R (y) (R is increasing monotone)
- if R(x) = R(R(x)) (R is idempotent)
  An element x of E is closed if and only if x = R(x).

The compositions h = Φ ° Ψ and h′ = Ψ ° Φ constitute closure operators of the Galois connection. The operator Φ ° Ψ generates closed subsets of objects while the operator Ψ ° Φ generates closed subsets of attributes.

In a maximum di-clique (A, B) (closed Galois lattice), B is the set of all attributes owned simultaneously by all objects in A, no other object possessing them all. A concept θ is determined by its extension and intension: (Extension (θ), Intension (θ)), where the extension represents all the objects that share the intension attributes, and vice versa, the intension consists of all the attributes that are common to the objects of the extension. These concepts are formal, which means that they are mathematical entities that do not necessarily correspond to a concept of the real world.

With this simple example (Fig. 4), the lattice representation of objects in relation to attributes is canonical. A single reading is allowed with this diagram. The scales of Gutman, can be exploited to exhibit Galoisian classifications. With this same structure and with generation methods association rules from data mining, it is possible to generate associations between attributes. These associations express hidden dependencies existing between the attributes.

Fig. 4 Galois lattice of the context given by Table 3

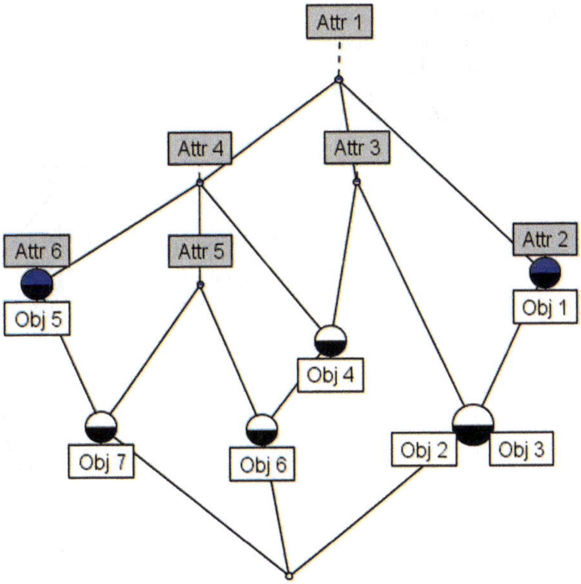

Table 3 Galois context

A	B	C	D	E	F	G
	Attr 1	Attr 2	Attr 3	Attr 4	Attr 5	Attr 6
Obj 1	×	×				
Obj 2	×	×	×			
Obj 3	×	×	×			
Obj 4	×		×	×		
Obj 5	×			×		×
Obj 6	×		×	×	×	
Obj 7	×			×	×	×

The Table 3 shows an example of context represented by C = (O; A; R) where O = {Obj 1, Obj 2, Obj 3, Obj 4, Obj 5, Obj 6, Obj 7} and A = {Attr 1, Attr 2, Attr 3, Attr 4, Attr 5, Attr 6}. This context expresses whether an object X has a Y attribute or not. It also gives, the binary matrix describing the correspondence R of the context C = (O, A, R).

ConExp [19] software constructs a concept lattice from finite formal contexts. A *formal concept* represents the node of the lattice. In the pair (O, A), O represents the subset of the object set and A is the subset of the attribute set. Each node x of the structure represents a formal concept (O, A). The extent O (resp. intent A) of this concept represents the collection of all objects attached to nodes reachable by descending (resp. ascending) mappings from this node x to the bottom (resp. top) element of this lattice (Fig. 5).

Fig. 5 Galois lattice exploration of the context given by Table 3

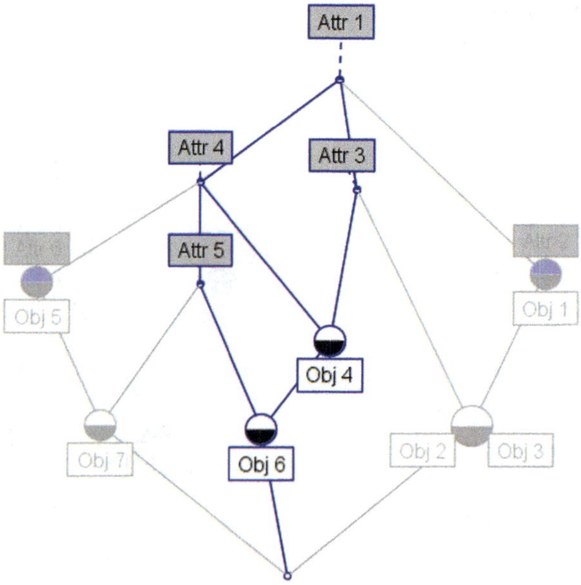

$$\{\text{Obj } 6\}/\{\text{Attr } 5, \text{Attr } 3, \text{Attr } 4, \text{Att } 1\}, \{\text{Obj } 6, \text{Obj } 4\}/\{\text{Attr } 4, \text{Attr } 3, \text{Att } 1\}$$

The base of association rules consist of two parts—that base of strict rules (Duquenne-Guigues base) and base of approximate rules (so called Luxenburger base). ConExp allows calculating base of association rules.

The display format of association rule is a small modification of the format for implications. It is: *No <Number of objects, for which premise holds> Premise = [Rule confidence]=><Number of objects, for which premise and conclusion holds> Conclusion.*

$$3 < 2 > \text{Attr } 1 \text{ Attr } 5 = [100\%] = > < 2 > \text{Attr } 4;$$
$$4 < 2 > \text{Attr } 1 \text{ Attr } 6 = [100\%] = > < 2 > \text{Attr } 4;$$
$$5 < 7 > \{\} = [100\%] = > < 7 > \text{Attr } 1;$$

4.2 *Fuzzy Galois Lattice*

Inaccurate and fuzzy values has led to a great deal of research in the field of fuzzy formal concept analysis [20–22], particularly with regard to the generalization of closure operators concerning fuzzy relations and the underlying problem of fuzzy formal concepts generation. The Formal Concept Analysis consists in inducing formal concepts <Objects, Attributes> from a formal binary context. These formal concepts are based on the following definition: "An object belongs to the Objects subset if it has all the attributes" [23]. Since the relation considered in the formal context is Boolean, an object has an attribute altogether or does not have it, this is a limit. Indeed, in real problems the formal concept analysis is brought to use formal contexts of various natures, obtained by measurements, observations, judgments, etc., where the relation between an object and an attribute can be uncertain, Gradual or imprecise. Various approaches have been proposed in the literature for the extension of the formal concept analysis to the fuzzy frame. The common point between these approaches is the notion of vague formal context. Classical Formal Context Analysis is a data mining method that, in the data, is in the form of a matrix with lines corresponding to objects, columns corresponding to attributes, and matrix entries containing values 1 and 0, depending on whether an object has or does not have an attribute. Many attributes are fuzzy rather than crisp. That is, it depends on the degree to which an object has an attribute (fuzzy degree). The degrees are taken at an appropriate scale of degrees of truth in a real interval [0, 1] or a subset of [0, 1], for more details see works given in [20–22].

Table 4 Formal fuzzy Galois context

	Attr 1	Attr 2	Attr 3	Attr 4	Attr 5	Attr 6
Obj 1	0.7	0.7	0	1	0	0
Obj 2	0.3	0.3	0.7	0	1	0
Obj 3	0.5	0.5	0.3	0	1	0
Obj 4	0.2	0	0.5	0.9	0	0
Obj 5	0.7	0	0.9	0.8	0.5	1
Obj 6	0.8	0	0	0.5	0	1
Obj 7	0.9	0	0	0	0	1

Table 5 Binary Galois context: threshold level 0.5

	Attr 1	Attr 2	Attr 3	Attr 4	Attr 5	Attr 6
Obj 1	1	1	0	1	0	0
Obj 2	0	0	1	0	1	0
Obj 3	1	1	0	0	1	0
Obj 4	0	0	1	1	0	0
Obj 5	1	0	1	1	1	1
Obj 6	1	0	0	1	0	1
Obj 7	1	0	0	0	0	1

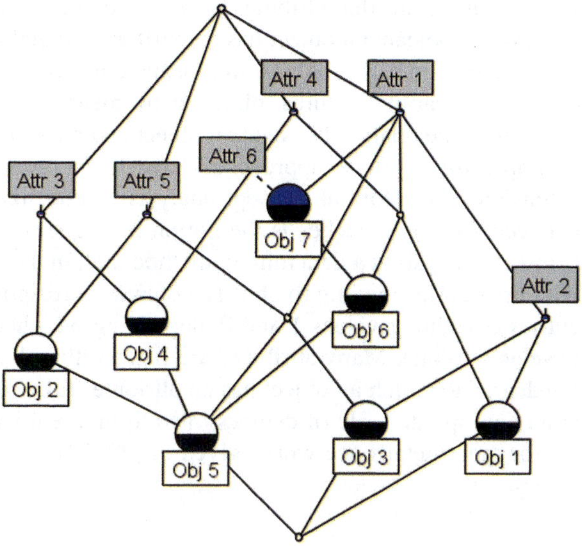

Fig. 6 Galois lattice of the context given by Table 5, threshold level 0.5

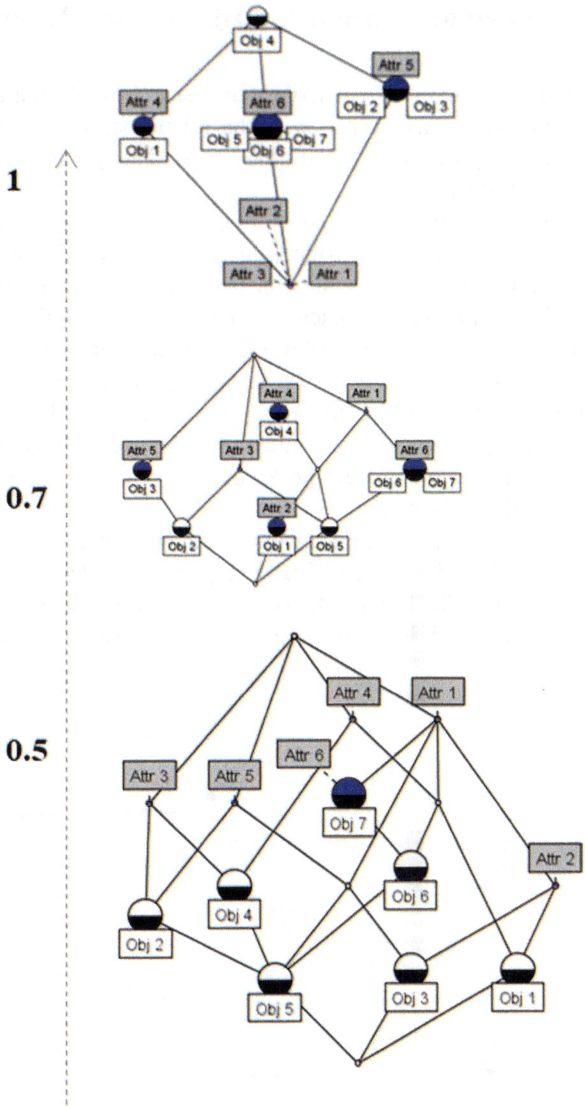

Fig. 7 Ordered Galois lattice of the context given by Table 5, thresholds level: 0.5, 0.7, 1

The data table below represents an example of fuzzy formal context (Table 4).

In this example, we take 0.5 as a threshold. So any element of the context C $(C_{ij)})$ are such that: $\begin{cases} 1; \mathit{if}\ C_{ij} \geq 0,5 \\ 0; \mathit{if}\ C_{ij} < 0,5 \end{cases}$

We obtain then the binary Galois context displayed below (Table 5).

The diagram below shows the concept lattice obtained with threshold 0.5 (Fig. 6).

We duplicated the Galois lattice with other thresholds. We took as an example: 0.5, 0.7 and 1 (Fig. 7).

5 Formal Concept for Discovering Knowledge Patterns

In the following, we describe our research work that relies on the use of the closed Galois lattice associated with the relationship expressing the correlations between banking processes and digital risks. The general architecture of the system is described in Fig. 8.

Figure 8 describes the architecture of the proposed exploration process. The process is initiated by collecting various data of activities and data exchanged in the banking system. This magma of data is then exploited to form matrix structures defining the dependencies or coupling between the elements put in relation. This step requires the choice of objects and attributes in the repository of the bank. These matrices materialize fuzzy contexts, which will be subsequently thresholded (binarized according to the choice of thresholds) to create binary Galois correspondences. Then these contexts will be used to generate Galois' lattices that permit the exploration according to the holistic exploration practice illustrated by an example in Sect. 4.

After receiving the answers of the specialists, we fill the correspondence matrix of the digital risks and banking processes. The impact is represented by values between 0 and 1. The analysis of the correspondence matrix is based on the Galois lattice method. It is therefore necessary to transform its elements into binary values.

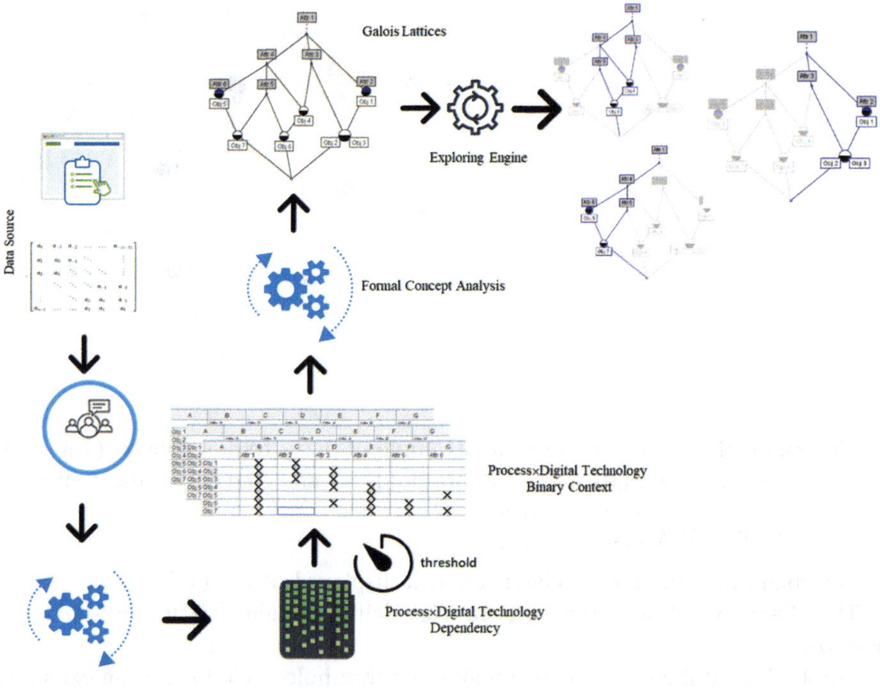

Fig. 8 Exploration process architecture

To do this, we have to determine the threshold from which the values of the matrix C ($C = \left[C_{ij}\right]$; $C_{ij} \in [0; 1]$) become binary:

$$C_{ij} = \begin{cases} m = average\left(C_{ij}\right) \\ 0;\, if\ C_{ij} \leq m \\ 1;\, if\ C_{ij} > m \end{cases}$$

In our analysis, the threshold represents the average of the values of the matrix C.

Below, the binary context obtained using the threshold 0.35 (which represents the average of the values of the matrix C).

$$avg\left(C_{ij}\right)_{1 \leq i \leq 20; 1 \leq j \leq 10} = 0.35$$

The lines of the binary context represent the banking processes:

- The realization processes: $R_{i;1 \leq i \leq 8}$
- The supporting processes: $S_{i;1 \leq i \leq 7}$
- The management processes: $M_{i;1 \leq i \leq 5}$.

The columns of the binary context represent the digital risks $RD_{ik;1 \leq i \leq 5; 1 \leq k \leq 3}$ (Table 6).

Table 6 Process digital technology binary context

	‵RD11	RD12	RD21	RD22	RD23	RD31	RD41	RD51	RD52	RD53
″R1	×	×	×	×	×	×	×	×	×	×
″R2		×				×	×	×	×	×
″R3		×				×	×	×	×	×
″R4										
″R5				×	×					
″R6					×					
″R7										
″R8										
″S1										
″S2										
″S3										
″S4										
″S5	×	×	×	×	×	×	×	×	×	×
″S6	×	×	×	×	×	×	×	×	×	×
″S7										
″M1	×	×	×	×	×	×	×	×	×	×
″M2										
″M3	×	×	×	×	×	×	×	×	×	×
″M4	×	×	×	×	×	×	×	×	×	×
″M5	×	×	×	×	×	×	×	×	×	×

Table 7 The intent and extent of Galois lattice

Intent (digital technology)	Extent (banking processes)
RD23	R5, R6, R1, S5, S6, M1, M3, M4, M5
RD22	R5, R1, S5, S6, M1, M3, M4, M5
RD11, RD21	R1, S5, S6, M1, M3, M4, M5
RD12, RD31, RD41, RD51, RD52, RD53	R2, R3, R1, S5, S6, M1, M3, M4, M5

Below, the result of the Galois lattice.

The ConExp software brings out the following rules below the intent and extent of Galois lattice (Table 7):

- 1 < 7 > "RD11 = [100%] => < 7 > RD12 RD21 RD22 RD23 RD31 RD41 RD51 RD52 RD53;
- 2 < 7 > RD12 RD23 RD31 RD41 RD51 RD52 RD53 = [100%] => < 7 > " RD11 RD21 RD22;
- 3 < 9 > RD12 = [100%] => < 9 > RD31 RD41 RD51 RD52 RD53;
- 4 < 7 > RD21 = [100%] => < 7 > " RD11 RD12 RD22 RD23 RD31 RD41 RD51 RD52 RD53;
- 5 < 8 > RD22 = [100%] => < 8 > RD23;
- 6 < 9 > RD31 = [100%] => < 9 > RD12 RD41 RD51 RD52 RD53;
- 7 < 9 > RD41 = [100%] => < 9 > RD12 RD31 RD51 RD52 RD53;
- 8 < 9 > RD51 = [100%] => < 9 > RD12 RD31 RD41 RD52 RD53;
- 9 < 9 > RD52 = [100%] => < 9 > RD12 RD31 RD41 RD51 RD53;
- 10 < 9 > RD53 = [100%] => < 9 > RD12 RD31 RD41 RD51 RD52;

Holistic exploration

For each δ in DT (Digital Technology Set)

1. *Let $\Theta(\delta)$ be the intention set containing digital technology in the MAP table <Intent, Extent>*
2. *Driving transformation for all processes in $\Theta(\delta)$.*

Results obtained by the exploration of the Galois lattice are complementary and concern the same holistic analysis. Objects materialize banking processes, attributes denote digital risks. Figures 9, 10 and 11 relate simply to the navigation in the lattice. However, the practice of holistic exploration is conducted to better determine in an exhaustive and optimal way the grouping of the processes to be digitized according to a group of digital risks. Exploration could be driven by risk or process. This optimal and holistic navigability is achieved thanks to the Galois structure.

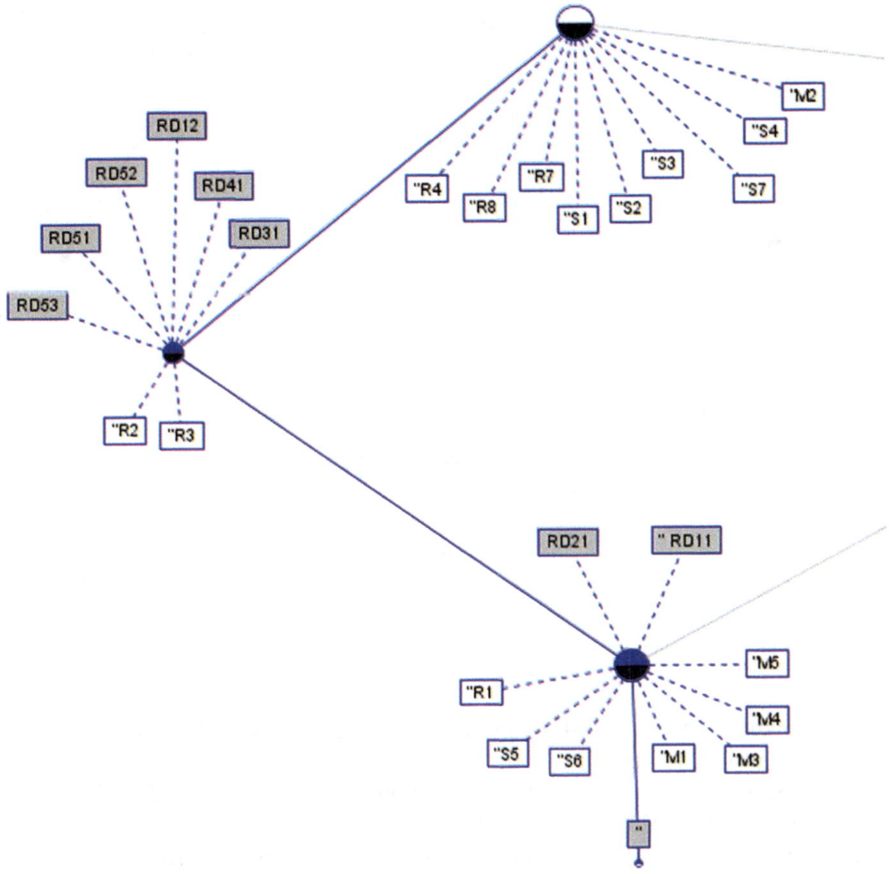

Fig. 9 Exploration process result 1

Among the results to be retained from this analysis, we quote:

- To treat the digital technology **RD23** (high frequency trading firm), we must transform the following processes at the same time **R5** (Execute and perform market transactions), **R6** (Realize security deals), **R1** (Evolution of Services and Products), **S5** (Develop and improve the information system), **S6** (Improve infrastructure and logistics), **M1** (Mentor, lead and execute the strategy), **M3** (Establish the politics and supervise the banking activity), **M4** (Control and manage risks), **M5** (Improve and manage growth drivers).
- To deal with the digital risk **RD22** (high frequency trading firm), we must transform the following processes simultaneously: **R5** (Execute and perform market transactions), **R1** (Evolution of Services and Products), **S5** (Develop and improve the information system), **S6** (Improve infrastructure and logistics), **M1** (Mentor, lead and execute the strategy), **M3** (Establish the politics and supervise

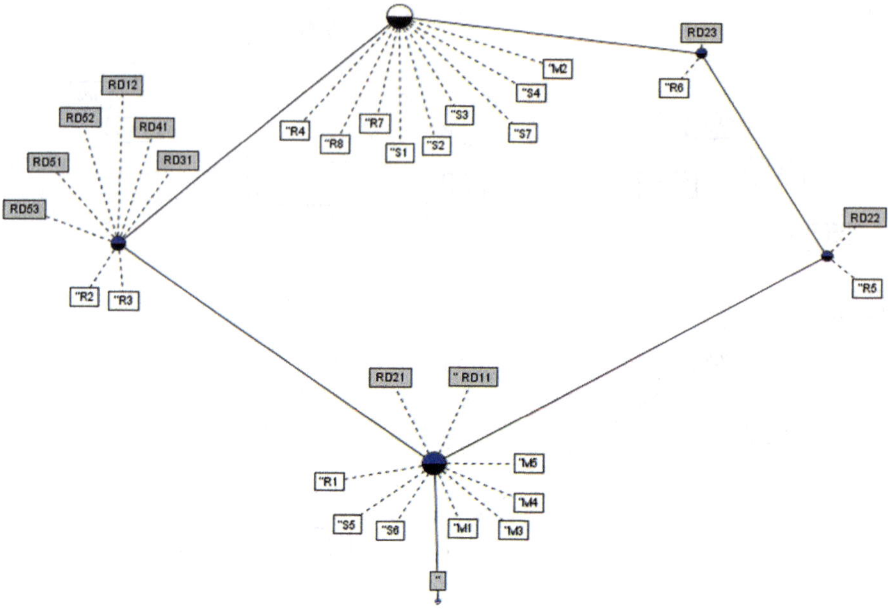

Fig. 10 Exploration process result 2

the banking activity), **M4** (Control and manage risks), **M5** (Improve and manage growth drivers).

- The processes involved in the transformation of the digital risks: **RD11** (Mobile payment) and **RD21** (Crowdfunding) are: **R1** (Evolution of Services and Products), **S5** (Develop and improve the information system), **S6** (Improve infrastructure and logistics), **M1** (Mentor, lead and execute the strategy), **M3** (Establish the politics and supervise the banking activity), **M4** (Control and manage risks), **M5** (Improve and manage growth drivers).
- We have to deal with all these digital risks: **RD12** (M-banking), **RD31** (Cybercriminality), **RD41** (Big data and IT analytics), **RD51** (Facebook), **RD52** (Twitter), **RD53** (others), if we want to transform these processes: **R1** (Evolution of Services and Products), **R2** (Sell products and services), **R3** (Optimize and develop the customer relationship), **S5** (Develop and improve the information system), **S6** (Improve infrastructure and logistics), **M1** (Mentor, lead and execute the strategy), **M3** (Establish the politics and supervise the banking activity), **M4** (Control and manage risks), **M5** (Improve and manage growth drivers).
- All the following processes are involved in the digital transformation: **R1** (Evolution of Services and Products), **S5** (Develop and improve the information system), **S6** (Improve infrastructure and logistics), **M1** (Mentor, lead and execute the strategy), **M3** (Establish the politics and supervise the banking activity), **M4** (Control and manage risks), **M5** (Improve and manage growth drivers).

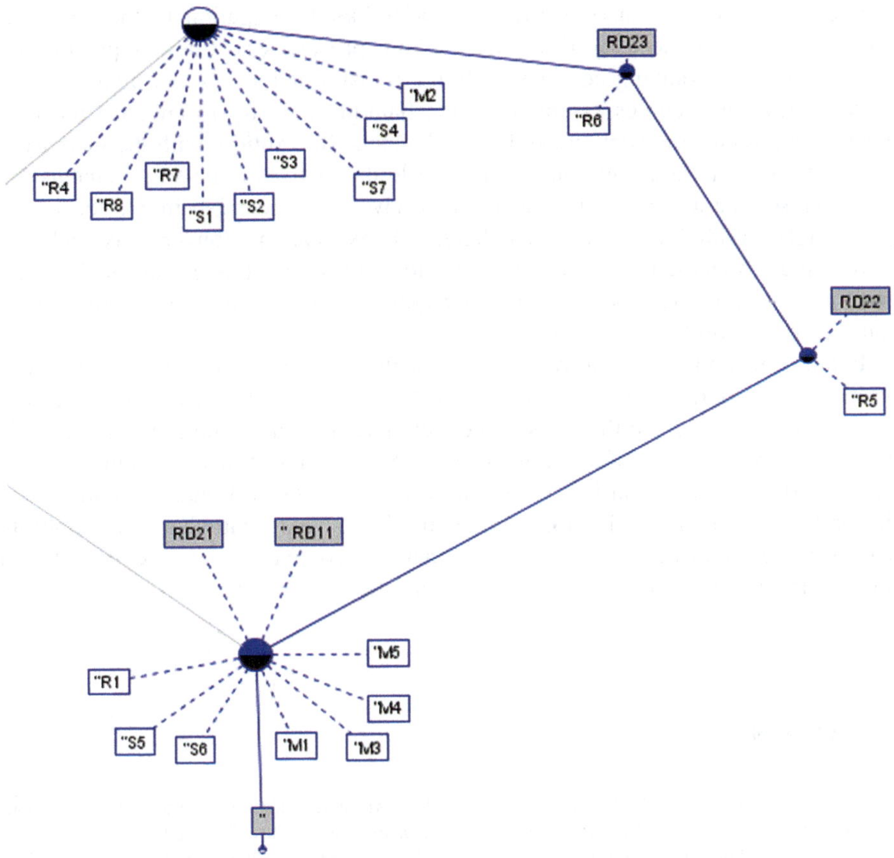

Fig. 11 Exploration process result 3

- For other processes that were not selected in the results of the Galois lattice, they will be treated secondly. It should not be forgotten that all processes are impacted by the digital transformation; the goal is to select those to prioritize.

6 Conclusion

In this chapter, we presented a new approach based on the application of Galois lattices on data captured from the problematic concerning digital transformation in the banking sector. We have shown how to use such a risk-driven approach to lead the digital transformation of the banking industry. Our proposal is holistic and guarantees the treatment of the impacts of a risk or a risk group according to the coherent grouping of banking processes. Risk-based navigation in the lattice of formal concepts induced by risks and banking processes is a new approach and promises the harmonizing conduct of digital transformation in other areas.

Nowadays, it becomes essential for the banking industry to enhance the new experience, as customers embrace digital channels. In addition, with the increasing competition from non-traditional players (digital natives), the requirements on customer service are increasing, which can have a considerable impact on banks profitability. With the growing challenges, banks can no longer make ad hoc incremental changes, it becomes essential to rethink the strategy and technology integration to provide a better customer experience, streamline operations and, in some case, reinvent the whole business.

Finally, beyond the seductive aspect that these technological innovations represent, it is important to remember that the Internet of Things is a mean at the service of a concrete use that must generate value. Leading such a development is above all a business transformation issue with its organizational and human components. In conclusion, an innovative project around the IOT must be carried out like any transformation. It is a long-term commitment and a path that raises multiple and complex challenges. It is essential that the company takes the time to go through the various stages and begins to reap the benefits of its IoT strategy from day one.

References

1. Dey, N., Wagh, S., Mahalle, P., Pathan, M. (eds.): Applied Machine Learning for Smart Data Analysis. CRC Press, Boca Raton (2019). https://doi.org/10.1201/9780429440953
2. Worldpay: Global Payments Report 2018. https://www.worldpay.com/us/insight/articles/2018-11/global-payments-report-2018 (2018)
3. Alan Goode: Goode intelligence: biometrics for banking: best practices and barriers to adoption (2018)
4. Dey, N., Hassanien, A.E., Bhatt, C., Ashour, A.S., Satapathy, S.C. (eds.): Internet of Things and Big Data Analytics Toward Next-Generation Intelligence. Springer, Berlin (2018)
5. Mckinsey Global Institute: Unlocking the potential of the Internet of Things. https://www.mckinsey.com/business-functions/digital-mckinsey/our-insights/the-internet-of-things-the-value-of-digitizing-the-physical-world (2015)
6. Juniper Research: Contactless to account for more than 1 in 2 POS transactions globally by 2022. https://www.juniperresearch.com/press/press-releases/contactless-account-more-than-1-in-2-pos-trans (2017)
7. GSMA Intelligence: The mobile economy sub-saharan Africa. https://www.gsmaintelligence.com/research/?file=809c442550e5487f3b1d025fdc70e23b&download (2018)
8. Mezghani, K., Aloulou, W.: Business Transformations in the Era of Digitalization. Advances in E-Business Research (1935–2700). IGI Global (2019). ISBN 9781522572633
9. Takalkar, V., Mahalle, P.N.: Trust-based access control in multi-role environment of online networks. Springer J. Wirel. Pers. Commun. (2018). 11277-017-5078-2
10. International Organization for Standardization (ISO): ISO 9000:2015: Quality Management Systems—Fundamentals and Vocabulary, International Organization for Standardization. https://www.iso.org/standard/45481.html (2015)

11. Debauche, B., Mégard, P.: BPM Business Process Management, Pilotage Métier de l'entreprise, Lavoisier (2004)
12. Ganter, B., Wille, R.: Conceptual scaling. In: Roberts, F. (ed.) Applications of Combinatorics and Graph Theory to the Biological and Social Sciences, pp. 139–167. Springer (1989)
13. Ganter, B., Wille, R.: Formal Concept Analysis: Mathematical Foundations. Springer, Inc., New York, Secaucus, NJ, USA (1997) ISBN: 3540627715
14. Wille, R.: Restructuring lattice theory: an approach based on hierarchies of concepts. In: Rival, I. (ed.) Ordered Sets, pp. 445–470. Reidel (1982)
15. Wille, R.: Concept lattices and conceptual knowledge systems. Comput. Math. Appl. **23**(6–9), 493–515 (1992)
16. Wille, R.: Line diagrams of hierarchical concept systems. Int. Classif. **11**(2), 77–86 (1984)
17. Barbut, M., Monjardet, B.: Ordre et Classification. Algèbre et Combinatoire, vol. 2. Hachette Université, Paris (1970)
18. Atkin, R.: Mathematical Structure in Human Affairs. Heinemann, London (1974)
19. Yevtushenko, S.A.: System of data analysis "Concept Explorer". In: Proceedings of the 7th national conference on Artificial Intelligence KII-2000, pp. 127–134, Russia (2000)
20. Burusco, A., Fuentes-Gonzales, R.: Concept lattice defined from implication operators. Fuzzy Sets Syst. (2000)
21. Georgescu, G., Popescu, A.: Concept lattices and similarity in non-commutative fuzzy logic. Fundam. Informaticae **53**(1), 23–54 (2002)
22. Yahia S.B., Jaoua, A.: Discovering knowledge from fuzzy concept lattice. In: Kandel, A., Last, M., Bunke, H. (eds.) Data Mining and Computational Intelligence, pp. 167–190. Physica-Verlag (2001)
23. Klir, G.J., Yuan, B.: Fuzzy Sets and Fuzzy Logic. Theory and Applications. Prentice Hall, Upper Saddle River, NJ (1995)

Unmanned Aerial Vehicle and IoT as Enabling Technologies for 5G: Frameworks, Applications and Challenges

R. Sokullu and M. A. Akkaş

Abstract From the very beginning the Internet was envisioned to provide connections—first between people and the documents they used, now not only among people but between machines as well. And with the current pace of technology development, Gartner's predictions rise to include 26 billion devices connected to the internet by 2020 with a global economic added value of $1.9 trillion. Among these the frontline is held by self-driving cars and connected cars. The focus of the chapter is how unmanned aerial vehicles, also simply called drones, can be integrated together with the IoT to become one of the major enablers of 5G and Beyond 5G (B5G) networks. A short introduction into the topic covering the main functionality, advantages and challenges of UAVs is followed by discussing the complementary roles that drones and IoT devices can have in the communication canopy. The second part will concentrate on specific architectural issues, discussing different frameworks and network models that can accommodate UAV and IoT integration. The role of UAVs as major entities in the 5G communication network, the promises and challenges are discussed in detail. Following is an overview of some of the most recent applications based on merging these two technologies, including two ongoing projects.

Keywords Drone communication · UAV · 5G and beyond 5G · Internet of things · Smart city

R. Sokullu (✉)
Department of Electrical & Electronic Engineering, Ege University, İzmir, Turkey
e-mail: radosveta.sokullu@ege.edu.tr

M. A. Akkaş
Department of Computer Engineering, Bolu Abant İzzet Baysal University, Bolu, Turkey
e-mail: alperakkas@ibu.edu.tr

N. Dey et al. (eds.), *Internet of Things, Smart Computing and Technology: A Roadmap Ahead*, Studies in Systems, Decision and Control 266,
https://doi.org/10.1007/978-3-030-39047-1_10

1 Introduction

The last decade has seen an unprecedented expansion in the range and variety of applications involving drones. While in the past they were primarily associated with military applications, the reduction in size and weight combined with enhanced communication abilities have given rise to an enormous number of civil applications [1–3]. In general, the term "UAV" or "unmanned aerial vehicle" is used to define a large category of flying devices that are remotely operated or supported from the ground. It covers a vast range including small planes, balloons and smart flying robots, varying in terms of operational altitude, flying and maneuvering abilities, endurance (physically and in terms of power consumption), weight and cost. Numerous classifications have appeared in literature but it is most common to group them based on the flight altitude they maintain and their size. UAVs covering altitudes above 10 km are considered HAP (high altitude platforms), while those below that are considered LAP (low altitude platforms). UAVs also vary greatly in size and weight but generally those for commercial use are below 10 kg, many off-the shelf models weighing less a kilogram. They also differ in their payload abilities which can vary from several hundred grams to 5–7 kg [4, 5]. These flying vehicles, commonly known as drones, have some similar major functional blocks: the set of sensors, the autopilot unit, the communication block, the ground control unit and the battery (Fig. 1). Most typical components of the sensor block include the barometer and the pilot tube, the distance, temperature and magnetometer sensors as well the GPS and an IMU (inertia measurement unit). The barometer sensor is used to gather information about the altitude by sensing the pressure; the pilot tube in turn, mounted in the fixed wing aircrafts helps measure air velocity to be used when flying at stall speed. To preserve its altitude a drone must maintain certain minimum speed called stall speed. The IMU sensor, considered one of the most important elements of the flight control block, is used to ensure stabilization (three or six axis). Distance sensor evaluates the obstacles in the surrounding environment, and together with the UAV controller ensures that no collisions occur. The role of the autopilot block, the next main component in the UAV, is to interpret the data from the other modules, send commands to the electronic speed control and to secure the balance of the aircraft. It also controls the servos and additional motors using instructions from the remote control unit. The communication module, the third major component, consists of a remote controller, telemetric module and video transmitter. Most common frequencies used for the communication link are 433 MHz, 2.4 GHz or 5.8 GHz, however there is ongoing research examining also the possibilities of the terahertz band as well as licensed bands for propriety and special purpose drones [6]. Two flying modes are common for UAVs: autonomous flying mode and remotely controlled flying mode. For safety reasons, law regulates that unlicensed UAVs should continuously be in the RC coverage area. Sensor data is sent by the telemetric module to the ground control station for evaluation. It can also be fed into a camera and transmitted to the GCS (ground control station), which is the fourth main component, involved in estimating the location and the

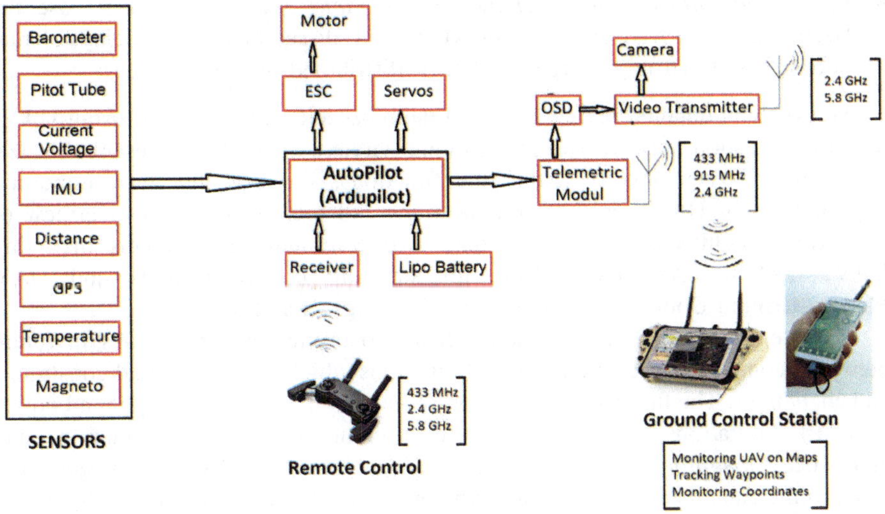

Fig. 1 Major functional blocks of flying unmanned vehicles

general performance of the UAV. Battery is the final component in this list. It is also very important and usually a LiPo (lithium polymer) battery is used because of its high capacity and low weight.

Drones have some important advantages which make them so desirable and easy to use in numerous civil applications. First of all, they are cost-effective and can be swiftly deployed, suitable for unexpected or limited-duration missions. Second, UAVs can be used to establish line-of-sight (LoS) communication links for short ranges which in many application scenarios leads to increased flexibility of deployment and better performance over direct communication between source and destination. Furthermore, the maneuverability of these small flying devices offers new opportunities for network performance enhancement, because it allows to dynamically adjust the state of the UAV to suit the communication and application environment. When a UAV experiences good channel conditions to a given ground terminal/station it can increase its transmitting rate and/or also lower its speed to sustain good wireless connectivity for a longer period.

Such benefits make UAVs a very integral part of future wireless communications which are envisaged to support an ever increasing set of diverse applications with orders-of-magnitude capacity improvement. In [7] the authors define the following 3 major roles for UAVs in the area of communications:

- *UAV-aided ubiquitous coverage*: in this case UAVs are deployed to facilitate and support already existing communication infrastructure;
- *UAV-aided relaying*: in this case the UAVs are employed to implement wireless connectivity between two or more remote users/groups of users, which do not have a direct and reliable links for communication;

- *UAV-aided information dissemination and data collection*: in this case the UAVs are deployed for the collection and dissemination of delay-tolerant information to (from) a large number of distributed wireless devices.

An excellent detailed review covering the most relevant aspects of using UAVs as communication network entities in future wireless networks is presented in [8]. The authors systematically define the functions, services and most important requirements of UAV based communications and present a concise overview of networking architectures. They propose two communication models involving UAV: UAV-to-UAV and UAV-to-Infrastructure communications and UAV Fleet-to-Ground Control communications (see Figs. 2 and 3).

However, these high expectations related to drones and their roles in future networks come at a cost. Drones have limitations which pose additional constraints and challenges. The first most obvious one comes from their size, weight and power (SWAP) limitations which restricts their communication, computation, and endurance capabilities. Furthermore, the high mobility of UAV systems generally leads to the establishment of highly dynamic network topologies, a fact that poses additional challenges for network management and operation. Also due to the mobility of UAVs and most often the lack of fixed backhaul links and centralized control, interference coordination among large numbers and formations of drones can become a serious problem. Managing groups of drones will require additional communication resources in terms of additional control and non-payload communications (CNPC) links. To ensure safety, real-time control, collision and crash avoidance much more stringent latency and security requirements are imposed on

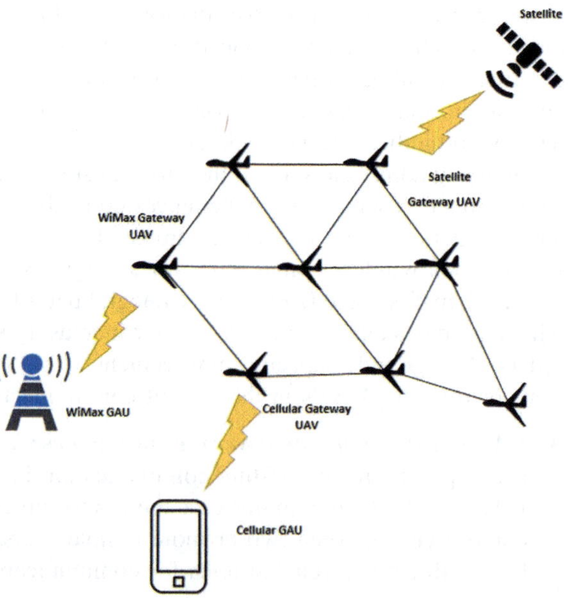

Fig. 2 UAV-to-UAV and UAV-to-infrastructure communications

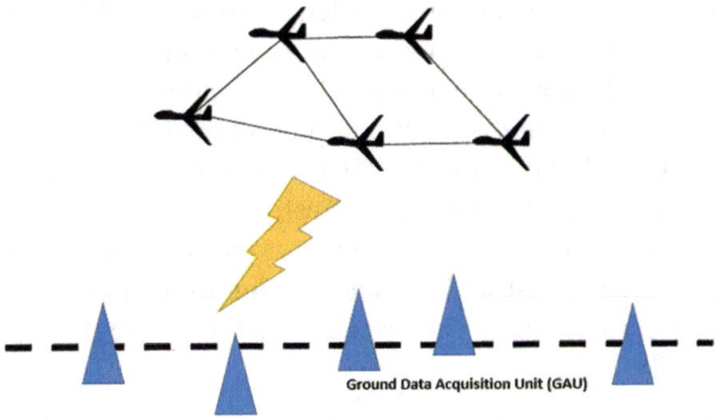

Fig. 3 UAV fleet communication with ground control

the network. Some researchers propose UAV based architectures containing two types of links: control and non-payload communications (CNPC) links and data links [7]. The CNPC links, due to their critical functions in controlling the drones' movement and operation, will require protected spectrum and immensely increased and effective security mechanisms (Fig. 4).

In literature the energy consumption issue of drones has been considered from different angles. A lot of research has focused on exploring optimal path trajectories and limiting the number of drones required to provide coverage of a given area of interest (AoI) [8]. Another important angle is determining the optimal operational altitude. The greater the altitude of the UAV, the greater the probability of establishing LOS high QoS links, however this leads to increased path loss caused by increased user-infrastructure distances. It also has additional environmental and managerial problems which mandate finding a good balance in terms of optimum flying height. Other aspects debated in the academics community relate to the procedures of how drones will access the wireless channel when they have network entity roles (i.e. participate in package delivery, be used for traffic surveillance and disaster management operations etc.) Cognitive radio technology (CRT) and spectrum sensing appear to be the most viable solutions for these problems so far.

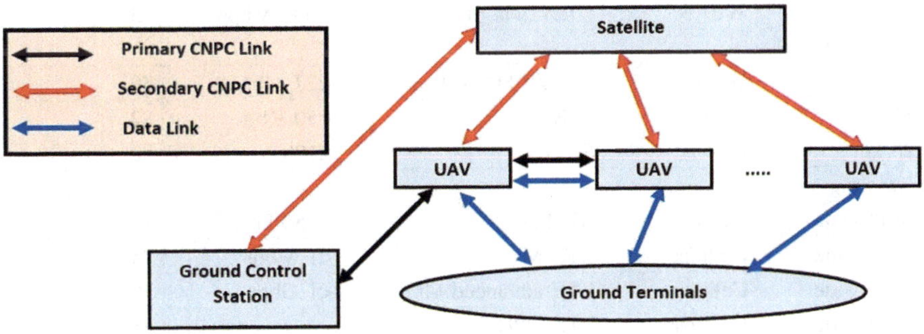

Fig. 4 Communication channels for drone aided networks

D2D communications complemented with UAVs are also both and opportunity and a challenge for the engineering community [9].

In addition to being used in a variety of applications today, drones are also expected to be one of the important enabling technologies for 5G and Beyond 5G (B5G) cellular networks [10]. UAVs have the inherent ability to complement and even substitute existing terrestrial cellular networks by serving network users experiencing strong shadowing or interference conditions, supporting overloaded or damaged terrestrial BSs, and serving users around idle BSs in ultra-dense networks and in rural areas. The primary benefits of the drone technology have been specified briefly in [10] as:

- Drones are able to operate in environments which are inaccessible, undesirable or dangerous for humans
- Drones are deployed on-demand and easy to relocate when required
- Drones can help improve coverage by establishing higher line-of-sight (LOS) connections for ground users, especially in terrains with crowded or varied landscape
- Drones' flying altitude can be adjusted to meet QoS based on user density, desired data rate, interference, shadowing effects etc.

A list, summarizing the range and delays of the communication technologies that can potentially be utilized with UAVs, is given in Table 1.

Another technology, which is considered a major enabler for 5G and Beyond 5G (B5G) networks is the Internet of Things (IoT). There are many different definitions of the term. One of the most detailed and encompassing ones is given in [11] as follows: the IoT defines a technological environment, which is continuously and rapidly developing based on cutting edge high-tech developments. Conceptually is

Table 1 Wireless technologies available to UAVs

Range	Category	Technology	Data rate	Latency (ms)
Short range				
60 m	WPAN	Bluetooth 4.0	<1 Mbps	50
<100 m	WPAN	Zigbee	<250 Kbps	50
<250 m	WPAN	802.11a/b/g/n/ac	<600 Mbps	75
Middle range				
<1 km	WPAN	WAVE 802.11P	<27 Mbps	50
<15 km	LPWA	LoRA	<50 Kbps	82
<20 km	LPWA	SigFox	<100 bps	82
Wide range				
World wide	Cellular	NB-IoT	<250 kbps	75
World wide	Cellular	LTE-M	<1 Mbps	75
World wide	Cellular	LTE advanced (4G)	<1 Gbps	50
World wide	Cellular	LTE D2D	–	25
World wide	Cellular	5G	<10 Gbps	3

envisages the integrated operation of an enormous number of devices, some very small in size, known as "smart objects", which will operate globally in unison. Even though some will still be directed by humans, the greater portion will operate autonomously, driven by highly sophisticated specified software, that will capture and control their behavior. Through the IoT environment, simple objects will become active participants in many daily human activities, using various communication technologies. Even though these two technologies, the UAV and IoT have started in answer to very different application challenges and requirements in the last several years there is a strong tendency of merging between them. UAVs large coverage, maneuverability and possibility to create good communication links complement the flexibility, application oriented nature and versatility of IoT. A number of isolated, quite different studies have been reported in the last several years related to this issue. In the following we try to provide a concise review of this topic covering most recent works on the subject of IoT and UAV.

2 Frameworks for Integrating IoT and UAVs

Fulfilling the main expectation from the IoT technology "to connect things anywhere, anytime providing any service" can be enhanced by UAVs thanks to their autonomy and flexibility, however currently existing network architectures do not accommodate this integration. That is why an important research area is determining the framework structure that can incorporate UAVs in the IoT vision. The authors in [12] propose and evaluate a communication framework that integrates UAVs in urban IoT environments. UAVs connect to the ground station, using novel multihop multipath network architecture. The authors conducted real-world experiments which illustrate the operation of the framework and prove that it can be effective and reliable when dealing with local congestion. A framework which considers UAVs as platforms to assist communication base stations in an IoT surges to UAVs in an IoT environment is proposed in reference [13]. The allocation of bandwidth to the UAVs is based on a game-theory approach and in modelled as a non-cooperative evolutionary game. Reference [14] introduces an optimization framework for aerial sensing also considering an IoT augmented virtual environment. The goal is to create an architecture which will allow users to navigate scenes of their choice based on augmented-reality or virtual-reality devices. The authors illustrate their idea using a scenario likened to virtual human teleportation. The work in reference [15] proposes a new mobile edge computing framework for IoT which is based on air-to-ground integration. To illustrate their framework, the authors discuss 4 different use cases and showing that that the proposal satisfies the most important requirements of 5G—low latency, high mobility and high throughput. UAvs can be used for efficiently collecting data from ground based IoT devices. The authors in [16] propose a framework that optimizes the deployment and mobility of UAVs. Their approach is based on clustering of IoT devices which allows minimizing their transmission power while preserving the required reliability. Reduction in the energy consumption for transmission purposes allows the UAVs to serve devices on the ground longer, i.e. the lifetime of the drone networks is increased.

In [9] the authors present a novel framework incorporating UAVs and IoT with a specific focus on the security and privacy issues (Fig. 5).

Their proposed framework aims to ensure security, safety, and privacy for mobile IoT while ensuring the fulfillment of required services by means of a set of distributed systems. Under this framework, users can deploy and query their "flying" objects (drones) securely within the limitations of privacy regulations. The framework consists of three main elements: the control center, the distributed systems, and the embedded solutions on the UAVs. For the drone-to-drone communication mobile routing algorithms are devised, while communication with the control unit is realized over the Internet. The whole operation of the ecosystems is controlled through the Control center (CC), which incorporates both traditional and novel elements specific to the IoT environment. The newly introduced functional block relate to searching operations, group instructing and information retrieval. The second major entity, Distributed Systems (DSs) is used to realize tasks like monitoring, management, and reputation, which include registration and authentication. The last major entity, represents the "flying" things.

To realize the potential of UAVs as network enablers for IoT two important factors have to be taken into consideration: how to ensure energy-aware

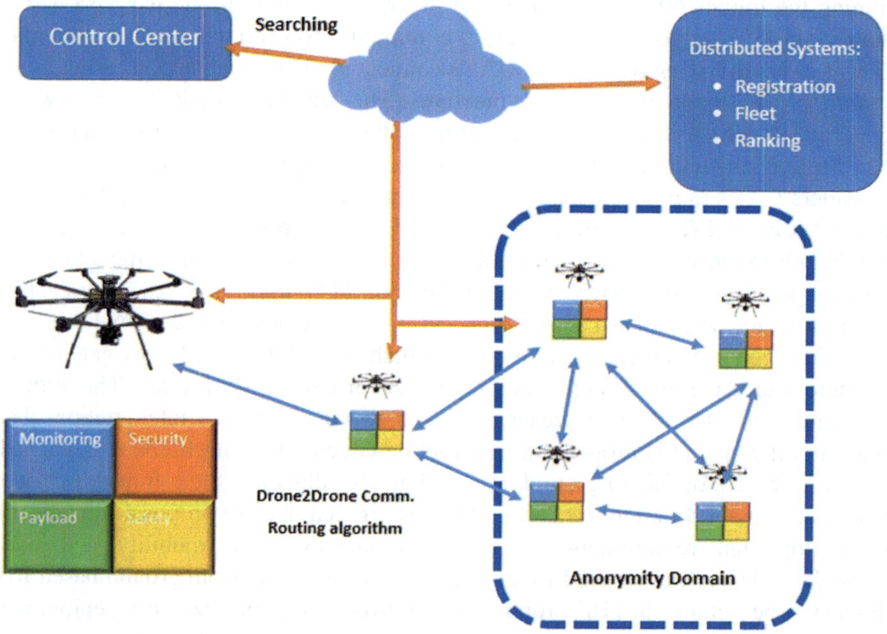

Fig. 5 Security and privacy oriented framework incorporating UAVs and IoT

deployment and how to provide operation in an energy efficient manner [5]. The deployment of UAVs is not a simple issue because it has to allow UAVs to replenish their energy supply without obstructing the network operation, while the second refers to the optimization of the UAVs operation to minimize energy consumption while achieving the desired goals. The authors in [16] considered a scenario for using a mobile swarm of drones to serve ground users, represented by discrete IoT devices. Their analytical study defines the optimal 3D trajectory of each UAV in such a way that it minimizes the total energy used for maneuvering the UAVs when serving the IoT devices. The authors show through simulation that the total transmission power of the IoT devices is reduced by 45% compared with a case, in which stationary aerial base stations are considered. In another work [17] the same team studies the more extended case, where they consider different classes of users on the ground: users which establish a link with the UAV, called downlink users (DUs), and directly communication users, called D2D users. The authors reported that, due to the interference between the two types of users, the total system sum rate first increases but then, with increasing the density of D2D users in the area it declines. Furthermore, with the increase of UAVs' altitude and the probability of establishing LOS (Line Of Sight) links there is increased interference between the D2D users.

Datta et al. [18] discusses how the Drone-IoT platform can be used to serve the needs of emergency services. The paper presents an advanced set of IoT and Web of Things (WoT) services which are planned to augment the EURECOM IoT Platform. These additional entities in the framework which will specifically target services such as early assessment of incidents in roads and/or structures, delivery of small medical equipment as early support to ambulance response, fast response to civil-triggered alerts for the emergency responders (e.g. police, firefighters, ambulance), search operations, and surveillance in crowded events. These extensions include UAV operations specific web services for real time video, defining UAV tasks and missions, and collecting telemetry data, given in the Fig. 6.

In this upgraded architecture UAVs are treated as a different type of IoT devices. They stream telemetry data and video data using the MQTT protocol, and store it locally for off-line processing. The authors claim that their work is a significant step towards realizing the so called "Drone-as-service" concept.

Extending the interplay between UAV-enabled connectivity and IoT functionality is considered also in [19], where the authors consider the case of establishing UAV aided communication in case of large scale disasters when infrastructure, buildings and power systems fail. The study is based on a detailed simulation using USARSim. It includes establishing an end-to-end communication network system for emergency management field workers, a Task Allocation and Coordination Strategies sub-system and a Localization and Navigation subsystem.

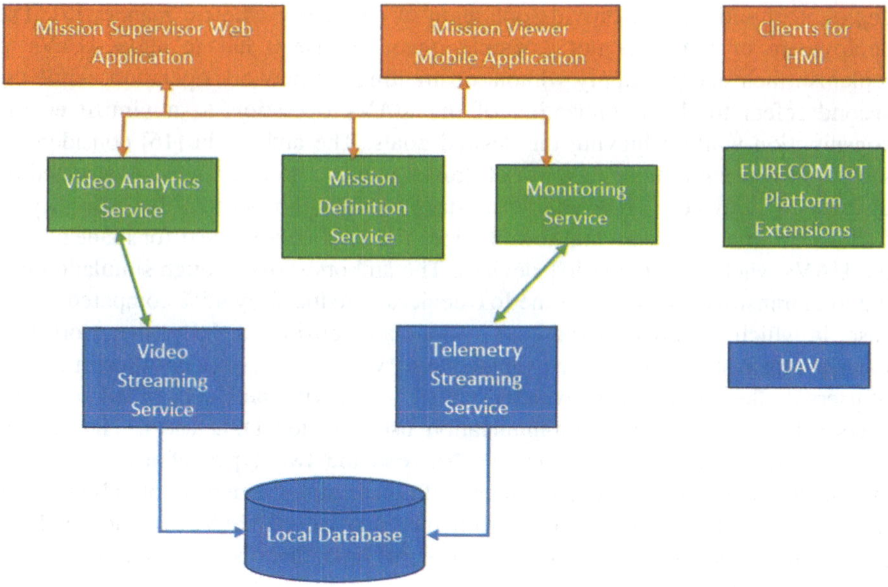

Fig. 6 Emergency services oriented extensions for the EUROCOM drone-IoT platform

3 IoT Applications Integrating UAVs

5G is expected to enhance mobile broadband, enable applications that require ultra-reliable, very low latency and very high availability networks. Without providing an exhaustive list we can name some of the most popular applications that are envisioned for 5G as improving traffic safety and control, logistics, tracking, and fleet management, extensive support for a variety of industrial applications and remote manufacturing. Other so called "smart applications" include smart agriculture, precision farming, smart buildings, smart meters, smart homes etc. However, until recently, the communication networks were mostly human-centered, i.e. their design and operation was aimed at supporting humans and their communication (Human-to-Human networks—H2H). However, this paradigm is bound to change drastically in 5G and B5G networks. More and more data will be passing through the network to connect non-human entities. Thus UAVs in 5G are expected to help close the communication gap between Machine-to-Machine networks and Human oriented networks and services (see Fig. 7) [20].

A large number of 5G use cases using drones involve also automation and robotics. These will pave the road to providing a variety of drone based 5G applications for a wide range of industries. Among the top use cases for 5G-enabled drones, we consider applications related to transportation and logistics, agriculture, communications and "smart" applications of different kinds and scope. A main

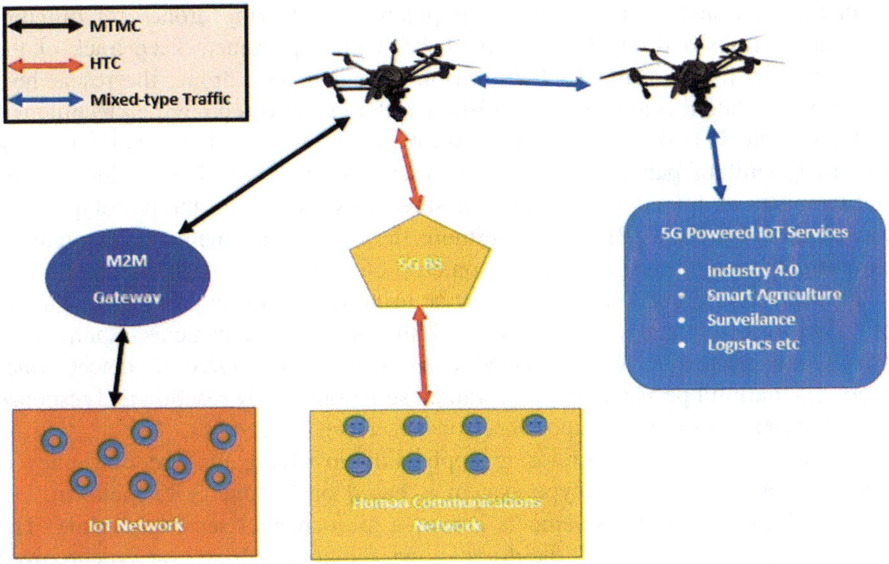

Fig. 7 UAVs closing the gap between H2H and M2M communications

factor that makes these applications feasible is the possibility to establish autonomous and beyond line-of-sight control. One of the most striking examples can be vendors such as Amazon that have started using UAVs for cargo transport on demand. Another example is the Aquila project, one of the most ambitious projects undertaken by Facebook, which plans to utilize "swarms of drones" to provide "WiFi in the sky" service to remote areas. Such projects can be viewed as major factors to unleash drones as 5G enablers and pave the way for their exploitation as vehicles of communication for future wireless networks.

4 Agriculture and Environment Applications

Agriculture is one of the commencing 5G candidate applications to benefit from merging UAV and IoT technologies. In [21] the authors discuss how drones equipped with appropriate cameras, sensors and integrating modules can help achieve efficient precision agriculture and present a short overview of available solutions.

The authors of [22] propose a new method for precision agriculture. The goal is to distinguish between sparse and a dense field formations based on data provided both from the drones and satellite systems. While satellite information may not be accurate enough drone images have high resolution and precision but drones have limited flying time. The suggested solution involves image processing and uses image statistics to minimize drone activity in order to achieve extend their operational time.

In [23], the authors point out the popularity of utilizing drones for different agricultural tasks and develop a method that will help farmers keep track of the work in their fields. Using RGB-D sensors mounted on a drone, the researchers propose a method that allows distinguishing different various ploughing techniques.

Filev et al. [24] discusses an IoT prototype device that allows monitoring of various agricultural parameters. A variety of sensors is networked to provide detailed soil temperature, humidity, moisture measurements. Tests with the prototype were carried out in Sao Paulo, Brazil. Furthermore, the system is augmented with reference climate data to help making decisions on crop life and its sustainability.

Reinecke and Prinsloo [25] proposes the using drones for improving the quality of monitoring different types of crops. Using specialized cameras mounted on drones and related software selected crops can be monitored to detect water shortages, harmful pests and other unwanted situations. Early reaction and response in such cases is of utmost importance for preserving high yields.

Another IoT enabled monitoring application involving the use of drones is described in [11]. The prototype system is based on Raspberry Pi platform. For water and air quality measurements it uses a multichannel sensing module. The collected data is first saved to the drone's own memory and later on synchronized with a remote server database. The server side uses advanced processing algorithms to evaluate the data. A mobile IoT application platform is developed for data visualization and statistical analysis.

5 Communications Related Applications

Another important aspect where UAVs can play a decisive role is assisting the communication functions of 5G IoT. In [26], the authors discuss the issue of drones as base stations (Drone Base Stations—DBS) for IoT services. Owing to the dynamic distribution of IoT users drone based stations, which can be flexibly deployed for hotspot areas, can potentially improve the wireless latency of IoT users by mitigating the heavy traffic loads of macro BSs. However, such a scenario is envisaged to pose two major challenges: the DBS deployment should address heavy traffic demands and the traffic load should be balanced among macro BSs and DBSs to avoid traffic overload and congestion. Therefore, the authors propose a traffic load balancing scheme called TALL, to minimize the wireless latency of IoT users. The proposed scheme acts at two stages where first DBSs are placed in locations with higher user densities and then the traffic loads among BSs are allocated to further minimize the latency ratios of DBSs.

Another proposed solution in this respect is resented in [27]. The authors investigate the feasibility and significance of creating a multi-tier drone network as compared to the case of a single tier of drones. The architecture and components of the proposed system are given in the figure below (Fig. 8). They address two major issues—the spectral efficiency of the downlink and the drone deployment for load balancing and conclude that for given load conditions the multi-tier solution is more efficient.

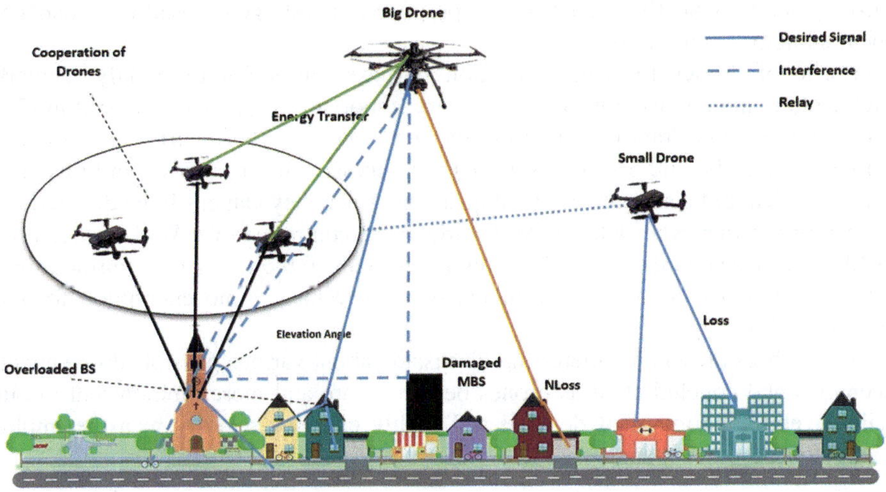

Fig. 8 Multi-tier drone integrated network architecture

In their work [28], the authors address a very interesting angle of UAV and IoT crossline—i.e. how can IoT devices embrace cognitive and deep learning functions. It has been long considered that IoT devices are small device, restricted both in size and computational abilities so cognitive applications are a far reach. The subject of "intelligence of IoT devices" has so far sounded quite out of place. However the authors in [28] argue that these devices will have to embrace some kind of cognitive functions in the next-generation networks. Their pioneering work focuses on ways to use drones together with IoT to realize the paradigm of what they call "Flying IoT". They consider an interesting application known as "Follow the leader", which automatically detects, tracks and follows a moving human target. State-of-the art image classification methods known in literature require gigaflops to manage a single image, computing in real time requires much more ALU power, much larger and heavier devices than small drones can afford. The challenge in solving this task is balancing between performance of image processing methods and energy consumption. Instead of going after an ideal most probably unachievable goal given today's technology super CPU, their novel approach focuses on designing a new sensor-cloud architecture which will allow to partition the process of data collection and processing between the edge and the cloud.

Another perspective of using UAVs to aid communication functions of IoT devices is considered in [29]. The authors consider cooperative UAV communications and model a swarm of micro UAVs (MUAVs) which are used for the 3D exploration of a hazardous area i.e. examining the situation after a nuclear power plant failure or a volcano eruption. They propose communication aware steering algorithms for the cooperative operation of MUAV swarms, which are optimized for spatial efficiency with controlled mobility. They especially focus on air-to-ground connectivity control and propose a novel agent based role management strategy for multihop links.

Usage of drones for communication purposed does not raise only network architecture and connectivity issues. A very important aspect is considered in [30] where the authors draw our attention the questions of cybersecurity, privacy and public safety. The authors point out that the pervasive use of drones will raise not only technical but also many societal questions since they can easily be directed for malicious use like cyber-attacks. Many drones are controlled over Wi-Fi which uses IEEE 802.11 standards, which is by its nature vulnerable to attacks. Furthermore, even professional drones can be hijacked because there is no encryption for the on-board chips.

The authors present an interesting discussion about various possible drone attack scenarios and conclude that as drones become more and more omnipresent in our daily lives the adoption of detailed and highly effective security measure implementation becomes mandatory.

6 Surveillance Applications

Another large group of applications which require the integration of UAVs and IoT are surveillance applications. An interesting example in that line is the work presented in [30] where a new method for discovering archeological sites combining satellite and UAV communication is suggested. The three step method includes satellite imagery, drone imagery and computer modelling to increase precision. First satellite images are pre-processed, followed by incorporating controlled information from drones with a mounted camera and finally a 3D model of the site is designed. UAV with onboard sensors is used first in a coarse search and then in a more refined search. Low-resolution image data is used to provide feed to the control algorithm of the drone to determine the trajectory for close inspection. The second phase, taking images from the most promising objects, results in a sequence of images which are later combined to create the 3D map of the area of interest. This three step procedure allows creating a good balance between resolution, accuracy and energy consumption of the drone.

In [31], the authors consider a surveillance application related to IoT data collection inside thick forests (wild animals, forest conditions etc.), called "IoT-Wild". Most IoT solutions suffer from strong restrictions in data transmission size and frequency due to the energy consumption limitations and need for extremely long network lifetime. To overcome these restrictions, the authors propose a forest monitoring system which incorporates low-cost, low-power IoT sensors and devices together with UAVs deployment in order to balance between system lifetime and extended data collection. (See Fig. 9.) A major challenge in this case is the establishment of a reliable link between the IoT devices placed in the forest, under a thick canopy and the drone which is hovering above the forest vegetation. Traditional network metrics, such as maximum range, transmission rate, and packet

Fig. 9 Integrating drones and IoT in the wild: "IoT wild"

loss are considered. Even though compared to an unobstructed scenario, the network performance is much reduced, the authors are optimistic. A continuation of this interesting work maybe to consider more sophisticated drone trajectory planning together with application specific routing among the IoT devices, which will ensure connectivity at specific locations with good channel conditions.

The fast technological developments, which result in reducing both the size and price of drones, while increasing their capabilities has made them so accessible that a new problem has been created for the community: "the amateur drone's problem" [32].

When the Federal Aviation Administration (FAA) started requiring owners to sign up in the beginning of 2016, only in the first 20 days of January there were over 200 000 drones registered in the US. While drones have very useful applications, operation of amateur drones in specific areas around airports, crowded event areas, privately owned territories etc. can actually cause serious problems invading personal privacy and creating hazardous even life endangering situations. Unfortunately legislation worldwide has been slow in defining precise rules and regulations for amateur drone operation. In many cases, even if available it has been slow or inadequate to provide a clear-cut solution. Furthermore, existing anti-drone technologies are not sufficient either. Amateur drones are generally very small and light, and appear quite unexpectedly, which makes them hard to detect. Even more, the addition of artificial intelligence and advanced materials into the latest designs of amateur drones make them very resilient to counter-capture and stealth. Most important however, existing anti-drone technologies are designed for specific drones or scenarios which greatly limit their general application. Thus, a group of researchers have focused their attention of providing a drone-detection a reaction system for amateur drones called DragNet. It will not only incorporate IoT and drones but will also introduce cognitive intelligence into the IoT system. They define the vision of Cognitive IoT as a new network paradigm, where (physical/

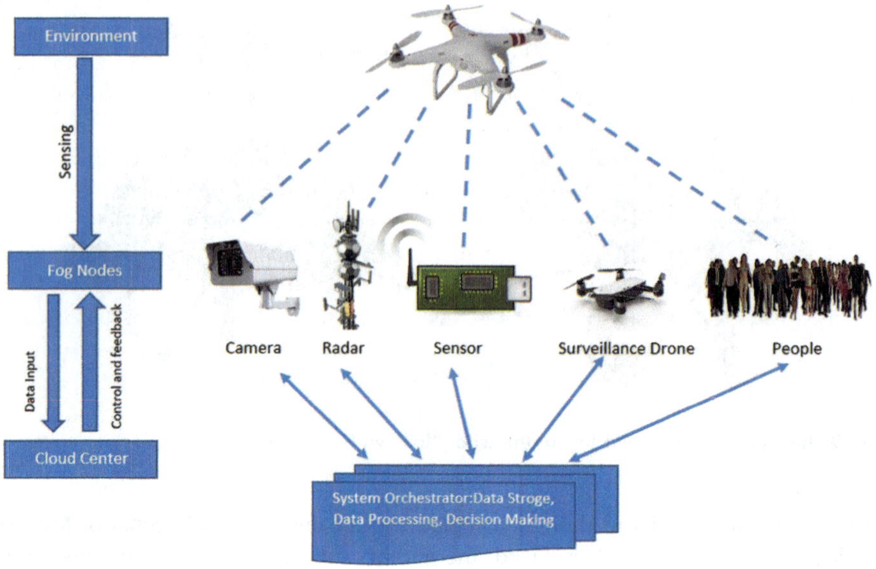

Fig. 10 Overall architecture of DragNet

virtual) things or objects interact following a context-aware perception-action cycle, thus allowing to bridge the physical world of objects and the social world of human demand and behavior, to form what the authors call "intelligent physical-cyber-social (iPCS) system". The overall architecture of the suggested DragNet system is given in Fig. 10.

The system architecture is based on the fog-to-cloud computing framework, where various IoT devices, passive cameras, radars as well as drones or people, serve as fog computing platform that provide information about the environment. Based on that local information the cloud center acts as a decision making center incorporating data from the whole variety of nodes. It is the global decision center and determines the actions to be taken be it capturing, jamming or directly destroying the amateur drone. They provide a short case study for a scenario in which unauthorized amateur drones are flying over a specific protected region. They consider a ternary hypothesis testing where H0, where no drone exists, H1 an authorized drone exists,, and H2 and an unauthorized amateur drone exists and provide results for the relationship between the number of sensors, the number of samples and the classification accuracy. Humbly stating that this work has only scratched the top of the problem, they are staring a very new interesting research direction.

7 Ongoing Projects

An interesting futuristic application called "AVUS" (Smart Flying umbrella) is discussed as an on-going work in [33]. The proposed Smart umbrella aims to perform all the traditional functions of a classical umbrella, (the design of which has survived little changes through the centuries) while adding new smart features to increase the comfort of the user. The umbrella, flying above the customer can also provide weather status information; can give notices before rain starts and report dangerous high levels of sun radiation using a smartphone dependent IoT environment. Apart from this, the paper mainly focuses on possible device and methods that can ensure some peculiarities in implementing smart following of humans. The authors detail their vision and discuss related problems that can be faced while carrying out these functions. They conclude by presenting ideas about possible solutions.

Our current ongoing project considers an interesting scenario relating IoT traffic management systems to emergency ambulance services. It aims to realize an IoT application for exchanging data between moving vehicles and/or moving vehicles and a ground station using an UAV as an intermediate connection point. In the connected world of today, not only self-driving cars, but person driven cars have a lot of reasons to exchange information.

We consider the following niche application: an ambulance is trying to proceed in heavy city traffic, where there are no side emergency lanes. If cars well ahead of the ambulance have information that it is approaching there is a possibility that they can open up a lane for its arrival. To realize this information from the ambulance (speed, direction, current coordinates, destination coordinates etc. in addition to the sound signal) has to be transmitted both to the cars travelling in the same direction (but well ahead of the ambulance) and to a traffic control center (TCC). However, due to heavy traffic, crowded landscape (high rising buildings, narrow streets) and over populated conditions cellular communication links are undesirable or unreliable to use. Thus an alternative will be to utilize V2V communication and/or vehicle-to-control center communication over an UAV hovering above the streets. From communication point of view this is a centralized system (decisions are taken by the TCC) based on the data from the IoT sensors from the ambulance, route maps and traffic maps from the he control center database. The architecture involves different communication links: point-to-point links between the ambulance and the TCC, between each vehicle and the TCC, and point-to-multipoint links from the TCC/UAV to the different vehicles. The TCC, based on the route and traffic information, can redirect vehicles and ensure opening the way for the ambulance. Sensors from the vehicles on the other hand will be providing information to the TCC related to speed and direction of travel. Thus the TCC computer based operation will be able to determine the most suitable traffic solution which will be transmitted back to the ambulance and the vehicles.

The first stage of this project was to realize the data exchange between moving vehicles (without human intervention) and a control center without relying on any infrastructure wireless network or cellular network. Each vehicle is equipped with a set of sensors, collects specific data from its immediate environment and transmits it to the central station, while the central station in its turn provides data to send the vehicles to a pre-defined point, specified by GPS coordinates.

To achieve this goal a client-server model ad hoc network is established between two moving vehicles and a control center. Vehicle movement along a given trajectory is remotely controlled by a computer. For a pre-defined time interval each vehicle attempts a connection to the server; if successful it transmits its data and awaits data from the server; if unsuccessful it searches for another connection possibility (open WiFi or cellular i.e. possibly UAV) to use for the transmission. The developed algorithm intentionally mimics heavy crowded location communication environment. The prototype system was tested in different environment conditions (without obstacles, with obstacles) for major network parameters like packet loss, delay, throughput and resilience (Fig. 11).

Three different types of data transmissions were tested: plain data packets less than 10 bytes; still image and video. While the central station is always fixed, vehicles were moving in an unobstructed or partially obstructed environment at different speeds, transmitting from different distances. Due to the short distances at the test site packet losses were negligible. Two different scenarios were examined: vehicles moving in an obstructed environment and in an environment with no obstructions. We also tested the prototype system with mobility of up to 5 m/s. Results for the case of short data packet transmission are given in Table 2 (without obstruction) and Table 3 (with obstruction).

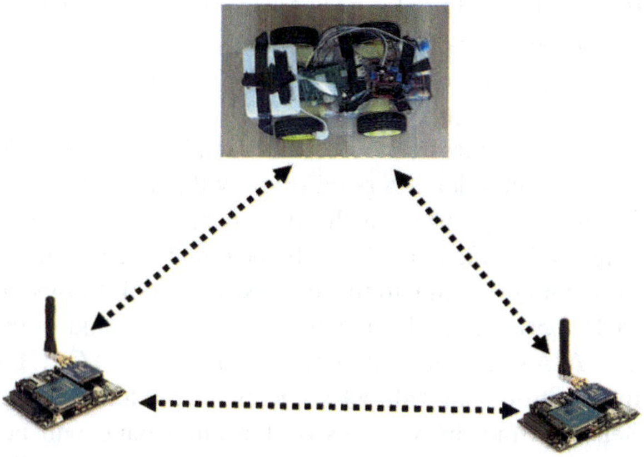

Fig. 11 Project scenario

Table 2 Data with no obstruction with and without mobility

Type of data	Distance (m)	Mobility	Bytes	Transmission duration (s)
Data packets	1	No	<10	0.0802
Data packets	1	Yes	<10	0.0808
Data packets	10	No	<10	0.0802
Data packets	10	Yes	<10	0.0802
Data packets	20	No	<10	0.0805
Data packets	20	Yes	<10	0.0802
Data packets	30	No	<10	0.0080
Data packets	30	Yes	<10	0.0801

Table 3 Data transmission with obstruction with and without mobility

Type of data	Distance (m)	Mobility	Bytes	Transmission duration (s)
Data packets	1	No	<10	0.0801
Data packets	1	Yes	<10	0.0796
Data packets	10	No	<10	0.00801
Data packets	10	Yes	<10	0.0801
Data packets	20	No	<10	0.0802
Data packets	20	Yes	<10	0.0802
Data packets	30	No	<10	0.0804
Data packets	30	Yes	<10	0.0805

The other experiments studied the transmission of larger pieces of information images of up to 1 Mbyte and video up to 15 Mbyte. Tests were carried out for obstructed and unobstructed environment with and without mobility similar to the small data packets case. Results were satisfactory and a comparison is presented in the Figs. 12 and 13, respectively for still images and video.

Even though this is still an ongoing project we believe the results are promising and UAVs can be used to provide emergency connections when there is traffic and communication congestion.

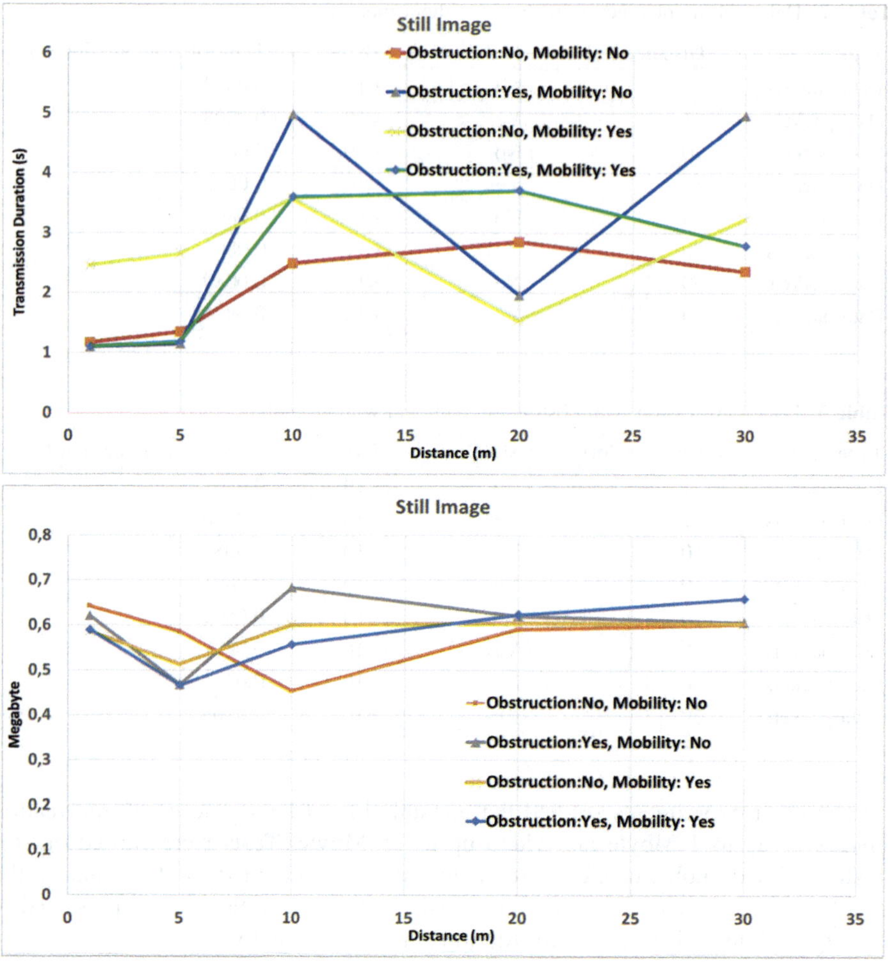

Fig. 12 Still image transmissions with and without obstruction and mobility

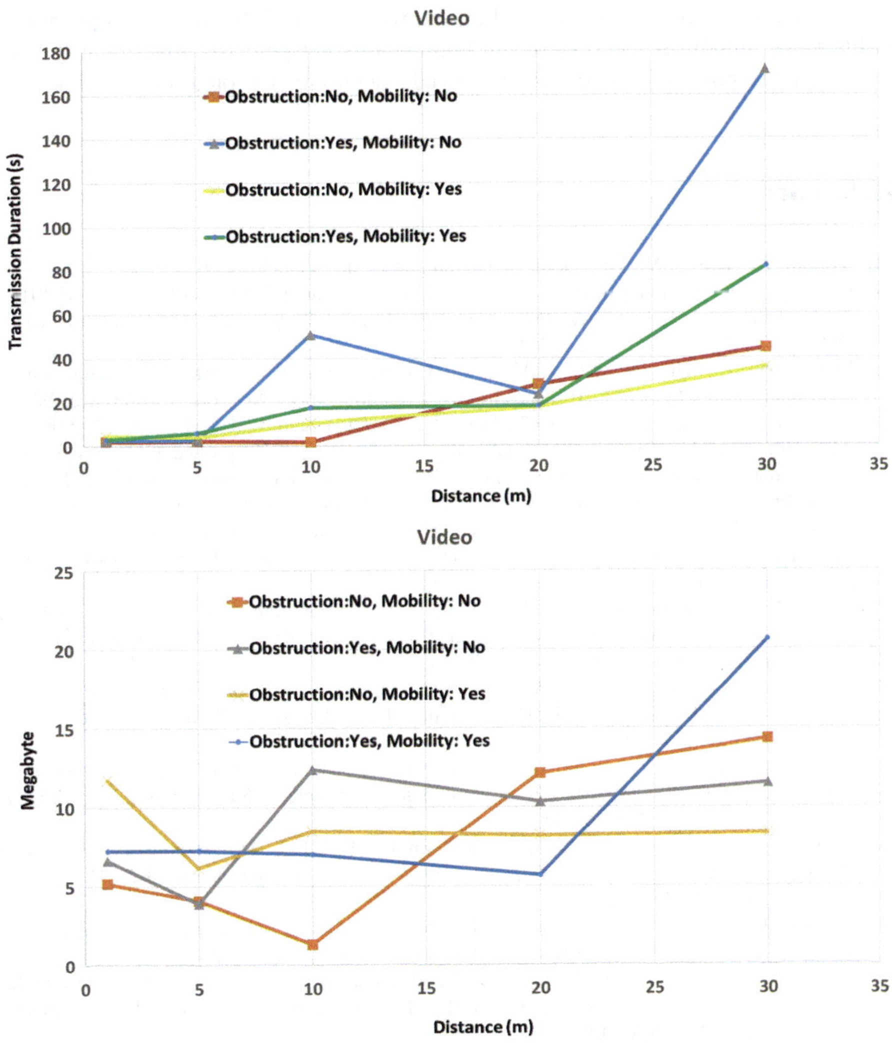

Fig. 13 Video transmission with and without obstruction and mobility

8 Conclusion

In this chapter we have discussed the topic how unmanned aerial vehicles (UAVs), known as drones, can be integrated together with IoT devices to become a major enabler for 5G and B5G networks. First a concise overview of UAVs structure and functionality is provided, followed by a summary of their main advantages and limitations and a discussion of how they complement the IoT paradigm. The second section deals with the possibilities of integrating UAVs and IoT in the overall wireless network architecture, providing an overview of different studies presented in literature. Finally some very recent applications merging IoT functionality with UAVs communication support are reviewed, including two ongoing projects.

References

1. Kimon, V., George, V.: Handbook of Unmanned Aerial Vehicles (2015)
2. US Department of Transportation: Unmanned Aircraft System (UAS) Service Demand 2015–2035: Literature Review & Projections of Future Usage. Technical report, v.0.1, DOT-VNTSC-DoD-13-01, Sept 2013
3. Merwaday, A., Guvenc, I.: UAV assisted heterogeneous networks for public safety communications. In: 2015 IEEE Wireless Communication Network Conference Work, WCNCW 2015, pp. 329–334 (2015). https://doi.org/10.1109/GLOCOM.2014.7037248
4. Al-Hourani, A., Kandeepan, S., Jamalipour, A.: Modeling air-to-ground path loss for low altitude platforms in urban environments. In: 2014 IEEE Global Communication Conference, GLOBECOM 2014, pp. 2898–2904. https://doi.org/10.1109/GLOCOM.2014.7037248
5. Naqvi, S.A.R., Hassan, S.A., Pervaiz, H., Ni, Q.: Drone-aided communication as a key enabler for 5G and resilient public safety networks. IEEE Commun. Mag. **56**, 36–42 (2018). https://doi.org/10.1109/MCOM.2017.1700451
6. Saeed, A., Gurbuz, O., Akkaş, M.A.: Terahertz-band communications for aerial vehicles at different altitudes. In: Third International Balkan Conference on Communications and Networking Skopje, North Macedonia (2019)
7. Zeng, Y., Zhang, R., Lim, T.J.: Wireless communications with unmanned aerial vehicles: opportunities and challenges. IEEE Commun. Mag. **54**, 36–42 (2016). https://doi.org/10.1109/MCOM.2016.7470933
8. Jawhar, I., Mohamed, N., Al-Jaroodi, J., et al.: Communication and networking of UAV-based systems: Classification and associated architectures **84**, 93–108 (2017). https://doi.org/10.1016/j.jnca.2017.02.008
9. Lagkas, T., Argyriou, V., Bibi, S., Sarigiannidis, P.: UAV IoT framework views and challenges: towards protecting drones as "things". Sensors (Switzerland) **18** (2018). https://doi.org/10.3390/s18114015
10. Zanella, A., Bui, N., Castellani, A., et al.: Internet of things for smart cities. IEEE Internet Things J. **1**, 22–32 (2014). https://doi.org/10.1109/JIOT.2014.2306328
11. Sekander, S., Tabassum, H., Hossain, E.: Multi-tier drone architecture for 5G/B5G cellular networks: challenges, trends, and prospects. IEEE Commun. Mag. **56**, 104–111 (2018). https://doi.org/10.1109/MCOM.2018.1700666
12. Shaikh, Z., Baidya, S., Levorato, M.: Robust multi-path communications for UAVs in the urban IoT. In: 2018 IEEE International Conference on Sensing, Communication Networking, SECON Work 2018, pp. 1–5 (2018). https://doi.org/10.1109/SECONW.2018.8396356
13. Yan, S., Peng, M., Cao, X.: A game theory approach for joint access selection and resource allocation in UAV assisted IoT communication networks. IEEE Internet Things J. **6**, 1663–1674 (2019). https://doi.org/10.1109/JIOT.2018.2873308
14. Chakareski, J.: Aerial UAV-IoT sensing for ubiquitous immersive communication and virtual human teleportation. In: 2017 IEEE Conference Computer Communication, 2017, pp. 718–723 (2017). https://doi.org/10.1109/INFCOMW.2017.8116465
15. Zhou, Z., Feng, J., Tan, L., et al.: An air-ground integration approach for mobile edge computing in IoT. IEEE Commun. Mag. **56**, 40–47 (2018). https://doi.org/10.1109/MCOM.2018.1701111
16. Mozaffari, M., Saad, W., Bennis, M., Debbah, M.: (2016) Mobile internet of things: Can UAVs provide an energy-efficient mobile architecture? In: 2016 IEEE Glob Communication

Conference GLOBECOM 2016—Proceedings. https://doi.org/10.1109/GLOCOM.2016.7841993

17. Mozaffari, M., Saad, W., Bennis, M., Debbah, M.: Unmanned aerial vehicle with underlaid device-to-device communications: performance and tradeoffs. IEEE Trans. Wirel. Commun. **15**, 3949–3963 (2016). https://doi.org/10.1109/TWC.2016.2531652
18. Datta, S.K., Dugelay, J.L., Bonnet, C.: IoT Based UAV platform for emergency services. In: ICTC 2018, pp. 144–147 (2018). https://doi.org/10.1109/ICTC.2018.8539671
19. Tuna, G., Mumcu, T.V., Gulez, K.: Design strategies of unmanned aerial vehicle-aided communication for disaster recovery. In: HONET 2012, pp. 115–119 (2012). https://doi.org/10.1109/HONET.2012.6421446
20. Sarigiannidis, P., Lagkas, T., Bibi, S., Ampatzoglou, A., Bellavista, P.: Hybrid 5G optical-wireless SDN-based networks, challenges and open issues. IET Netw. **6**, 141–148 (2017)
21. Saha, A.K., Saha, J., Ray, R., et al.: IOT-based drone for improvement of crop quality in agricultural field. In: 2018 IEEE 8th CCWC 2018, pp. 612–615, Jan 2018. https://doi.org/10.1109/CCWC.2018.8301662
22. Murugan, D., Garg, A., Singh, D.: Development of an adaptive approach for precision agriculture monitoring with drone and satellite data **10**, 5322–5328 (2017)
23. Tripicchio, P., Satler, M., Dabisias, G., Ruffaldi, E., Avizzano, C.A.: Towards smart farming and sustainable agriculture with drones. In: Proceedings of the 2015 International Conference on Intelligent Environments IE, Prague, Czech Republic, pp. 140–143, 15–17 July 2015
24. Filev, R., Ieee, M., Netto, I,: Precision agriculture using remote monitoring systems in Brazil. In: 2017 IEEE Global Humanitarian Technology Conference (GHTC) (2017). https://doi.org/10.1109/GHTC.2017.8239290
25. Reinecke, M., Prinsloo, T.: The influence of drone monitoring on crop health and harvest size. In: 1st International Conference on Next Generation Computing Applications (NextComp) (2017)
26. Matos, J., Postolache, O.: IoT enabled aquatic drone for environmental monitoring. In: EPE 2016, pp. 598–603 (2016). https://doi.org/10.1109/ICEPE.2016.7781410
27. Fan, Q., Ansari, N.: Towards traffic load balancing in drone-assisted communications for IoT. IEEE Internet Things J. **6**, 3633–3640 (2019). https://doi.org/10.1109/JIOT.2018.2889503
28. Genc, H., Zu, Y.: Ultra-low-power processors flying IoT: toward low-power vision in the sky. IEEE Micro **37**(6) (2017). https://doi.org/10.1109/MM.2017.4241339
29. Goddemeier, N., Daniel, K., Wietfeld, C.: Role-based connectivity management with realistic air-to-ground channels for cooperative UAVs. IEEE J. Sel. Areas Commun. **30**, 951–963 (2012). https://doi.org/10.1109/JSAC.2012.120610
30. Vattapparamban, E., Güvenç, I., Yurekli, A.I., et al.: Drones for smart cities: Issues in cybersecurity, privacy, and public safety. In: IWCMC 2016, pp. 216–221 (2016). https://doi.org/10.1109/IWCMC.2016.7577060
31. De Lima, C.M.A., Da Silva, E.A., Velloso, P.B.: Performance evaluation of 802.11 IoT devices for data collection in the forest with drones. In: GLOBECOM 2018—Proceedings, pp. 1–7 (2018). https://doi.org/10.1109/glocom.2018.8647220
32. Ding, G., Wu, Q., Zhang, L., et al.: An amateur drone surveillance system based on the cognitive internet of things. IEEE Commun. Mag. **56**, 29–35 (2018). https://doi.org/10.1109/MCOM.2017.1700452
33. Ahmadhon, K., Al-Absi, M.A., Jae Lee, H., Park, S.: Smart flying umbrella drone on internet of things, AVUS. In: ICACT 2019, pp. 191–195, Feb 2019. https://doi.org/10.23919/ICACT.2019.8702024

Multi-criteria Decision Making for Routing Process in MANET

M. Deshmukh, S. N. Kakarwal, R. Deshmukh and D. V. Kurmude

Abstract Next-generation computing paradigms are expected to support wireless networking with high data transfer rates and autonomous decision-making capabilities. The self-organized, decentralized and infrastructure-less features of mobile ad hoc networks (MANETs) established their importance for communication in indoor and outdoor environments. In MANET, a broadcast storm causes network issues as there are redundant broadcasts and packet collisions. Traditional broadcast schemes have focused on avoiding broadcast storms by inhibiting some rebroadcasts. Another issue is the link failures caused by node mobility and their energy exhaustion. In this research, we propose two schemes to address these network challenges. The first approach refines the counter threshold based on neighborhood, mobility and energy parameters of the node and makes use of the refined thresholds to make the broadcasting decision. The second proposed scheme exploits TOPSIS multiple criteria decision-making model to make broadcast decisions emphasizing on neighborhood, mobility and energy criteria of the mobile node. The proposed schemes perform best in various network densities as compared to ad hoc on-demand distance vector (AODV) routing by increasing lifetime of network and decreasing delay, packet dropping and energy consumption, while achieving a good packet delivery ratio.

Keywords MANET · Broadcasting · Multi-criteria decision making · TOPSIS, etc.

M. Deshmukh (✉)
Computer Engineering Department, Pillai College of Engineering, Mumbai, India
e-mail: mdeshmukh@mes.ac.in

R. Deshmukh
Computer Science Department, Dr. B. A. M. University, Aurangabad, India
e-mail: rrdeshmukh.csit@bamu.ac.in

S. N. Kakarwal
Computer Engineering, PES College of Engineering, Aurangabad, India

D. V. Kurmude
Physics Department, Milind College of Science, Aurangabad, India

N. Dey et al. (eds.), *Internet of Things, Smart Computing and Technology: A Roadmap Ahead*, Studies in Systems, Decision and Control 266,
https://doi.org/10.1007/978-3-030-39047-1_11

1 Introduction

In communications, MANETs continuously evolve since every node communicates with other nodes by establishing one-hop and multi-hop communication paths. Furthermore, MANET nodes move, causing alternative paths to be established among nodes over a period of time. These features of MANETs encouraged their adoption for indoor and outdoor communications [1]. MANETs have been applied in indoor environments including conference and class rooms, police operations, theme parks and in fire disaster relief teams. MANETs have been applied outdoors for disasters near beaches, communications among business people near railway stations, sport venues, marketplaces and in military operations such as automated battlefields. Furthermore, through the advent of sensor-enabled smart mobile devices, MANETs have become a crucial element in the framework of Smart City and Internet of Things (IoT) scenarios [2, 3]. The framework of internet of things (IoT) with the keystone as identity of wireless mobile computing devices have become foundation for incorporating security methods [4, 5]. Hence, many researchers have worked toward creating efficient MANETs. The dynamic topology of MANETs forced routing protocols to use multicasting and broadcasting mechanisms. Moreover, most MANET routing protocols have used flooding as a basic broadcasting mechanism. In flooding, after a node acknowledges a broadcast packet, it sends out the packet to all of its neighbours. In turn, these neighbours receive the packet and transmit it to their neighbours. This transmission process continues until the packet has been broadcasted throughout the entire network. Flooding is very simple to implement, but causes redundant broadcasts, collisions and network contention. Such severe problems are known collectively as a broadcast storm [6]. A broadcast storm causes network issues as there are redundant broadcasts and packet collisions. Traditional broadcast schemes have focused on avoiding broadcast storms by inhibiting some rebroadcasts.

Another issue is the link failures caused by node mobility and their energy exhaustion [7]. Since mobile nodes are battery-driven, they suffer from limited energy availability. In a MANET, some nodes may run out of energy if involved in retransmission, causing network partitioning. Therefore, extending battery life has become a key objective of researchers working in the MANET domain. The other dominant cause of link failure is the mobility of nodes. The continuous movement of some nodes induces network retransmission instability and causes partitioning. Thus, there is also a need to focus on the speed of mobile nodes to reduce link failure. As a result, multi-criteria decision-making is needed in the real world to achieve quality-of-service (QoS) routing in MANETs.

Multi-criteria decision-making is a method that finds out the best alternative among a set of possible options. Many multi-criteria decision-making models exist, including the well-known MAXMIN, MAXMAX, Analytical Hierarchy Process (AHP) and Technique for Order Preference by Similarity to the Ideal Solution (TOPSIS) models. In MAXMIN, the overall performance of an alternative is decided by its weakest attributes. In contrast, in MAXMAX the overall performance of an alternative is decided by its best attribute value. AHP performs pairwise comparisons of alternatives to find a preference. The TOPSIS chooses the alternative closest to the ideal solution and farthest from the negative ideal alternative [8]. Non-traditional broadcast schemes, such as speed-and energy-based schemes, have also been exploited to increase the lifetime of a network. In this paper, we aim to bring together the advantages of both traditional and non-traditional broadcast schemes.

The remainder of this paper is arranged as follows. Section 2 provides general idea of related work through a classification of broadcast schemes. Section 3 proposes adaptive, mobility-aware, energy-efficient, counter-based broadcasting scheme (AMECBB). Section 4 presents novel counter-based broadcasting using multi-criteria decision-making scheme (TCBB). Section 5 summarises the performance analysis through simulation. Section 6 concludes the paper.

2 Background and Related Work

Both broadcast storms and link failure affect network lifetimes. Many researchers have used traditional, non-traditional and hybrid broadcast schemes to address issues in MANETs, thus establishing three categories, each of which can be further classified into subcategories.

2.1 *Traditional Broadcast Schemes*

Traditional approaches can be classified into two categories, heuristic-and topology-based schemes [9].

In heuristic schemes, the decision to retransmit is made using heuristics such as probabilities, counter values, positions and distance. In topology-based approaches, topological information such as neighbouring nodes, the number of network nodes, or the type of network are used instead.

Probabilistic broadcast schemes have been proposed to alleviate broadcast storms associated with blind flooding. These inhibit some nodes from forwarding broadcast packets in order to reduce redundancy. They allow a node to rebroadcast the packet with a probability p, but are not concerned with density of network nodes. In fixed probabilistic schemes, all nodes initially receiving a broadcast packet will rebroadcast it to all other nodes in the network with a probability p. A smaller value of p reduces the impact of the storm.

The theme behind counter-based broadcast schemes is that the broadcast packet may be blocked by a busy medium or other queued messages. Thus, it is possible that the node may receive the same packet repeatedly from other broadcasting nodes before it actually starts transmitting the packet. This is made possible by inserting a small random delay before packet broadcasting which causes the timing of broadcasting to differentiate. This not only allows nodes sufficient time to receive redundant packets but also prevents collisions. In this scheme, each node is associated with a counter for each broadcasted packet, which keeps trace of the number of times a packet is acknowledged by a node. When this counter c reaches a predefined threshold, the packet is dropped or else the packet is transmitted. In fixed-counter-based broadcasting, a smaller threshold will cause reduce broadcasts and avoid collisions, again minimising storms.

In position-based schemes, each node can establish its position in order to precisely estimate the additional coverage that can be achieved. Position-based schemes can be supported by a global positioning system (GPS). The challenge with a position-based scheme is the cost of computing extra expected coverage regions.

Distance-based schemes use the distances between source and destination nodes to make the broadcasting decision. Euclidean distance is often used, but received signal strength (RSS) and hop count can also be employed. To avoid redundant retransmission, nodes situated further away from senders are preferred. In Euclidean distance-based schemes, the support of a positioning system like GPS is required to obtain the Euclidean distance between two nodes. In RSS-based schemes, nodes compute the relative distance as a function of signal power. Alternatively, many routing protocols use hop count to compute the distances between the source and destination nodes.

In topology-based broadcast schemes, topological information is used decide whether to broadcast. These approaches can be categorised into neighbour-knowledge-based and cluster-based schemes. The former makes use of information regarding neighbouring nodes, while in the latter, the network is partitioned into several groups of clusters. Each cluster has a head that acts as a gateway and is responsible for rebroadcasting the message and selecting forwarding nodes on behalf of the cluster.

2.2 *Non-traditional Broadcast Schemes*

Non-traditional broadcast approaches include those based on energy, mobility and others.

Mobility-based schemes aim to eliminate redundant broadcasts by choosing low-speed nodes to discover more stable paths [10–12]. When network node mobility is high, link breakages can occur affecting the network lifetime. Hence, mobility-based schemes achieve network lifetime reduction by retransmissions. Energy-based schemes aim to eliminate redundant broadcasts by selecting high-energy nodes to discover routes that bypass low-energy nodes [13]. As each mobile node in a MANET is responsible for routing packets during route discovery, link failure is possible due to the low energies of nodes. Specifically, in densely connected regions of a network, nodes may not be available, causing link failures that affect network lifetime. Hence, energy-based schemes can also achieve higher network lifetimes by reducing retransmissions.

K. Egoh and S. De introduced a multi-parameter mapping function that aggregates all decision parameter into one in order to rank potential relay candidates [14]. In order to optimise the delay and distance from a source to a destination node, they minimised a weighted sum to address greedy-forwarding.

2.3 Hybrid Broadcast Schemes

Hybrid broadcast schemes combine the advantages of two or more approaches to optimise broadcasting in MANETs. Fixed counter-based schemes use a static threshold to condense redundant rebroadcasts. Setting a low threshold allows broadcasts to be reduced, but does not provide significant reach. When a high threshold is set, broadcasts are not reduced but reachability improves. Tseng et al. [15] attempted to dynamically adjust thresholds based on information on node neighbours in order to reduce broadcasts and reachability simultaneously without causing delays.

In another approach, Yassein et al. [16] combined probability-based and neighbour-knowledge-based schemes that dynamically adjust rebroadcasting probabilities depending on the region of the network. In dense regions, rebroadcast probabilities were set to low values and achieved considerable broadcast savings, and in sparse regions the thresholds were set high. However, Yassein et al. [17] showed that conventional probabilistic flooding frequently does not achieve a high degree of reachability, partly because each network node has the same rebroadcasting probability regardless of the number of neighbours. They incorporated a new probabilistic, ad hoc on-demand distance vector (AODV) algorithm to demonstrate a high delivery ratio of data packets.

Mohammed et al. [18] proposed a scheme that combines probability, counter, and neighbour-knowledge-based approaches in order to mitigate broadcast storms. The algorithm considers the variation of network node density, and assigns two different broadcast probabilities to the sparse and dense regions of the network to improve reach and reduce broadcasts. Khalaf et al. [19] suggested a dynamic probabilistic scheme for mitigating broadcast storms by reducing the dependency on thresholds and determine the density near a given node. Through extensive simulation, their algorithm outperformed other existing protocols by reducing network overhead.

A new, counter-based scheme was introduced by Yassein et al. [20] that uses three dynamic thresholds to improve broadcasting efficiency. Through simulation, the packet delivery rate and network throughput were enhanced. A hybrid broadcasting scheme was proposed by Reina et al. [21] that aimed to reduce broadcast storms in mobile networks using probabilistic, neighbour-knowledge and position methods. To decide the forwarding probability, a forwarding zone criterion was used. It was implemented using the position of the node as obtained from GPS or using RSS, and demonstrated a reduction in broken links.

Yang and Wu [22] exploited directional antennas for broadcasting based on network coding and to reduce the volume of broadcasts and energy consumption. They used the antennae to forward nodes' broadcasts, either original or those coded to restricted regions that were locally selected. They also extended the directional antenna approach with a dynamic code. Via simulations, they showed that static efficient broadcasting using network coding and directional antennae (EBCD) conserves considerable energy with smaller overhead, and performed well despite collisions.

Finally, Yassein and Al-Dubai [23] proposed inspired-counter-based broadcasting for dynamic source routing (DSR) in mobile networks, a hybrid of counter and neighbour knowledge schemes. In this paper, five counter thresholds were assigned based on the information collected from the nodes' neighbourhood, and classified with counter thresholds as low, medium, high, very high or ultra, leading to the name DSR-5C. Simulations were performed using the Qualnet simulator, and performance results showed that DSR-5C improved DSR performance load.

3 Adaptive Mobility-Aware Energy-Efficient Counter-Based Broadcasting Scheme

Due to the dynamic nature of MANETs, many problems can occur during packet transmission. Basic routing protocols such as AODV and DSR induce issues such as broadcast storms or link failure. MANET topologies are often random and dynamic with varying degrees of node density in various regions of the network. In MANETs, some nodes are highly mobile and others not. In sparse networks, shared coverage is lower since very few nodes act as intermediate nodes. If these intermediate nodes are highly mobile, then link failure can occur, which decreases packet delivery. To improve it, node mobility must be considered. Moreover, if the network topology is too dense, nodes may quickly exhaust their energy, further increasing the likelihood of link failure. Link failure can be inhibited by considering node energies. The adaptive, mobility-aware energy-efficient, counter-based broadcasting (AMECBB) scheme proposed here is a hybrid of approaches using counters, probability and neighbour knowledge, and focuses on mobility and node energies to make broadcasting decisions.

In a simple counter-based scheme, a node maintains a counter to trace redundant broadcast packets. Upon acknowledgment of a packet, a counter is incremented by one. Then, a delay is introduced to enhance broadcast reduction. During this delay period, the counter is updated, and after expiry, the counter associated with a node

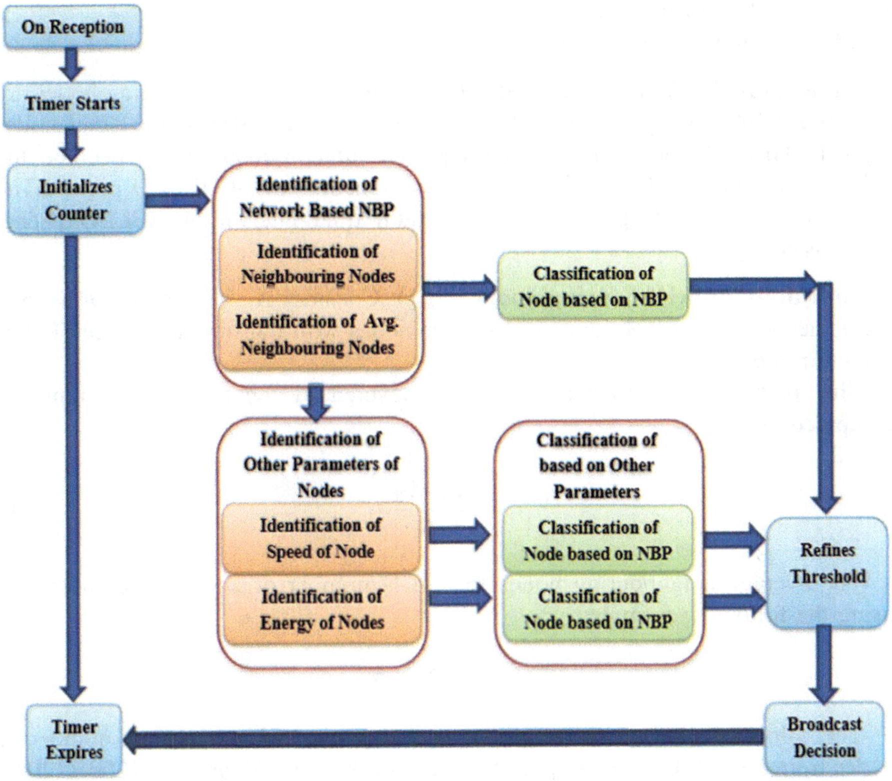

Fig. 1 Framework of adaptive mobility-aware energy-efficient CBB scheme

is compared with the counter threshold. If the count has reached the threshold, then the node suppresses retransmission. Otherwise, it retransmits the packet.

The framework of the AMECBB scheme consists of four important modules viz., identification of network structure, classification of network structure, and rebroadcast threshold estimation, and rebroadcast probability estimation as shown in Fig. 1.

3.1 Identification of Network Structure

When a node receives a packet, it starts the timer and initializes counter (C) to 1 to keep track of redundant transmissions. Subsequently, it starts the process of identifying the network structure, following these steps:

1. Identification of neighbouring nodes.
2. Identification of average neighbouring nodes.

3. Identification of node mobility.
4. Identification of node energy.

Identification of neighbouring nodes (N_{nbg}): At first, the neighbouring nodes are determined by sending the HELLO packet to all other nodes. If the other nodes reply to HELLO packet during Hello packet interval those nodes become the neighbouring nodes of the node who started transmission of HELLO packet. The Hello packet interval specifies the maximum time interval of HELLO packet transmission.

Identification of neighbouring nodes (N_{avg}): Consider A_N be the area of Mobile adhoc network having N number of mobile nodes with transmission range (T_R) as shown in the Fig. 2.

The fraction of the total network area covered by the transmission range is computed as:

$$\alpha = \frac{\pi * (T_R)^2}{A_N} \tag{1}$$

The average neighbouring nodes for the N number of nodes is computed by using the following formula:

$$N_{avg} = N \times \alpha \tag{2}$$

The values obtained for average neighbouring nodes for the varied number of nodes from 50 to 150 with a step of 25 nodes for the network area of (1200×1200), (1000 × 1000) and (800 × 800) is shown in Table 1. The graph is drawn in Fig. 3 regarding the number of neighbours and average neighbouring nodes computed.

The average number of neighbors increases as the number of nodes increases in the network. Intuitively, the average number of neighbors is less in the sparse network and an average number of neighbors are more in the dense network. More

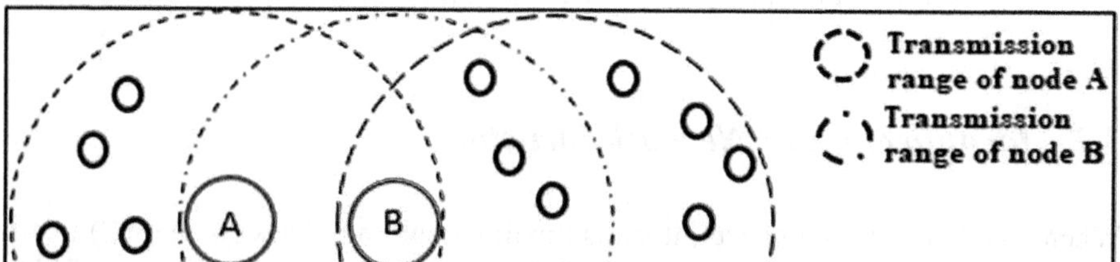

Fig. 2 Transmission range over the area of the network

Table 1 Average neighboring nodes

No.	No. of nodes	Avg. neighboring nodes (800 * 800)	Avg. neighboring nodes (1000 * 1000)	Avg. neighboring nodes (1200 * 1200)
1.	50	6	9	15
2.	75	10	14	23
3.	100	13	19	30
4.	125	17	24	38
5.	150	20	29	46

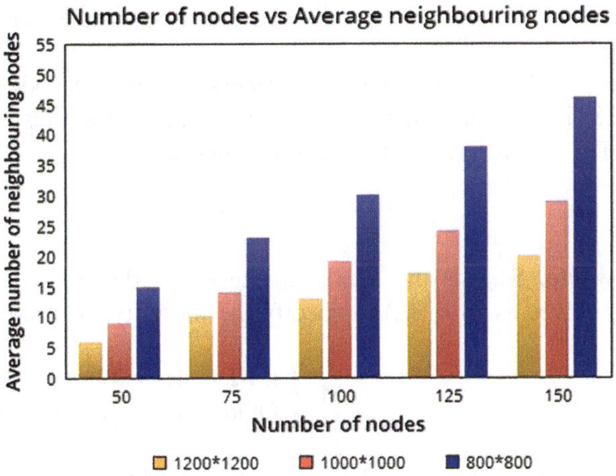

Fig. 3 Average neighboring nodes

interestingly, in the large area of network nodes are scattered too far to form sparse networks. As a result, the average number of neighbors is less in large area network. On contrary, in the small area of network nodes are placed too near to form dense networks. As a result, the average numbers of neighbors are more in a small area network. Hence the AMECBB considers the type of network as one of the parameters to decide to rebroadcast probability.

Identification of node mobility (S): It determines the speed of the node (S) that received the packet. The speed of node aids to decide stabilized routing path from source to destination to reduce link failure due to the high mobility of mobile nodes.

Identification of node energy (E_N): It determines the energy of the node (E_N) that received the packet. The energy of node aids to decide long-lasting routing path from source to destination to reduce link failure due to high energy exhaustion of mobile nodes. The fraction of energy node (E) is computed as the ratio of the energy of the node (E_N) to the initial energy of node (E_{ini}). Consequently, AMECBB focuses on mobility and energy of mobile nodes to have stabilized long lasting paths from source to destination.

3.2 Classification of Network Structure

After identification, classification of the network structure begins, and occurs at three different levels. At first level, the algorithm classifies the node as sparse if the number of neighbouring nodes is less than the average number of nodes in the network, and classifies the node as dense if the number of neighbouring nodes is more than the average. The conditional attribute (F1) evolves as ($N_{nbg} < N_{avg}$) to specify the type of mobile network (MN).

$$\text{MN:} \begin{Bmatrix} \text{if F1} = 1; & \text{MN} = \text{Sparse} \\ \text{Otherwise} & \text{MN} = \text{Dense} \end{Bmatrix} \tag{3}$$

At the second level, the process classifies a node as stable if its speed of node (S) is less than a speed threshold (S_{th}), and classifies the node as unstable otherwise. The speed threshold S_{th} is computed as:

$$S_{th} = S_{max} \div 2 \tag{4}$$

where, S_{max} is the maximum speed of the node. The conditional attribute (F2) evolves as ($S < S_{th}$) to specify the mobility of network (M).

$$\text{M:} \begin{Bmatrix} \text{if F2} = 1; & \text{M} = \text{Stable} \\ \text{Otherwise;} & \text{M} = \text{Unstable} \end{Bmatrix} \tag{5}$$

During the third level, a node is classified as strong if its energy fraction (E) exceeds the energy threshold (E_{th}), and as weak otherwise. The conditional attribute (F3) evolves as ($E > E_{th}$) to specify the energy efficiency of the network (E).

$$\text{E:} \begin{Bmatrix} \text{if F3} = 1; & \text{E} = \text{Strong} \\ \text{Otherwise;} & \text{E} = \text{Weak} \end{Bmatrix} \tag{6}$$

3.3 Rebroadcast Threshold Estimation (R_{th})

During decision-making, the counter threshold is set as per the classification of the network structure. This step determines how many retransmissions will be allowed for a node.

Observation 1: If the network is sparse, then rebroadcasting a packet is quite significant for three causes. First, the redundancy of rebroadcasts is minor because nodes are scattered and individually responsible for less shared coverage. Second, rebroadcasting may cause network partitioning because of less shared coverage. Third, in a sparse network there is much less shared coverage, since very few nodes

Table 2 Characteristic decision table

No.	F1	F2	F3	F1 ^ F2 ^ F3	Threshold (R_{th})
1	1	1	1	d = 1	High Sparse
2	1	1	0	d =2	High Moderate
3	1	0	1	d = 3	Low Moderate
4	1	0	0	d = 4	Low Sparse
5	0	1	1	d=5	High Dense
6	0	1	0	d = 6	Low Moderate
7	0	0	1	d = 7	High Moderate
8	0	0	0	d = 8	Low Dense

act as intermediate nodes. If these intermediate nodes are highly mobile, then link failure can occur.

Observation 2: If the network is dense, then packet rebroadcasting is quite significant for three causes. First, rebroadcast redundancy is major because nodes are responsible for more shared coverage. Second, coverage is complete since nodes are not scattered. Third, in a dense network, nodes may quickly exhaust their energy and induce link failure.

The AMECBB refines counter threshold to d decision counter thresholds (R_{th}) as High Sparse, Low Sparse, High Dense, Low Dense, High Moderate and Low Moderate from three conditional attributes viz., F1, F2, and F3. The AMECBB is methodized as follows with Table 2.

If the node is classified as the sparse stable strong node then the High Sparse counter threshold is set to allow more broadcasts and if is classified as the sparse stable weak node then the High Moderate counter threshold is set to allow high moderate transmissions. If the node is classified as the sparse unstable strong node then the Low Moderate counter threshold is set to allow low moderate transmissions. If the node is classified as the sparse unstable weak node then Low Sparse counter threshold is set to allow fewer broadcasts. If the node is classified as the dense stable strong node then the High Dense counter threshold is set to allow more broadcasts and if is classified as the dense stable weak node then the Low Moderate counter threshold is set to allow low moderate transmissions. If the node is classified as the dense unstable strong node then the High Moderate counter threshold is set to allow high moderate transmissions and if is classified as the dense unstable weak node then the Low Dense counter threshold is set to allow fewer broadcasts. The classification tree of all possible conditional attributes is shown in Fig. 4.

Intuitively, the counter threshold C is a function dependent on the neighbourhood (MN) and can be refined on the basis of node mobility (M) and energy (E).

Fig. 4 Classification tree

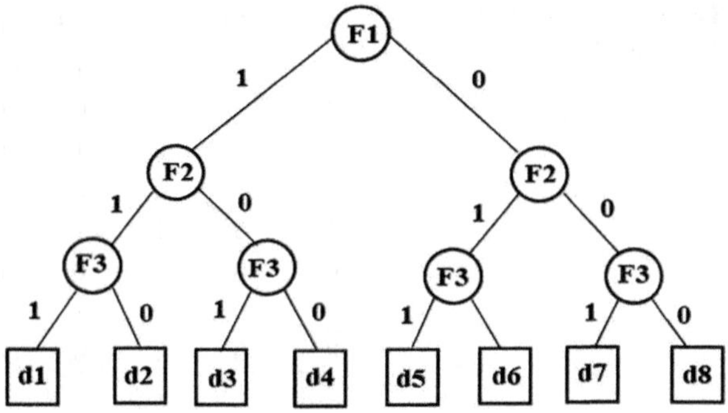

3.4 Rebroadcast Probability Estimation (P)

The selection of the counter threshold is most important in AMECBB as rebroadcast probability depends on the rebroadcast threshold. These refined counter thresholds adjust the rebroadcast probability dynamically. Assume C_{Th} be the estimated broadcasting threshold from the total Nth broadcasting thresholds. The maximum-likelihood estimate calculates the rebroadcast probability (P) as (C_{Th}/ Nth). Table 3 lists the rebroadcast thresholds and their corresponding rebroadcast probabilities and Fig. 5, shows the linear dependency of rebroadcast probability on rebroadcasting threshold.

Finally, AMECBB computes a random probability (P_{rand}) in between values zero and one [0, 1] and compares with the rebroadcast probability (P). If the rebroadcast probability is more than or equal to random probability then the packet is transmitted, otherwise, the packet is dropped. The pseudo code of AMECBB protocol is described in detail as shown in Fig. 6.

Table 3 List of rebroadcast probabilities

No.	Rebroadcast threshold	Rebroadcasting probability
1.	1	0.16
2.	2	0.33
3.	3	0.5
4.	4	0.66
5.	5	0.83
6.	6	1

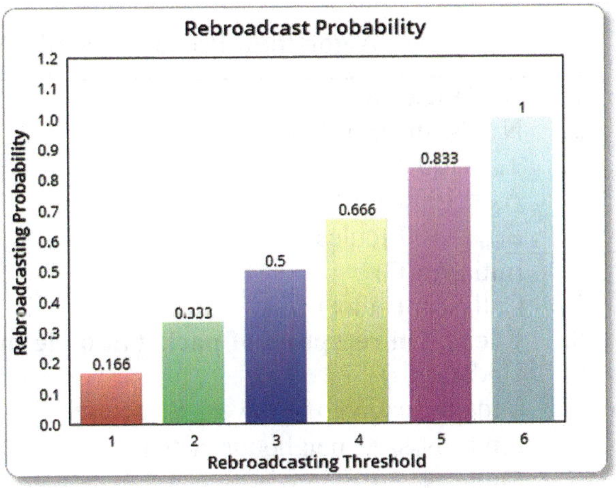

Fig. 5 Rebroadcast probability

4 Novel Counter-Based Broadcasting Using TOPSIS Multi-criteria. Decision Making (TCBB)

Here, we describe the novel counter-based broadcasting scheme (TCBB) using adaptive multi-criteria decision-making that determines whether to broadcast and is a good approximation of a quality model according to a set of parameters. In order to address the issues of MANET, it considers the three main decision parameters of the mobile node, the neighborhood (MN), mobility (M) and energy (E). The weights of these criteria reflect their relative importance in the decision-making process. As all the decision attributes are equally important, they are equally weighted. In this scheme, each time a node receives a broadcast packet it does not broadcast the packet immediately. It waits for Random assessment time to reduce the storm effect and initializes the three attribute weights. Next, it constructs a table of neighboring nodes (N_{nbg}) by periodic transmission of HELLO packets. The algorithm computes the average neighboring nodes (N_{avg}) by using the Eqs. (1) and (2). The neighborhood criteria (MN) is estimated as the ratio of neighboring nodes (N_{nbg}) to the average neighboring nodes (N_{avg}). In spite of estimation of mobility criteria (M) and energy criteria, it finds its speed (S), energy (E_N), maximum speed (S_{max}) and initial energy (E_{ini}) of the mobile node. To choose low-speed nodes to discover more stable paths, the mobility criteria (M) is estimated by taking the ratio of the speed of node (S) to the maximum speed (S_{max}) of the mobile node. Likely, to choose strong nodes to have the long-lasting routing path the energy criteria (E) is estimated by taking the ratio of the energy (E_N) of node to the initial energy (E_{ini}) of the node. The TCBB scheme constructs a rebroadcasting probability function by way of a multiple parameter mapping functions that aggregates the neighbourhood (MN), mobility (M) and energy (E) parameters into a single parameter. The TCBB computes a random probability (P_{rand}) in between values zero and one [0, 1] and compares with the rebroadcast probability (P). If

Algorithm: Pseudo-code of AMECBB Protocol

Initialization :
N = Number of Nodes
$T_R = 250$ m
$A_N = 1000 \times 1000$
$E_{init} = 100$ Joules
Initialize(C)
Call SendHello()
Event: On reception of packet at node (n_i)
Update(C)
\# Identification of Network Structure
Finds $N_{nbg} \leftarrow$ neighbour_list( )
Finds N_{avg}
Finds $S_N \leftarrow$ speed( )
Finds $E_N \leftarrow$ Energy( )
\# Classification of Network Structure
$E = E_N / E_{init}$
If $N_{nbg} > N_{avg}$ then MN = Dense
Else MN = Sparse
End
If $S_N > S_{Th}$ then M = Stable
Else M = Unstable
End
\# Rebroadcast Threshold Estimation
If MN = Dense && M= Stable && E = Strong then C_{Th} = High_Dense
Else if MN = Dense && M= Stable && E = Weak then C_{Th} = Low_Moderate
Else if MN = Dense && M= Unstable && E = Strong then C_{Th} = High_Moderate
Else if MN = Dense && M= Unstable && E = Weak then C_{Th} = Low_Dense
Else if MN = Sparse && M= Stable && E = Strong then C_{Th} = High_Sparse
Else if MN = Sparse && M= Stable && E = Weak then C_{Th} = High_Moderate
Else if MN = Sparse && M= Unstable && E = Strong then C_{Th} = Low_Moderate

Fig. 6 Pseudo-code of AMECBB protocol

28. Else if MN = Sparse && M= Unstable && E = Weak then C_{Th} = Low_Sparse
29. End
Rebroadcast Probability Estimation
30. Finds Rebroadcast Probability (P) = C_{Th}/N_{Th}
31. When RAD Expires if $P \geq P_{rand}$ [0,1] then
32. Broadcast (n_i)
33. Else
34. Drop (n_i)
35. End

Fig. 6 (continued)

rebroadcast probability is more than or equal to random probability then the packet is transmitted, otherwise, the packet is dropped. The pseudocode of TCBB protocol is described in detail as shown in Fig. 7.

Algorithm: Pseudo-code of TCBB Protocol

1. **Initialization :**
2. N = Number of Nodes
3. T_R = 250 m
4. $A_N = 1000 \times 1000$
5. RAD = 0.0001
6. E_{ini} = 100 Joules
7. [w1,w2,w3] =[0.35,0.3,0.35]
8. **Event:** On reception of packet at node (n_i)
9. Get $N_{nbg} \leftarrow$ neighbouring nodes( )
10. Compute $N_{avg} = \frac{N \times \pi \times (T_R)^2}{A_N}$
11. Get S, $S_{max} \leftarrow$ speed( )
12. Get $E_N \leftarrow$ energy( )
13. Compute MN (Neighborhood fraction)
14. Compute M (Mobility fraction)
15. Compute E (Energy fraction)
16. Compute P = TOPSIS_WEIGHT(MN, M, E)
17. When RAD Expires if $P \geq P_{rand}$ [0,1] then
18. Broadcast (n_i)
19. Else
20 Drop (n_i)
21. Exit

Fig. 7 Pseudo-code of TCBB protocol

4.1 Computation of Rebroadcast Probability Using TOPSIS

Graphics Multi-criteria decision-making with m alternatives and n criteria can be expressed in matrix form, where A_1, A_2, …, A_m are feasible alternatives evaluated for an optimal selection, and C_1, C_2, …, C_n are evaluation criteria that can impact the selection of alternatives. The weight vector $W = [w_1, w_2, \ldots, w_n]$ is composed of individual weights W_j for each criterion C_j satisfying $\sum_{j=1}^{n} W_j = 1$.

Step 1: Compute the normalised decision matrix. The normalised value r_{ij} is calculated as follows: The rebroadcast probability function consists of three main criteria, the neighbourhood (MN), mobility (M) and energy (E). Graphics Multi-criteria decision-making with m alternatives and n criteria can be expressed in matrix form, where A_1, A_2, …, A_m are feasible alternatives evaluated for an optimal selection, and C_1, C_2, …, C_n are evaluation criteria that can impact the selection of alternatives. The weight vector $W = [w_1, w_2, \ldots, w_n]$ is composed of individual weights W_j for each criterion C_j satisfying $\sum_{j=1}^{n} W_j = 1$.

$$r_{ij} = X_{ij} \sqrt{\sum_{(i=1)}^{m} X_{ij}^2} \; i = 1, 2, \ldots, m \text{ and } j = 1, 2, \ldots, n \tag{7}$$

Step 2: Compute the weighted, normalised decision matrix. The weighted normalised value v_{ij} is calculated as follows

$$v_{ij} = r_{ij} \times W_j \quad i = 1, 2, \ldots, m \text{ and } j = 1, 2, \ldots, n \tag{8}$$

where, W_j is the weight of the jth criterion or attribute and $\sum_{j=1}^{n} Wj = 1$

Step 3: Compute the positive (A^+) and negative ideal (A^-) solutions.

$$A^+ = \{(\max v_{ij} | j \in C_b)_2 (\min v_{ij} | j \in C_c)\} = \left\{ v_j^+ | j = 1, 2, \ldots, m \right\} \tag{9}$$

$$A^- = \{(\min v_{ij} | j \in C_b)_2 (\max v_{ij} | j \in C_c)\} = \{ v_j^+ | j = 1, 2, \ldots, m \} \tag{10}$$

Step 4: Compute the separation measures using an m-dimensional Euclidean distance. The separation measures of each alternative from the positive and negative ideal solutions, respectively, are:

$$S_i^+ = \sqrt{\sum_{j=1}^{m} \left(V_{ij} - V_j^+ \right)^2}, j = 1, 2, \ldots, m \tag{11}$$

$$S_i^- = \sqrt{\sum_{j=1}^{m} \left(V_{ij} - V_j^- \right)^2}, j = 1, 2, \ldots, m \tag{12}$$

Step 5: Compute the relative closeness (RC_i^+) to the ideal solution. The relative closeness of alternative A_i with respect to A^+ can be estimated from positive and negative ideal solutions of each attribute which forms the rebroadcast probability (P).

5 Performance Evaluation

The aim of this section is to evaluate the performance of our two proposed counter-based broadcasting schemes, viz., the adaptive, mobility-aware, energy-efficient counter-based broadcasting scheme (AMECBB) and counter-based broadcasting using a TOPSIS multi-criteria decision-making scheme (TCBB) with AODV using the NS-2 Simulator (NS-2.34).

The parameters used in the following simulation experiments for Mobile Ad-hoc network are listed in Table 4. The network has a constant transmission range of 250 m. The MAC layer scheme follows the IEEE 802.11 MAC specification. The movement pattern of each node follows the random waypoint model. The radio propagation model used is the two-ray ground reflection model which considers both the direct path and a ground reflection path. The transceiver of MANET is configured with Omni directional antenna. The mobiles nodes move with a speed ranging from 1 m/s to a maximum of 4 m/s over the terrain size of 1000 * 1000. The simulation was performed for 225 ms with different network densities (50, 75, 100, 125, 150).

The following measures were used to evaluate the effectiveness of the schemes over AODV. Packet over varying network densities of (50, 75, 100, 125, 150) for our proposed schemes and AODV.

Table 4 Simulation parameters

No.	Parameter	Value
1.	Network area	1000 * 1000
2.	Transmission range	250 m
3.	Simulation time	225 ms
4.	Number of nodes	50, 75, 100, 125 150
5.	Traffic pattern	CBR (UDP)
6.	Max. speed of nodes	4 m/s
7.	Pause time	25 s
8.	Routing protocol	AODV, AMECBB, TCBB
9.	MAC protocol	IEEE 802.11
10.	Mobility model	Random way point mobility model
11.	Propagation model	Two ray ground
12.	Antenna	Omni directional

5.1 Packet Delivery Ratio

The packet delivery ratio is the ratio of the number of received data packets to the number of sent data packets. A higher packet delivery ratio means better performance. Figure 8 shows the number of nodes versus the packet delivery ratio. For the AODV protocol, the packet delivery ratio decreased from a low-density network to a high-density network as more number of packets are dropped from a sparse network to dense network. However, for the proposed schemes, the packet delivery ratio was constant due to the reduction in packet dropping. AMECBB caused an average incremental growth of 30% in the packet delivery ratio compared with the AODV routing protocol, while TCBB incurred an average incremental growth of 32%. TCBB performed better than AMECBB by increasing the packet delivery ratio by 2% more.

5.2 Total Consumed Energy

Total energy consumption is defined as the difference between the current and initial energies of nodes. A lower total energy consumption means better performance for a proposed scheme. Figure 9 shows a graph of the number of nodes versus total energy consumed, highlighting that total energy consumption increased from the low-density network to the high-density network when using the proposed schemes as well as for the AODV routing protocol. AMECBB reduced the total energy consumption by more than 24% compared to the AODV routing protocol, while TCBB reduced it by more than 32%. TCBB performed better than AMECBB by reducing the total energy consumption by 11% more.

Fig. 8 Packet delivery ratio

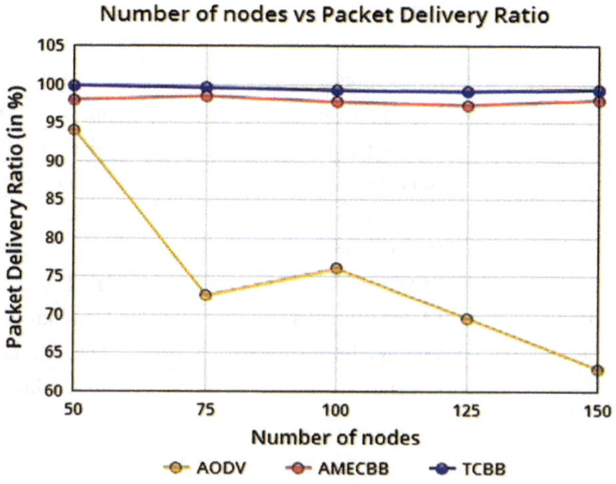

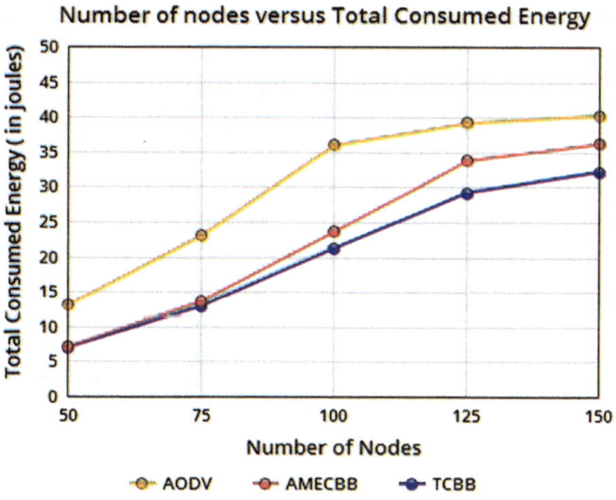

Fig. 9 Total consumed energy

5.3 *Delay*

Delay is the total time required by a packet to travel from the source to the destination. A lower delay means better performance. Figure 10 presents a graph of the number of nodes versus delay. Delay increases linearly from a sparse network to a dense network in the proposed schemes as well as for the AODV routing protocol. AMECBB reduced more than 47% of the delay, and TCBB more than 65%, compared to the AODV routing protocol. Indeed, TCBB outperformed AMECBB in reducing the delay by an additional 34%.

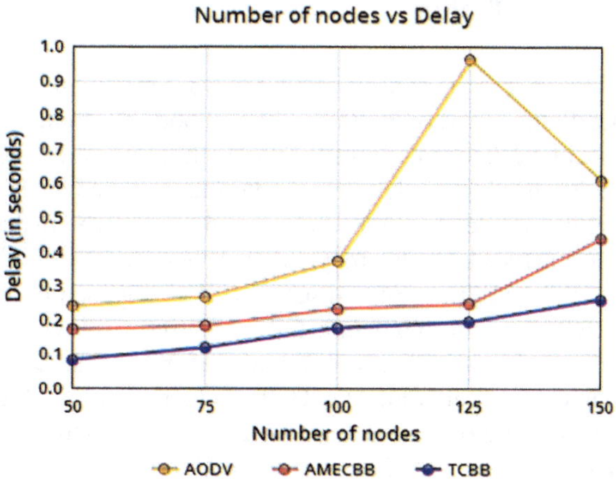

Fig. 10 Delay

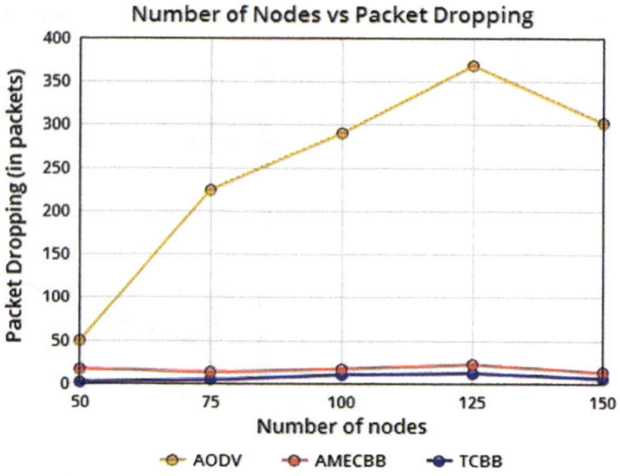

Fig. 11 Packet Dropping

5.4 *Packet Dropping*

The Data packets are considered dropped if they travel through a network and fail to reach their destination. A smaller number of packets means better performance. Figure 11 shows the number of nodes versus packet dropping. The number of dropped packets increased from the low-density network to the high-density network for the AODV protocol. For proposed schemes, the inverse was true because of reduction in retransmission of packets. Indeed, TCBB performed better than AMECBB by reducing dropped packets by an additional 55%.

6 Conclusion

A basic counter-based method was modified to dynamically refine counter thresholds based on node neighbourhood information and node mobility and energy to achieve better network lifetime. In a sparse network, shared coverage is lower since very few nodes act as intermediate nodes. If these intermediate nodes are highly mobile then link failure can occur which decreases packet delivery. In a dense network, nodes may exhaust their energy quickly, which in turn also causes link failure during route discovery. We proposed two schemes, viz., AMECBB and TCBB that can avoid link failure caused by low node energy and stability. Both schemes focus on the mobility and energy of nodes while forwarding packets from source to destination. The TCBB scheme exploited TOPSIS multi-criteria to make broadcasting decisions. The AMECBB approach reduced dropped packets by more than 93% compared to the AODV routing protocol, while the TCBB method reduced packet drops by more than 97%. As packet loss was reduced, the packet delivery ratio increased while delay was reduced by more than 47% for AMECBB and 65% for TCBB when compared with AODV. Further, energy consumption was reduced by 24% with AMECBB and by 32% with TCBB.

References

1. Conceicao, L., Curado, M.: Clustering for indoor and dense MANETs. In: Proceedings of 5th International Conference on Internet of Things, Smart Spaces, Next Generation Computing, ruSMART 2012, St. Petersburg, Russia, pp. 225–236, 27–29 Aug 2012
2. Dey, N., Ashour, A.S., Bhatt, C.: Internet of things driven connected healthcare. In: Internet of Things and Big Data Technologies for Next Generation Healthcare, pp. 3–12. Springer, Cham (2017)
3. Dey, N., Hassanien, A.E., Bhatt, C., Ashour, A.S., Satapathy, S.C. (eds.): Internet of Things and Big Data Analytics Toward Next-Generation Intelligence. Springer, Berlin (2018)
4. Kimbahune, V.V., Deshpande, A.V., Mahalle, P.N.: Lightweight key management for adaptive addressing in next generation internet. Int. J. Ambient Comput. Intell. (IJACI) **8**(1), 50–69 (2017)
5. Railkar, P.N., Mahalle, P.N., Shinde, G.R.: Access control schemes for machine to machine communication in IoT: comparative analysis and discussion. In: 2018 IEEE Global Conference on Wireless Computing and Networking (GCWCN), Nov 2018
6. Ni, S.-Y., Tseng, Y.-C., Chen, Y.-S., Sheu, J.-P.: Broadcast storm problem in a mobile ad hoc network. In: Proceedings of 5th Annual ACM/IEEE International Conference on Mobile Computing and Networking (MobiCom), pp. 151–162 (1999)
7. Fazio, F., Guerriero, F., Fazio, P.: Link Stability and Energy aware routing protocol in distributed wireless networks. IEEE Trans. Parallel Distrib. Syst. **23**(4), 713–726 (2012)
8. Brzostowski, J., Roszkowska, E., Wachowicz, T.: Multi-criteria decision making models by applying the TOPSIS method to crisp and interval data. Multiple criteria decision making. International Scientist Journal issued by University of Economics, Multiple Criteria Decision Making, vol. 6, pp. 200–230 (2011)
9. Yi, Y., Gerla, M., Kwon, T.: Efficient flooding in ad hoc networks: a comparative performance study. In: Proceedings of the IEEE International Conference on Communications (ICC), Anchorage, USA, pp. 1059–1063 (2003)
10. Khamayseh, Y., Darwish, O., Wedian, S.: MA-AODV Mobility aware routing protocols for mobile ad hoc networks. In: Proceedings of Fourth International Conference on Systems and Networks Communications. IEEE, Porto, Portugal (2009)
11. Yassein, M.B., Al-hassan, A.A., Taye, Z.A.: Performance analysis of the effects of network density and network mobility on velocity based scheme in MANET. In: Proceedings of 7th International Conference on Systems, Signals and Devices (SSD), pp. 1–7. IEEE, Amman, Jordan (2010)
12. Mustsfa Bani Khalaf, D., Al, A.Y., Abed, M.: New velocity aware probabilistic route discovery schemes for MANET. In: Proceeding of 20th Conference on Software, Telecommunications and Computer Network (SoftCOM), Split, Croatia, pp. 1–6. IEEE (2012)
13. Shivashankar, H.N., Varaprasad, G., Jayanthi, G.: Designing energy routing protocol with power consumption optimization in MANET. IEEE Trans. Emerg. Top. Comput. **2**(2), 192–197 (2013)
14. Egoh, K., De, S.: A multicriteria receiver-side relay election approach in wireless adhoc networks. In: Proceedings of the 2006 IEEE Conference on Military Communications, MILCOM 06, pp. 2389–2395. IEEE Press, Piscataway, NJ, USA (2006)

15. Tseng, Y.-C., Ni, S.-Y., Shih, E.-Y.: Adaptive approaches to relieving broadcast storms in a wireless multihop mobile ad hoc network. IEEE Trans. Comput. **52**(5), 545–557 (2003)
16. Yassein, M.B., Khaoua, M.O., Mackenzie, L.M.: Improving the performance of probabilistic flooding in MANET. In: Proceeding of International Workshop on Wireless Ad Hoc Networks. London, UK, pp. 1–6 (2005)
17. Yassein, M.B., Khaoua, M.O., Mackenzie, L.M., Papanastasiou, S.: Performance analysis of adjusted probabilistic broadcasting in MANET. Int. J. Wirel. Inf. Netw. **13**(2), 127–140 (2006). Springer Netherland
18. Mohammed, A., Ould-Khaoua, M., Mackenzie, L.: An adjusted counter-based broadcast scheme for mobile ad hoc networks. In: Proceeding of 10th International Conference on Computer Modeling and Simulation, Cambridge, UK, pp. 441–446 (2008)
19. Khalaf, M.B., Al, A.Y., Buchanan, W.: A new adaptive broadcasting approach for mobile ad hoc networks. In: Proceedings of 6th Conference on Wireless Advanced (WiAD), pp. 1–6. IEEE (2010)
20. Yassein, M.B., Nimer, S.F., Al-Dubai, A.Y.: A new dynamic counter-based broadcasting scheme in MANET. Simul. Model. Pract. Theory **19**(1), 553–563 (2011) (Elsevier)
21. Reina, D.G., Toral, S.L., Johnson, P., Barrero, F.: Hybrid flooding scheme for mobile ad hoc networks. IEEE Commun. Lett. **17**(3), 592–595 (2013)
22. Yang, S., Wu, J.: Efficient broadcasting using network coding and directional antennas in MANET's. IEEE Trans. Parallel Distrib. Syst. 148–161 (2010)
23. Yassein, M.B., Al-Dubai, A.Y.: Inspired counter based broadcasting for dynamic source routing in mobile networks. In: Proceedings of International Conference on Computer and Information Technology, pp. 1455–1459. IEEE (2015)

Development of an IOT-Based Atmospheric Fine Dust Monitoring System

N. Kavitha and P. Madhumathy

Abstract The major perilous type of air pollution is due to particulate matter since it is the primary leading factor in affecting the human health as well as it is having major impact even on the earth's typical weather condition and precipitation levels. The proposed methodology yields in developing a system which is user friendly, low cost module used for measuring the fine dust particle, amount of CO (carbon monoxide), in addition to that temperature and humidity is also been estimated for weather forecasting. The information about the pollutant from the environment is collected through fine dust sensor, humidity sensor and temperature sensor. The collected statistics about the environment is been forwarded to the Node MCU. ESP8266 is been used to operate Node MCU. The cost-effective Wi-Fi microchip ESP8266 is used to processes the data. The IoT server collects the processed data. IoT server is used to fetch data whenever there is a request from mobile application for the data. This proposed system after measuring the dust particle it analyses the dust levels in real world scenario in addition it tests and collects the pattern change of dust at different location. The analysed data is given to the users in the form off instant alerts to the subscribers. Preventive measures can be taken in prior based on the instant alert of the analysed data given to the subscribers. This preventive measures yields in providing better environment and leading a healthy life.

Keywords Node MCU (ESP8266) · Temperature sensor · Fine dust sensor · Gas sensor · Humidity sensor

N. Kavitha · P. Madhumathy (✉)
Electronics and Communication, Dayananda Sagar Academy of Technology and Management, Bengaluru, India
e-mail: sakthi999@gmail.com

N. Kavitha
e-mail: kavithaprashanthn@gmail.com

N. Dey et al. (eds.), *Internet of Things, Smart Computing and Technology: A Roadmap Ahead*, Studies in Systems, Decision and Control 266,
https://doi.org/10.1007/978-3-030-39047-1_12

1 Introduction

Internet of things (IoT) is an integrated communication technique, it connects physical devices and everyday objects from anyplace, anytime. Internet of things (IoT) communicates with sensors [1], devices and social contexts by using intelligent interfaces. There are numerous applications on different domains of IoT, such as retail environment [2], safety, security, healthcare, bio sciences automation, smart grid and many others. Here the cloud is used to store data [3]. IoT is one of the growing technologies [4]; it uses single architecture to connect two or more devices.

To establish communication with other devices microprocessors and microcontroller are connected with other devices so that they establish a connection to communicate with each other. To communicate between connected devices using internet a specific protocols are being used. Web socket, http are some of the specific protocols used for communication between devices. IoT is platform through which communication between different physical devices can take place and also it supports communication with itself.

IoT with embedded includes devices like microcontrollers, microprocessors and other physical devices connected by the internet communicate by means of MQTT, web sockets, https which are some of the specific protocols used for communication through internet [5]. These devices were able to transmit statistical data not only to the cloud server and a server in addition to that they were also able to intercommunicate with each other. In IoT infrastructure the devices can exchange data by getting connected with the cloud services. Here the data can be visualized by visualizing services and also analyzed by machine learning service. However increase in number of devices will not affect the utilization of a cloud because cloud manages the entire load and it will also see to it that all devices performance will be extremely up to the mark.

In recent years the air Pollution has surfaced globally [6] mainly due to eruptive industrial growth [7] and road vehicle transportation which is contributing towards weather change that has hazardous domestic and global consequences [8, 9]. Presently fine dust particles are increasing considerably [10, 11] suspended in air. The human health has been adversely affected due to these fine dust particles. The production of fine dust particle can be either natural origin or created by human activity [12]. Fine dust particle is classified into two types such as primary fine dust and secondary fine dust, primary fine dust is one in which dust particles [13] arise directly from a combustion process. Secondary fine dust is one wherein a complex chemical reaction from gaseous substances in the atmosphere creates fine dust.

Here motor vehicles, waste incineration plants, stoves and heating systems in residential buildings, bulk materials handling, animal husbandry, and certain industrial processes [14, 15] are categorized as man-made sources of fine dust. In urban areas, the main source of particulate matter is due to road traffic. An additional vital source contributing towards secondary dust pollution [16] is due to emissions of gaseous precursors from animal husbandry. Hence the fine dust particles are mainly classified as PM2.5 and PM10 based on the particle size. PM2.5 [17, 18] is a particulate matter whose dimension is equal to or less than 2.5 μm. PM10 is also one of the particulate matter whose diameter is equal to or less than 10 μm [19, 20]. These fine dust particle size is very minuscule [8, 21, 22] i.e., even slighter than the size of a human hair. The physical property of these particulate matter is particles size in the range 10 or 2.5 [5, 23] will yield in large proportion of dust that can penetrate easily deep into the lungs and yields to serious health problems, whereas particle size of larger diameter likely be trapped in the throat, mouth and nose.

Hence to address the above problems this paper aims at designing a mobile application which is flexible, reliable and provide approximate measure of dust particle [24], volume of CO (carbon monoxide) [12] to the users. Here arduino and Wi-Fi module has been inbuilt in fine dust sensor, MCU node. In addition we integrated gas sensor, humidity sensor, and temperature sensor for additional functionality of this system. Here dust levels in real time scenario are been estimated and also it examines and analyses the variation in dust pattern such as floating dust, blowing sand, and dust storm conditions at various scene. After the analysis the analysed data is provided to users in the form of instant alerts to the subscriber. On the basis of alert information people can take preventive measures in prior. This results in better environment and to lead healthy life.

2 Literature Survey

Kaivonen and Ngai [1] proposed an air quality monitoring system in real-time scenario by installing sensors on bus. This system was developed and tested in Uppsala Sweden to monitor air quality with movable sensors. Bus is been installed with sensor for air quality monitoring. The sensor measured NO_2, CO, temperature, humidity and ambient pressure. It worked well for CO sensor probe, but the NO_2 sensor had very less success rate in estimating the non-zero concentration. The data was successfully delivered without re-sending at a success rate of 70% by the usage of 4G mobile network. Storage and re-transmission function was implemented to reach maximum success rate. Here the system is less sensitive. The major drawback is the sensor just collect the data but analysis of data quality is not carried out. Case studies on how sensor measurements are affected by the presence of cooling system of the bus, moving speed and wind speed affect the measurement reading was not studied.

Binsy and Sampath [2] has proposed IoT based self tuned air quality examining system using a technique of data mining. Here data mining technologies and IoT technique is used for air pollution monitoring system. The device can vary the frequency of monitoring by itself on the basis of pollutant predictions. Air quality will be examined by accumulating all concentration pollutants which includes harmful gases, Carbon monoxide, dust level and meteorological parameters which provides temperature and GPS location. Further humidity information is predicted through monitored data. The level of pollutant is predicted through regression and data mining technique. Depending on the values being predicted consecutively it decides in which approach the device has to be operated. MATLAB tool is been used to analyse the monitored data that comes from ThingSpeak. In GPS map of the location pollution level is been updated where higher level of pollutant is marked in red and moderate level of pollutant is marked in green. Through ThingSpeak the predicted pollutant levels along with data can be sighted by the public. The system lags in regress analysis.

Zhalgasbekova et al. [3] proposed an Opportunistic collection of data to monitor indoor air pollutant on the basis of IoT. Here advance sensing methods such as movement of data mules, the accuracy closeness of sensor devices are used for collection of IoT data and low cost devices decides in cost effective manner IoT data will be accumulated and delivered it to a sink. It has implemented CollMule algorithm and in addition to that it has evaluated the algorithm further it extends the 3D kNN approach for determining, negotiating, accumulating and sending the sensed data at a faster rate and cost-effective manner. Android platform is used to handle indoor air quality using CollMule software prototype data from heterogeneous IoT devices. Based on amount of power consumed and usage of CPU for single algorithm cycle CollMule is evaluated. The usage of CollMule algorithm testing shows the practicability in the usage of mobile smart devices.

Lai et al. [7] proposed a methodology to predict and examine air quality by IoT based system using Kalman Filter for improving the accuracy. Here Kalman Filter algorithm is used for predicting the various air pollutants concentration. The proposed air quality examining system is cost effective and application of edge computing is based on Raspberry Pi and smart agriculture on IoT, Kalman Filter algorithm is executed using Raspberry Pi. In this paper delay in data transmission due to the limitations in bandwidth and network connection and the agricultural environment problem can be avoided by using machine learning algorithm, and it also improvises in taking the decision under real-time conditions by the implementation of Kalman Filter algorithm on the Raspberry Pi, it also exhibits a power computation strongly and it is been utilized on the edge devices, the major six type pollutants in air are SO_2, NO_2, CO, O_3, PM2.5 and PM10 [24, 25] are comprehended. Usage of Kalman Filter algorithm has improved the predicted value accuracy by 27% when compared to the observation of data from sensors, the accuracy of the sensors has improved from the aspect of the algorithm. The proposed system for air quality monitoring equipment is cost effective when matched up to the previous air quality monitoring equipment; the proposed system cost value is reduced by at least 10% having the equivalent examined accurateness. The major drawback of this paper in predicting air quality and monitoring is it has not considered the change in concentration of pollutants based on external environment influence. External environment includes factory emissions and wind speed which indirectly affect the pollutant concentration. In addition to that other factors also will directly affect the current concentration of pollutants. The monitoring and precautionary alerting means for edge estimating layer itself is not up to the mark.

Lu and Young [10] proposed a digital imaging technique for measuring particle fine dust size On-line. Numerous industrial applications inclusive of drug delivery, particulate emission monitoring and so on mainly uses on-line continuous particle sizing. But developing and implementing a cost effective, non-invasive instrument to monitor particulate emission continuously on-line still remains a challenge. This paper proposes a methodology to address the implementation of cost effective non-invasive instrument to measure the fine dusts through digital imaging technique. A digital imaging system is designed and implemented to collect the particle size information continuously. The equivalent particle size is obtained from the captured images by implementing ellipse-fitting algorithm. Thresholding processing is used to accomplish image segmentation; further 4-connectivity is used for accurate identification of connected regions in the binary images. Here evaluation of distribution in the particle size and the imaging system performance is gauged on the basis of statistical experimental results. The natural calcium carbonate narrow classified fractions is used for comparison and testing of particle sizing system. Here the design and operation of the system is simpler with low maintenance cost. But mainly the system lags in not monitoring of environmental data in addition to that the system doesn't even alert for bad environmental condition.

Kim [16] proposed a technique for developing atmospheric environment monitoring system using IoT. As we know that the main source for air pollution are fine dusts, industries etc. which is drastically increasing in recent years; henceforth it is necessary to develop cost effective air quality examining system which accurately estimates the pollutant concentration present in air and gives the pollutant information to public conveniences or homes. It proposes an effective system to monitor atmospheric environment on the basis of IoT (Internet of Things), this system effectively gathers the information of air pollutant without having any limitation of a particular place or space. This system mainly composed of two parts first part of the system include a device to measure atmospheric environment and second part of the system include an analyzer which performs the analysis of atmospheric environment measured in the previous step. Overall the system has been developed in such a way that it will be able to perform both the task such as collection and analysis of atmospheric environment. LTE communication network is been used to store the measurement results in a server. For estimating the performance of a system the proposed system atmospheric environment information was being evaluated with the study results of National Ambient Air Quality Monitoring Information System (NAMIS). Here the low cost, eco-friendly device has been developed for real time monitoring of gas sensor data through IoT technology, but the system monitors only gas particles not the particulate matter.

Wang et al. [21] proposed an methodology for measuring the fine particles size from industrial stacks using digital imaging technique. The estimation of particle size is mainly useful in various applications in industries and also in diverse fields such as observing particulate emission, drug delivery, atmosphere examining. Here electric power plant is the main parameter to affect combustion efficiency hence information about velocity, size and pulverized fuel concentration plays major role in potentially increasing safety risks and atmospheric emissions. Hence there is an immediate requirement for designing and implementing an efficient and low cost instrument for measurement of particulate emission on-line which is still remains as a challenge in this area. This paper primarily aims at tiny particles size distribution measurement such as PM2.5, PM10 and so on. Here new methodology is been proposed for addressing the problem for deriving the distribution of particle size. The main goal is to propose a system for designing and implementing image processing algorithms on the test images for the analysis of on-line particle size in future. The system was tested with calcium carbonate fractions. The statistical analysis data showed that correspondence between coefficients and the standard value measured value are above 90%. Analysis result shows the methodology is efficient and precise in estimating the particle size under static conditions. The proposed system mainly lags in implementing low cost instrument this area. The proposed system has a drawback of it is only focusing on tiny particles, such as PM2.5, PM10 the size distribution measurement.

3 Motivation

Conventional methods for monitoring air pollution generally require high investment costs for laboratory analytical instruments [26, 27], extensive maintenance, training and highly specialized personnel [24, 28]. Hence, there is an increasing demand for monitoring and also controlling systems for real-time environmental pollution [16, 28, 29]. In this regard toxic chemicals are the main source for ever-increasing pollution; hence the systems should have an ability to facilitate in detecting and quantifying the sources rapidly. The usage of Internet of Things (IoT) based system can make implementation of air pollution monitoring system less complex and other direct readings can be obtained at that instant. IoT based air pollution monitoring systems has an added advantage of easy installation, convenient maintenance and cost effectiveness. Moreover, they are advanced, lithe with respect to information is concerned [13] and also scrutiny is more as compared to traditional monitoring system. IoT based monitoring will helps us to provide unrefined measurements of air pollutant concentrations [29], which can be further used for analysis and based on the application defined it can be interpreted. Monitored data analysis is helpful in assessing day by day how the pollutant is increasing in air, it also help us to assess which areas air pollutant is worse than others and whether the levels of air pollutant are rising or falling. It can also be observed how there is an interaction between the pollutants and how they can be related to traffic levels or industrial activity is also been estimated [21]. The sensors and node MCU specification offers flawless interoperability between smart things without including the necessitate of complicated local mechanism [20, 30] and offer back the free will to communicate with the user, developers and enabling with devices connected. This paper proposes air pollution monitoring system on IoT basis that enables sensor data to be appropriately distributed within the stipulated time so that proper interpretation [31] can be made and actions are taken accordingly [32]. Obtaining these accurate real-time results allows regulatory agency to take necessary action whenever pollution occurs [33]. The regulatory agencies in India are the key inspiration for following our study in the development of a system. These regulatory agencies in India to ensures manufacturing industries which does not encompass a provisions for incorporating a means for treating the wastes they generate and there are many factories being built without including these technologies [34], These technologies mainly help us to reduce the waste they do so. These challenges have motivated us to design and implement a system which is of low cost, speedy response, low maintenance and can facilitate in generating measurements continuously. The major objective of this proposed system is to examine the, hazardous gases and air pollution in addition to monitoring it should also improve in creating awareness with reference to pollution by means of air pollution monitoring system [12].

4 Proposed Methodology

Few discontinuities such as monitoring air pollution is costlier, extensive maintaining and training is required in conventional technique and few system monitors only gas particles not the particulate matter are spotted with the existing system. The implementation of a well-performed instrument is very expensive and the system focuses on measurement of particulate matter only it could be enhanced more. If the system is cost-effective it only monitors only gas particles not the particulate matter. Furthermore many system lags in monitoring of environmental data in addition to that the system will not even alert for bad environmental condition and sensor data quality is been analysed.

To overcome the above mentioned difficulties a new method is proposed IoT with embedded system has been proposed wherein we develop user-friendly and cost-effective system for determining amount of CO (carbon monoxide), dust particle in addition temperature and humidity is also been determined for weather forecasting. Here determined values from temperature sensor, dust sensor, gas sensor and humidity sensor data collected from the environment will be fed to the MCU node running on ESP8266 which is one of the open source IoT platform. Further the Processed data is given to the IoT server, which will be fetched by the mobile applications when required. In addition to the data collected by that system it also analyses variation in the dust levels for real time scenario taken at different location and that analysis is helpful in instantaneously alerting the subscriber.

In IoT mobile and physical world connecting devices such as light, fan, temperature sensor, dust sensors, etc. can get connected with the cloud, here the communication between mobile and physical world connecting devices in a cloud they get connected with the cloud and communicate. Alike to PCs working mobile can also send information into the cloud which is been linked to devices in IoT and it can also retrieve the information from the cloud it is also a cloud of connecting devices. The most common IoT devices considered are raspberry pi, Intel Edison.

These IoT works for several version of Linux and also they can communicate the messages with web services like calling the web services, responding to a web services by getting connected to internet, usage of common IOT protocols like MQTT they get connected in which two machines can network and communicate the data between them. In IOT infrastructure they can obtain the commands from mobile and act accordingly to convey some data in return into the mobile. In IOT infrastructure there is variety of number of IOT devices like Intel Edison [35, 36] which is connected with relays sensors, and it puts data in return into the cloud. Based on tremendous processing ability of the cloud these data will be further processed. The outcome of processed data can be made available by integrating to a data visualization framework or machine learning to the mobile users.

The proposed system is based on Arduino development. It is having a Node MCU board. This board has 1 standard non concurrent serial ports (convention: 8 information bits, 1 stop bit and no equality), and it does not withstands increased Baud rate wherein a USB programming ports fit for transmitting information in 9600 Baud rate. The main ground for choosing this MCU board is due to its effectiveness in prototyping with minimal effort. Moreover, it is programmable with Embedded C language (Fig. 1).

Sensor is one of the subsystems used for events detection or environmental changes it forwards the data to the other electronics. Due to huge progression in micro machinery and microcontroller platforms, the uses of sensors have been

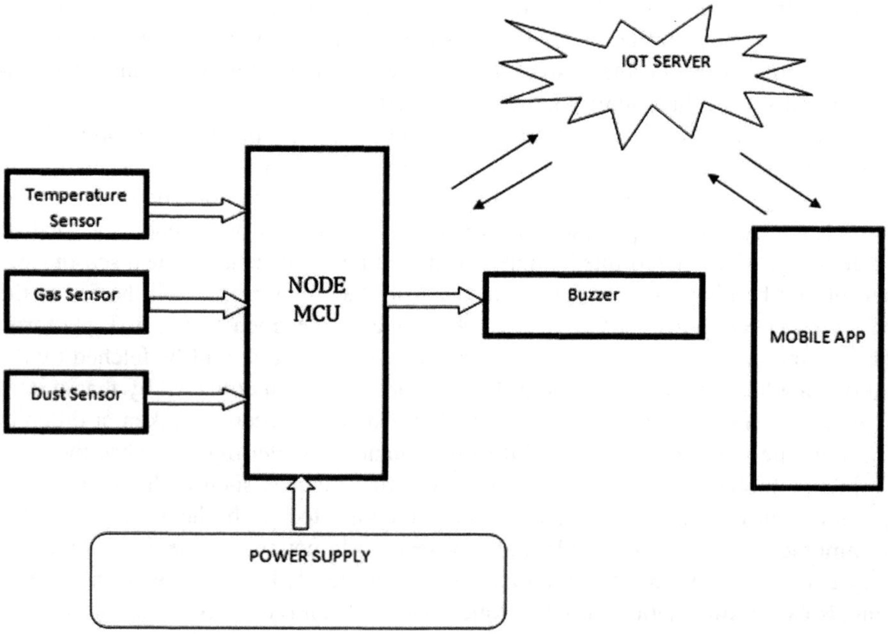

Fig. 1 Block diagram

expanded. Sensor detects analog values from the environment and sends to the microcontroller. Dust sensor, temperature sensors and gas sensor are connected to microcontroller for taking environment data from an environment.

- Dust Sensor—It will sense Dust particle from environment.
- Temperature Sensor—It will detect environment temperature.
- Gas sensor—It will detect hazardous gas (Like CO_2, NO_2) from Environment.

Node MCU: It is a board in which microcontroller and wifi module is present inbuilt. Node MCU runs on the ESP8266 Wi-Fi SoC which is an open source IoT platform. The ESP8266 is a microcontroller with full TCP/IP stack which is a Wi-Fi microchip with lower cost. The ESP8266 is a small computer microcontroller (or MCU for microcontroller unit) on a single integrated circuit. The different applications of **IoT servers** include management, monitoring and gathering data and analysis of data. They are open-source enterprise platforms which are modular based and it provides the required capabilities which are needed for the server-side to connect to devices. Cloud server has online storage capabilities but it is analogous to a Wi-Fi router. Here server is been remotely accessed through cloud.

Since peers are connected online through cloud server the data is directly fetched from the cloud whenever data is required. IOT server is established by creating one account in IOT page. Microcontroller will send sensor data to IOT server using Internet. One Mobile app will be created for alert system. All devices are connected with DC power supply.

5 Experimental Setup

Control supply is a reference to a wellspring of electrical compel. A contraption provides electrical or diverse sorts of essentialness to a yield load or assembling of weights is known as constrain PSU. This is generally associated with electrical essentialness supplies, less much of the time to mechanical ones, and once in a while to others. This power supply segment is required to change over AC flag to DC flag furthermore to decrease the plenitude of the flag. From the mains the available voltage motion is 230 V/50 Hz which is an AC voltage, but we require a DC voltage with the sufficiency of +5 and +12 V for a range of applications. Node MCU is WI-FI enabled and it is an Open-source, Interactive, Programmable, Low cost, simple and Smart (Fig. 2).

High Level of Integration: ESP8266EX is one of the major Wi-Fi integrated chips in the industry. It has a dimension of 5 mm × 5 mm. A 32-bit Tensilica MCU is integrated with minimal external circuitry. The temperature ranges from −40° to 125° in the industry which is one of the widest operating temperature range.

32-bit Tensilica MCU: Tensilica L106 is a 32-bit Micro Controller (MCU) it is been integrated with ESP8266EX. It posses 16-bit RSIC architecture, it consumes

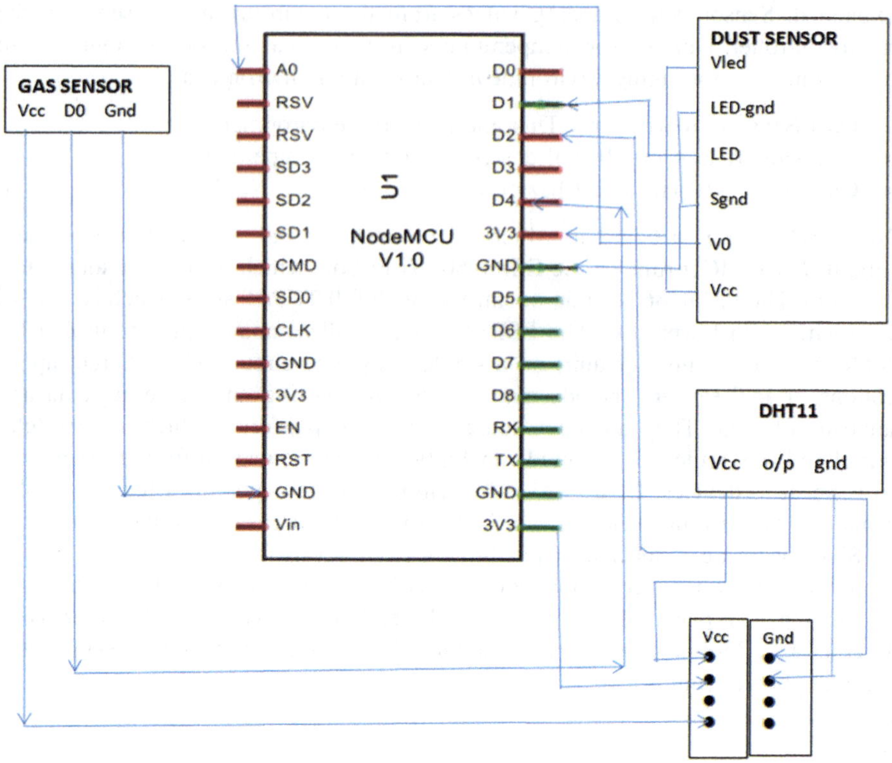

Fig. 2 Circuit diagram of proposed system

extra low power, it works at a maximum clock frequency of 160 MHz. It is Wi-Fi stack functional and Real Time Operation System (RTOS) enabled.

The DHT11 is a basic digital humidity sensor and temperature sensor which consists of digital signal output of the temperature and humidity. The temperature and humidity of neighbouring air is calculated by capacitive humidity sensor and a thermistor, and digital signal is been splited. It has long-term stability and higher reliability. The sensor is an inclusive of a resistive sense of wet components and NTC temperature measurement devices which is connected to 8-bit microcontroller.

Interfacing temperature and humidity sensor with Arduino UNO board:

- The first pin of the DHT11 is the VDD power supply 3.5–5.5 V DC.
- The second pin of the DHT is Data pin which supports serial data.
- The third pin is not used.
- The fourth pin of the DHT sensor is ground.

First the ground and the VCC of DHT11 here temperature sensor and humidity sensor is connected to the ground and 5 V of the Arduino. Then data pin of the

DHT11 sensor is been connected to the pin 2 of the Arduino. Finally fourth pin is ground to which negative of supply is been connected.

Gas sensor is a very sensitive to clean air and combustible gas; the conductivity is lower in clean air and conductivity increases when exist target combustible gas. As the gas concentration increases the sensor's conductivity will also increase. Potentiometer is been used for adjusting the sensitivity of the sensor. Henceforth any change in conductivity can yield to output signal of gas concentration correspondingly. The gas leakage detection in home and industry will be done by Grove—Gas Sensor module. It is mainly used for detecting Smoke, Propane, H_2, LPG, CH_4, CO and alcohol. Due to its elevated sensitivity, lower cost and speedy reaction, it is been preferred for taking precautionary measures at the early stage itself.

GP2Y1010AU0F is one of the optical sensors for air quality measurement it is mainly designed to sense dust particles. The reflected light of dust in air is been detected by diagonal arrangement of an infrared emitting diode and a phototransistor in the device. It mainly aims at determining very fine particles like cigarette smoke from home dust effectively which is commonly used for air purification. It consumes very less power (MAX: 20 mA) and to interface dust sensor with arduino board is connected to its pin6 by using its matching connector.

An audio signalling device is a buzzer which can be mechanical, electromechanical, or piezoelectric. Piezoelectric buzzers are mainly used for making beeps tones and alerts. It drives 3–30 V peak-to-peak square wave. Here either anyone of the pin is connected to the ground terminal or the other pin to the square wave output from a timer. It works quite well for the range 2–10 kHz. Typically buzzers and beepers are used in alarm devices, timers, and it requires confirmation from user input such as a mouse click.

6 Hardware Implementation

Here four sensors are used such as Dust sensor (GP2Y1010AU0F), Gas sensor (MQ 6), Temperature sensor and Humidity sensor (DTH11). Circuit diagram and hardware connection between sensors and node MCU is shown in Fig. 3.

For embedded software implementation arduino platform is chosen, here Arduino is used for managing lights and other actuators by sensing the environment in the form of input being received from a variety of sensors and it can estimate how it affects its surroundings. Arduino programming language is based on wiring and the Arduino development environment is based on processing is been used for programming microcontroller on the board. Arduino projects are stand-alone projects or it can be combined with software running on computer for communication purpose. We can Plug the Arduino to your PC via USB cable once we downloaded the arduino IDE. From which we can control the devices remotely.

Thing Speak is one of the IoT analytics platform service by MathWorks, the architect of MATLAB and Simulink. The online data streams in the cloud can be

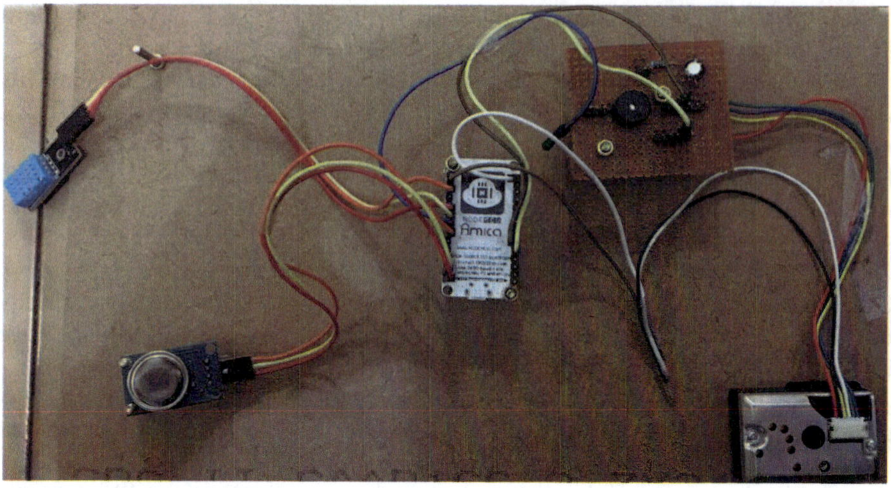

Fig. 3 Hardware implementation

visualized, summative and analyzed by Thing Speak. The data posted by your devices can be instantly visualized by Thing Speak. Analysis of data online and processing data is performed by executing MATLAB code in Thing Speak. The development of proof-of-concept in IoT systems through analytics is been carried out by Thing Speak. IoT systems can be built without building up of servers or develop web software. Thing Speak provides a hosted solution for small- to medium-sized IoT systems which can be used for production. Thing Speak permit us for aggregating and analyzing online data streams in the cloud. Our information will be stock up in channels by Thing Speak.

Internet-connected device is required for connecting hardware to Thing Speak. A specific embedded hardware prototyping platforms like Arduino, ESP-8266 and Raspberry Pi are required for transferring the information from our devices are been commonly used. REST API or an MQTT API are been used for sending data to Thing Speak from machines or local gateways.

Data will be written to Thing Speak and can also be fetched via Thing Speak Simulink blocks, it is being used in this models by Simulink user for embedded devices including iPhone, Android, Arduino, Raspberry Pi.

The data can be accessed online as well as offline through Thing Speak. For online or offline analysis all the information we send will be stored in one primary location in the cloud so that we can access our data through Thing Speak easily. The data stored in cloud also helps in protecting our private data with an API key that we control. We can make use of the option in internet for downloading the information securely and store in cloud when we login to your Thing Speak account. REST API and the appropriate API key is been used to read our data programmatically in JSON or CSV formats. Subscribing to an MQTT topic will allow our devices to read data from a Thing Speak channel. Third-party web services inclusive of climate data, public utility data and stock data from financial providers are being imported. The imported data and data collected from your devices and equipment can be used together for investigating correlations and to develop predictive algorithms.

Visualization of sensor data can be done remotely in real time through Thing Speak. This visualization of sensor data will be routinely charted, so that we can distantly supervise our devices from anyplace and also view our data from any internet browser or mobile device. Read-only views our data will be shared between the clients and colleagues. On the other hand we can manage our data through Thing Speak and we can also build our front end to our clients and for our customers to log into.

7 Results and Discussion

IoT embedded system and arduino platform is used for developing atmospheric Fine Dust Monitoring System. Here five different case studies are been carried out for analysis of different sensor and how the data which is been processed will be processed to the IoT server.

Case 1:

In case 1 the IoT server (ThingSpeak) will receive the processed data which will be sent. To stock up and get back the information by Thing Speak using the protocols over the internet. Thing Speak is being used which is an open source IoT platform for application and API. The sensor application includes logging in, location tracking and a social network of things with status updates are being enabled by Thing Speak. When the subscriber wishes for data it is being obtained from the server and it shows the result in mobile application. This is how mobile application communicates with IoT server.

Output

See Figs. 4 and 5.

Case 2:

In case 2 dust sensors is been tested. Dust sensor senses dust particles from the environment and that data is sent to the controller to process. A threshold value will be set, if the dust particles increase beyond that threshold value then there will be a buzzer sound. When the output pin of the dust sensor is connected to input pin of controller i.e., Vcc (5–7 V), due to its low power consumption capability it is easily interfaced with node MCU. Node MCU enters deep sleep power less than 10uA and wakes up when the dust sensor starts consuming the power. Typically it measures up to 0.9 V and sensitivity ranges up to 0.5 V/(0.1 mg/m^3). When it crosses the above the mentioned range, the buzzer starts making sound specifying that the dust levels have been increased above the specified levels. Furthermore the cloud server

Fig. 4 Output of sensors

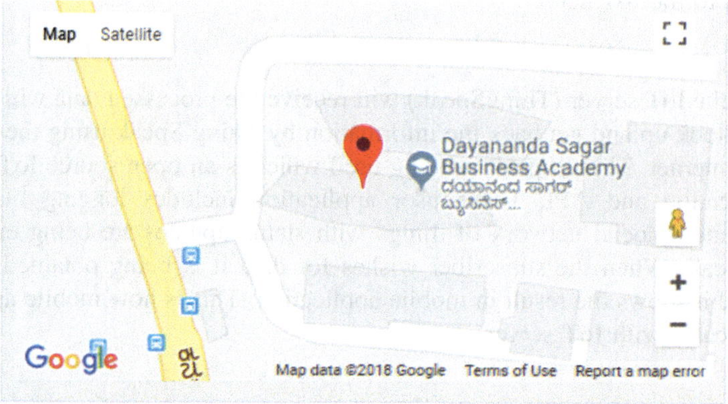

Fig. 5 Location

will be updated with processed data and this data is shown in the mobile application. Variation in dust density, it will be increased or decreased and output is shown in Fig. 4.

Case 3:

In case 3 gas sensor is been tested. Determining the gas present in the environment and yielding to the consequent changes are being observed by gas sensors. Resistance of MQ-6 is dissimilar to different kind and various concentration gases. There is no specific threshold value as such. MQ6 has tin Dioxide (SnO_2) its sensitive layer predicts the gas other than oxygen concentration(minimum value over 21%). Hence the sensitive layer measuring electrode and heater senses the gas and provides the sound as this gas is different from the oxygen layer. When digital pin of controller i.e., D0 and the output pin of the gas sensor is connected, then the data will be processed. Further this data is updated in the server and the same statistical data is shown in the mobile application. CO Gas is been sensed and output is shown in Fig. 4.

Case 4:

In case 4 temperature sensors is been tested. Temperature varies in the environment. Temperature sensor in addition to monitoring it also observes the increased temperature. Here the threshold value is set, if the temperature increases beyond the threshold value then alert comes through buzzer sound. It can transmit signal up to 20 m. Further this data will be updated in the cloud server and this data is shown in the mobile application. Increased temperature output is shown in Fig. 4.

Case 5:

In case 5 humidity sensor is been tested. Humidity sensor senses the humidity in environment and observes the change in humidity. Here the threshold value is set, if the humidity increases beyond the threshold value then alert message comes through a buzzer sound. The threshold value here is of the range 55–80%. Further humidity data is been updated in the cloud server and this data is shown in the mobile application. Increased humidity output is shown in Fig. 4.

8 Conclusion and Future Work

Indoor and outdoor air quality observing system is been designed by means of diverse sensors for monitoring fine dust. Various parameters of the environment are monitored by using three sensors in the system for monitoring. The temperature sensor, Gas sensor and fine dust senor are used for estimating temperature and humidity and dust density from environment. Precautionary measures can be taken by the subscribers when real time data taken from sensors are being updated in IOT server for yielding better environment and leading a healthy life. In this system subscriber can obtain all real time environmental data from various area and they can plan the measures for their life according to it. Under different environmental change every unit is being tested and appropriate results are been obtained from all modules. The traced data is taken and every information are made available in IOT server. Hence this system is very helpful in taking precautionary measures for betterment of the environment and also for leading a healthy life so that both human being and society is being benefited. This system can be extended by taking into consideration additional factors that grounds the pollution mainly by the industries, since our system is only monitoring four parameters and hence it can be improvised. The sensor outputs during the internet connectivity can be viewed by this system. Toxic emissions can be controlled by giving commands from distance.

Numerous pollutant sensors are very expensive and some do not have sensors that sense them. Hence building a cost effective sensors for different parameters is one of the future and very challenging task.

References

1. Kaivonen, S., Ngai, E.: Real-time air pollution monitoring with sensors on city bus. Digit. Commun. Netw. (2019) (Science Direct)
2. Binsy, M.S., Sampath, N.: Self configurable air pollution monitoring system using IoT and data mining techniques. In: ICICI: International Conference on Intelligent Data Communication Technologies and Internet of Things (ICICI), pp. 786–798 (2018)
3. Zhalgasbekova, A., Zaslavsky, A., Saguna, S., Mitra, K., Jayaraman, P.P.: Opportunistic data collection for IoT-based indoor air quality monitoring. In: Internet of Things, Smart Spaces, and Next Generation Networks and Systems. Springer (2017)
4. Park, T.J.: LPWA IoT network technology trends. Electron. Telecommun. Trends ETRI **32** (1), 46–53 (2017)
5. Eric Wang, Y.-P., Lin, X., Adhikary, A.: A primer on 3GPP narrowband internet of things. IEEE Commun. Mag. **55**(3), 117–123 (2017)
6. Amann, M.: Health Risks of Ozone from Long-Range Transboundary Air Pollution. Copenhagen WHO Regional Office Europe (2008)
7. Lai, X., Yang, T., Wang, Z., Chen, P.: IoT Implementation of Kalman filter to improve accuracy of air quality monitoring and prediction. Appl. Sci. **9**(9), 1831 (2019). https://doi.org/10.3390/app9091831
8. Brar, S.K., Verma, M.: Measurement of nanoparticles by light-scattering techniques. TrAC Trends Anal. Chem. **30**, 4–17 (2011)
9. Oh, J.S., Park, S.H., Kwak, M.K., HaePyo, C., et al.: Ambient particulate matter and emergency department visit for chronic obstructive pulmonary disease. J. Korean Soc. Emerg. Med. **28**(1), 32–39 (2017)
10. Lu, Z., Young, Y.: On-line size measurement of fine dust through digital imaging. In: 2015 IEEE 3rd international conference on smart instrumentation, measurement and applications (ICSIMA) (2015)
11. Kang, D., Kim, J.-E.: Fine, ultrafine, and yellow dust: emerging health problem in Korea. J. Korean Med. Sci. **29**(5), 621–622 (2014)
12. Li, J.F., Xu, L.H., Cai, X.S.: Smoke dust emission monitoring system of more integrated measurement methods. China Powder Sci. Technol. **15**, 24–27 (2009)
13. Compact Optical Dust Sensor. GP2Y1010AU0F, Sharp [cited Dec 2006]. https://www.sparkfun.com/datasheets/Sensors/gp2y1010au_e.pdf
14. Korea has the worst environment among OECD countries, The Dong-Ailbo. http://english.donga.com/List/3/04/26/535532/1. Accessed 20 Oct 2016
15. Air Quality Standards and Air Pollution Level. Ministry of Environment [cited 2013]. http://eng.me.go.kr/eng/web/index.do?menuId=253. Accessed 20 Oct 2016
16. Kim, S.H.: Development of an IoT-based atmospheric environment monitoring system. In: 2017 International Conference on Information and Communication Technology Convergence (ICTC) (2017)
17. Ghafghazi, S., Sowlati, T., Sokhansanj, S., Bi, X., Melin, S.: PM2.5 in China measurements sources visibility and health effects and mitigation. Renew. Sustain. Energy Rev. **15**, 3019–3028 (2011)
18. Carter, R.M., Yan, Y.: An instrumentation system using combined sensing strategies for online mass flow rate measurement and particle sizing, vol. 54, pp. 1433–1437 (2005)

19. Zhang, J.Q., Yan, Y.: On-line continuous measurement of particle size using electrostatic sensors. In: Proceedings of the I2MTC International Instrumentation and Measurement Technology Conference, Vail, CO, USA, pp. 877–880 (2003)
20. Zhang, H., Cai, X.S.: The measurement of size velocity and flow angle of coarse water with image method. Therm. Turbin **37**, 26–29 (2008)
21. Wang, L., Zhang, L., Yan, Y.: Imaging-based size measurement of fine particles from industrial stacks. In: 2014 12th International Conference on Signal Processing (ICSP), pp. 19–23 (2014)
22. Zhang, J.Q., Yan, Y.: On-line continuous measurement of particle size using electrostatic sensors. In: Proceedings of the 20th IEEE Instrumentation and Measurement Technology Conference, pp. 164–168, 20–22 May 2003
23. Mazumder, M.K., Ware, R.E., Yokoyama, T., Rubin, B.J., Kamp, D.: Measurement of particle size and electrostatic charge distributions on toners using E-SPART analyzer. IEEE Trans. Ind. Appl. **27**, 611–618 (1991)
24. Carter, R.M., Yan, Y.: On-line particle sizing of pulverized and granular fuels using digital imaging techniques. Meas. Sci. Technol. **14**, 1099–1109 (2003)
25. Carter, R.M., Yan, Y.: The effect of illumination wavelength on the measurement of size distribution of very small particles using a novel imaging based system. Part. Part. Syst. Charact. **25**, 298–305 (2008)
26. Yan, Y.: Recent advances in imaging based instrumentation for combustion plant optimization. In: Proceedings of 2010 IEEE International Conference on Imaging Systems and Techniques (IST), pp. 148–151 (2010)
27. Holoubek, J.: Some applications of light scattering in materials science. J. Quant. Spectrosc. Radiat. Transf. **106**, 104–121 (2007)
28. Carter, R.M., Yan, Y.: A novel imaging system for concurrent measurement of particle velocity and size distribution in a pneumatic suspension. In: Proceedings of Instrumentation and Measurement Technology Conference, pp. 2050–2054, 12–15 May 2008
29. Yamamoto, N., Tomita, K., Sugita, K.: Measurement of xenon plasma properties in an ion thruster using laser Thomson scattering technique. Rev. Sci. Instrum. (2012)
30. Carter, R.M., Yan, Y.: An instrumentation system using combined sensing strategies for online mass flow rate measurement and particle sizing. IEEE Trans. Instrum. Meas. **54**, 1433–1437 (2005)
31. Schaller, A., Mueller, K.: Motorola's experience in designing the internet of things. Int. J. Ambient Comput. Intell. (IJACI) **1**(1), 75–85 (2009)
32. Carter, R.M., Yan, Y., Cameron, S.D.: On-line measurement of particle size distribution and mass flow rate of particles in a pneumatic suspension using combined imaging and electrostatic sensors, Flow Meas. Instrum. **16**, 309–314 (2005)
33. Muller, C.L., Chapman, L., Johnston, S., Kidd, C., Illingworth, S., Foody, G., Overeem, A., Leigh, R.R.: Crowd sourcing for climate and atmospheric sciences: current status and future potential. Int. J. Climatol. **35**, 3185–3203 (2015)
34. Dey, N., Wagh, S., Pathan, M.S.: Applied Machine Learning for Smart Data Analysis. CRC Press, Boca Raton (2019). https://doi.org/10.1201/9780429440953
35. Durst, F., Melling, A., Whitelaw, J.H.: Principles and Practice of Laser-Doppler Anemometry, 2nd edn. Academic Press, London (1981)
36. Gao, L., Yan, Y., Lu, G., Carter, R.M.: On-line measurement of particle size and shape distributions of pneumatically conveyed particles through multi-wavelength based digital imaging. Flow Meas. Instrum. **27**, 20–28 (2012)
37. Black, D.L., McQuay, M.Q., Bonin, M.P.: Laser-based techniques for particle-size measurement: a review of sizing methods and their industrial applications. Prog. Energy Combust. Sci. **22**, 267–306 (1996)
38. Bhatt, C., Dey, N., Ashour, A.S.: Internet of Things and Big Data Technologies for Next Generation Healthcare (2017)

Internet of Things Security

Toward Smart and Secure IoT Based Healthcare System

Smita Sanjay Ambarkar and Narendra Shekokar

Abstract The protection of a patient's data is the prime concern in the healthcare sector. With the escalation in the adoption of Internet of Things (IoT) technology for the smart healthcare system, incidences of the revelation of privacy data also upswings hence it becomes necessary to devise a secure smart healthcare system. The requirement of the secure healthcare system is based on a critical survey and this year's Thales India Data Threat report. The report discloses the percentage of data breaches in past years and emphasizes the need for a tightening of patient data privacy regulation. As a result, the secure smart healthcare system has been recognized as a high priority goal to improve the sustainability of society. However, to concoct a legitimate secure smart healthcare system, threat triggered by integrating multiple devices and protocols need to be curtailed. In addition, a big challenge is to achieve accuracy despite the generation of a colossal amount of data per unit time. Encryption is the top choice for satisfying data privacy laws. Still, only encryption cannot impede data breach activities. It doesn't always make sense to lush low constraint IoT devices on an algorithm encrypting every data, because it will impose a substantial burden on the system. It is imperative to develop techniques that will detect and prevent threats that vex the security of a healthcare system. Here authors attempt to analyze the smart healthcare architecture, its threats, vulnerabilities and the security measures to provide a secure smart healthcare system.

S. S. Ambarkar (✉) · N. Shekokar
Computer Department, DJSCE, University of Mumbai, Mumbai, India
e-mail: smita.ambarkar27@gmail.com

N. Shekokar
e-mail: narendra.shekokar@djsce.ac.in

S. S. Ambarkar
Computer Department, LTCoE, University of Mumbai, Mumbai, India

N. Dey et al. (eds.), *Internet of Things, Smart Computing and Technology: A Roadmap Ahead*, Studies in Systems, Decision and Control 266,
https://doi.org/10.1007/978-3-030-39047-1_13

1 Introduction

IoT is a revolutionary technology capable of transforming the world to new technological heights by transferring information from every connected entity. This ubiquitous term mainly uses sensors and actuators that are interconnected via the Internet. During the past decade, IoT has gained popularity and acceptance due to its low-power lossy networks [1]. Nevertheless, IoT faces challenges such as multi-hop routing, unattended deployments; high heterogeneity, low power devices, etc. In addition to these challenges, secure communication is required for the real-time realization of IoT sensors. It is essential for the receiver to verify that the received data must be generated from the trustworthy sensor node.

In any country, the healthcare sector is a significant need for every person. IoT in healthcare is emerging as a newfangled technology impacting the entire business world. Stakeholders of the healthcare system can directly use the IoT based application with ease. However, the difficulty is non-transparency in medical system that emanates ambiguity in terms of cost and treatment for each patient [2]. The paucity of medical amenity, redundancy of medical data poses another problem with the health care system that adversely affects the efficiency of the patient's treatment. Therefore, it requires optimizing the healthcare system in an effective manner [3]. Along with optimization and efficiency, security is also equally important. People are nowadays more cautious about privacy of their medical data and many countries have given legal provision for the data privacy [4]. However, medical privacy faces the big challenge of data breach incidences [5]. Data breach in healthcare sector results in leakage of patients confidential data by unauthorized means and hence patient demented. Suppose if someone is having any ignominious disease and the information about that disease is leaked in his/her workplace or a social circle then it will be difficult for him/her to work or to communicate therewith a stable and constructive mind. Hence, the paper concentrates on the security of the healthcare IoT system.

The main contributions of this paper can be summarized as follows:

1. Present the precise architecture of the healthcare IoT system by analyzing various architectures.
2. Analyze and compare the existing research on threats in the IoT system and healthcare IoT systems precisely.
3. Investigate the security measures incorporated in the existing healthcare IoT system. Investigation lead to the conclusion that though there is security provided for the healthcare IoT system still it suffers from severe data breach incidences.
4. Paper further discusses potential directions for open research along with the recommendations for enhancing the security of the healthcare IoT system.

The rest of the paper is organized as follows: Sect. 2 surveys the literature to get an insight into the healthcare IoT architectures and the security measures incorporated in it. Section 3 proposes the IoT based healthcare architecture in detail.

Further Sect. 4 put forth the details of routing attacks along with the method of operation of routing protocol, afterward; this section gives critical analyses of the threats and attacks in the existing system implementations. Section 5 discloses open research issues and challenges. Further recommendations for future work are discussed in Sect. 6. Finally, Sect. 7 summarizes the paper and provides concluding remarks.

2 Related Work

The scarcity of automated and convenient healthcare system leads to many undetected healthcare problems. IoT digitized the whole healthcare system and reforms it to provide better medical facilities. Healthcare IoT applications can be classified into three categories [6] as shown in Fig. 1. It also describes applications in respective areas, which mainly focuses on to provide facilities to the patients like continuous monitoring, preventive care, enhancing patient satisfaction and engagement, advancement in care management, etc.

The interconnection of medical system with network makes the MCPS (medical cyber physical system) which are prone to many security challenges like cyber physical attacks, interoperability and threats [7].

Among the existing smart healthcare systems wireless sensor network (WSN), Radio Frequency Identification (RFID) and IoT are the most auspicious techniques for implementation.

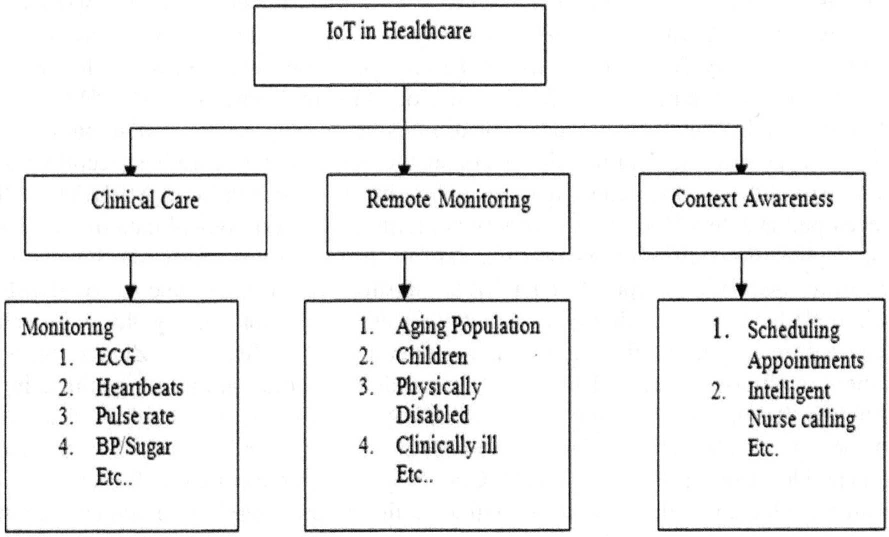

Fig. 1 The services provided by healthcare IoT

WSN is a flexible, scalable, dynamic, cost-effective ad-hoc network of analog or digital sensor devices which are communicated using an IEEE 802.15.4 radio signals. The battery-powered sensor nodes used in this multi-hop network consumes more power hence the overall lifetime of the WSN networks are highly get affected [8]. Conversely RFID is low power, a low-cost system consisting of RFID tag, RFID reader and server. RFID reader identifies the tag. RFID tag collects the information and after simple processing transfers it to the server. In the medical system, RFID tags attached to the patient to collect the physical health parameters. RFID is low cost, provides automatic identification and easy monitoring. RFID has a protracted lifetime, as tag does not require battery backups; therefore, RFID is well suited for healthcare systems. This is a promising technology comes with some limitations like feeble security and the reader's poor detection range only up to 15–20 m. Remote monitoring is difficult to implement using RFID techniques. Only a few endeavors have been there to integrate RFID and WSN for implementing the healthcare system [9]. Likewise, very fewer contributions presented in literature based on the IoT perspective to implement the healthcare system.

At first, the research work states the existing proposals for WSN technology for satisfying the healthcare systems. This exhaustive research survey delineates many architectural designs and security loopholes. In 2004 CodeBlue [10] developed at Harvard; it is a wireless infrastructure base sensor network lab, developed using PDA and PC for the deployment of exigency medical help, amalgamating low-power, and wireless significant sign sensors. CodeBlue uses a publish/subscribe model for data delivery. CodeBlue's authors realize the need for security in healthcare applications but authors did not address the same hence this architecture falls prey to DoS (Denial of service) attack, modification attacks, grey-hole attack, routing loop attack, masquerading attacks, sybil attack, snooping attacks (details of these attacks are given in Sect. 4). Other researchers in 2006 propose the architecture of Alarm Net at the University of Virginia [11] is a wireless sensor network specially designed for assisted living and home monitoring. It has been implemented as the network of MICAz sensor, stargate gateways, iPAQ PDAs, and PCs. Researchers address the authentication using an only secure remote password (SRP). Therefore the data transfer is not secure hence architecture is susceptible to confidentiality attacks, snooping, and replay attacks. Later in 2010 MEDiSN [12] developed at Johns Hopkins University is monitoring physiological data of patients using physiological monitors (PM) which are driven powers from the battery to measure patient's corporal data uses during catastrophe event. MEDiSN acknowledged data confidentiality, but the authors do not specify the approach required to attain confidentiality and authentication. Afterward, Zigbee based patient monitoring system [13] was recommended as a solution to assuage the aging problem. It emphasizes on the system including the delivery of first aid information in early 6 min after injury. The system provides services for monitoring of medical data and location tracking of patients. Generally, the system consists of a gateway, a router and handheld devices. Data confidentiality is addressed but no authentication mechanism is specified. Zigbee supports AES (Advanced Encryption Standard) 128-bit encryption to provide security. Another Zigbee based health monitoring system [14] is proposed which monitors the ECG of home based patients. It is also required by the patients who need regular health (ECG) parameter monitoring, analysis and interpretation.

Various research proposals are studied to perform a comprehensive survey of RFID based healthcare system. In 2014 night-care system based on ambient and passive RFID tags [15] was proposed. The system will help to monitor the elderly and disable people's state. The immediate help will be provided to them if any abnormal event gets detected. The authors failed to discuss the security aspect, the authentication problem will continue to pursue in the system.

In 2017 Anonymous Authentication protocol for RFID [16] was developed which has emphasized authentication, proposes the asymmetric design principle for lightweight authentication based on the generation of dynamic tokens. But in this architecture tag does not authenticate the reader so anonymous authentication is not abolished and also it falls prey to replay attack.

Another Internet of thing based lightweight RFID protocol [17] was proposed. These lightweight RFID protocols protect medical privacy using a secure authentication mechanism. The protocol is not factual and man in the middle attack could breach the security of the system. As all the existing security architectures for the healthcare system discussed above are prone to severe attacks summarised in Table 1, hence they are not providing adequate security to protect healthcare data. ALZ-MAS is a healthcare monitor [18] developed specifically for monitoring and assisting the patients suffer from Alzheimer disease. System consisting of three agents user agent, super user agent, scheduler user agent, admin agent and devices agent for automating the patient's activities.

Finally, IoT based healthcare architectures are analyses in detail, the researchers of remote patient monitoring architecture [19] propose a prototype in 2016 for continuous monitoring of patients when the patients shifted in the general ward from ICU. The system includes sensors that communicate wirelessly with Gateways. The overall functionality of these devices is made available through a RESTful based Web Service. The security aspect acknowledged but kept for future implementation. The smart healthcare system (SHS) [9] was proposed with the integration of 6LBR (6LoPAN border router), 6LowPAN router readers (6LRR), and 6LowPAN host tag (HT). The architecture combines the benefits of RFID and sensor network but the security aspect is considered only while implementing the mobile application. The real-time healthcare monitoring system [20] presents to measure the patient's vital signs. This real-time healthcare system discusses hardware and software used in the system along with different sensors required for the healthcare system. However the proposal failed to discuss the exact architecture of the healthcare IoT system, the security measures are left untouched.

Table 1 Security analysis of existing healthcare IoT systems

Healthcare application	Attacks/Threats	Security goals not achieved by the system
CodeBlue [10] (Harvard Sensor Network Lab)	DOS, snooping attacks, modification attacks, routing loop attack, Grey-hole attack, sybil attack, and masquerading attacks	Confidentiality, authentication, non-repudiation
Alarm Net [11] (University of Virginia)	Confidentiality attacks, reply snooping attacks	Confidentiality, authentication (As the only secure remote password is provided for authentication), non-reputation
MEDiSN [12] (Johns Hopkins University)	Confidentiality attacks, reply attack, DOS	Confidentiality, authentication (Mechanism used for checking authentication is not mentioned)
Remote Patient Monitoring [19]	Confidentiality attacks, reply attack, DOS	Confidentiality, authentication, non-repudiation. Security aspect is consider as future scope
Lightweight RFID Protocol [17]	Man in the middle attack, anonymity	Confidentiality, authentication is there but it will not eliminate the chances of anonymity
Anonymous Authentication protocol for RFID [16]	Confidentiality attack, anonymity	Confidentiality, authentication is there but anonymity is not present which leads to a serious security threat
Zigbee based patient monitoring system [13]	Anonymity	Authentication, confidentiality is provided by AES hence it will create overhead in the system
Secured Hospital System [9]	Confidentiality attack, authentication is there for mobile application	Confidentiality, authentication is only for mobile applications

The above critical analysis of RFID, WSN and IoT based healthcare system reveals that IoT based healthcare system carries very sensitive and personal data. The security measures incorporated in the existing systems are not adequate to annihilate various attacks hence the patient's data security is at high risk. Therefore this work proposes a novel IoT based architecture, the details are given in the following section.

Healthcare application	Advantages	Disadvantage
CodeBlue [10] (Harvard Sensor Network Lab)	The scalability of the network is very high, that it can scale to form very dense network and also ensures reliable data transfer	Communication challenges, programming challenges and computational challenges are not addressed in CodeBlue [10]

(continued)

(continued)

Healthcare application	Advantages	Disadvantage
Alarm Net [11] (University of Virginia)	Alarm Net is prominently design for health monitoring of assisted living and home monitoring patients. The core idea of Alarm Net is implementation of is to protect privacy of patients data and power management	Alarm Net prone to adversarial confidentiality attack [21], that attacker may get patients residential location addresses. Hence, the patient's data privacy is at risk. Another drawback of the system is as it is having open system architecture hence transactional confidentiality attack may also take place [21]
MEDiSN [12] (Johns Hopkins University)	MEDiSN especially designed for patient monitoring system during disaster. Architectural design includes sensors, relay point (RP) and Back-end servers	Authentication mechanism implemented in the system is not discussed by authors of [12]. Any authenticated user can get access for sensor and backend database server; hence security of system is at risk
Remote Patient Monitoring [19]	Remote patient monitoring system developed for patients who get discharge from ICU and shifted to general ward. These patients need equal monitoring. System uses RESTful web based service that makes the system flexible	The security of the system is not addressed at all. The sensor node is not able to discover another sensor node. Hence author [19] not specified anything about sensor node discovery
Lightweight RFID Protocol [17]	The system implements the robust lightweight mutual authentication scheme where RFID tag, RFID reader, and server authenticate each other. Author [17] claims that it is highly secure protocol, suitable for IoT implementations	The tag anonymity is not achieved in the system hence the mutual authentication is not achieved. Further, the system is prone to man in the middle attack
ALZ-MAS [18]	The system provides monitoring for Alzheimer patients by using different agent. The system automates the medical staff's work and reduces time consumed for routine work	The security parameter is considered while implementation of the system but it requires enhancement to provide strong authentication. Further more features need to be consider to improve performance
Anonymous Authentication protocol for RFID [16]	The system design based on the asymmetric design principle for lightweight authentication. Each entity of RFID system generates the dynamic tokens to achieve the authentication	The system may prone to serious security threat, as anonymity not achieved in the system. The generation of dynamic token causes congestion in the network
Zigbee based patient	The system claims to provide the faster medical facilities to the patients. The system will track	The AES algorithm will provide the confidentiality of the system as AES is complex algorithm and

(continued)

(continued)

Healthcare application	Advantages	Disadvantage
monitoring system [13]	location of patient and first aid will reach to patient within 6 min. The system will transfer data in confidential manner	requires high computational power the overall system performance is degraded
Zigbee based health monitoring system [14]	The system monitors the ECG of home based patients. It provides regular monitoring for patients who require periodic ECG monitoring	ECG is a vital sign measure for any patients if the ECG data is not delivered properly may cause a death threat for patient. The reliability of data transfer is not consider in the current standard of zigbee
Secured Hospital System [9]	The architecture integrates the 6LoWPAN router to provide smart architecture	The exact architecture of the system is not discussed in detail; the security parameter is not addressed

3 System Architecture Overview

The proposed architecture of the healthcare IoT system envisioned to collect the real-time patient's health data and deliver to central control. The data need to be analyzed based on different metrics [22] (source ID, Source Address, number of packets lost, source Type, etc.) and then generates the alert if any data breach activity is observed. To provide the services mentioned in Fig. 1, the proposed IoT architecture is shown in Fig. 3 and the architectural elements that are required to build the system are illustrated in Fig. 2. The architecture mainly includes three components as follows.

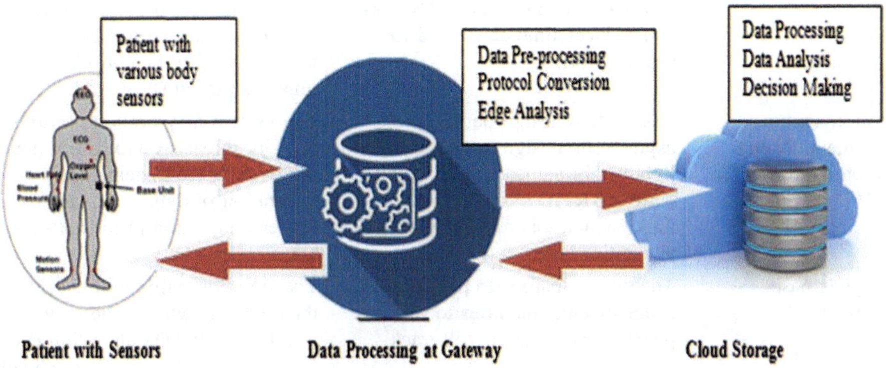

Fig. 2 Architectural elements of healthcare IoT systems

Body Area Sensor Network: The sensor nodes connected to the body of patients form the body sensor network (BSN). The patient's data collected by the sensor nodes, and then it is transferred towards the gateway using the 6LoWPAN protocol. The channel available to transfer this collected data is lossy hence BSN incurred high bit error ratio.

Therefore, the sensor node with low or degraded channel quality can transfer its data through some other efficient channel. This proposed architecture (Fig.3) uses a multi-hop routing technique which forms the clusters of sensor nodes. The energy efficiency of multi-hop routing is more energy efficient than single hop routing [23]. The proposed architecture uses the IPv6 routing protocol for low power and lossy networks (RPL) to route data towards the gateway. RPL [15] forms DODAG (destination-oriented directed acyclic graph) and RPL is an unsecured protocol that is prone to many attacks [15].

Internet-Connected Gateway: The gateway is the intermediary node that connects the BSN with the internet or cloud [3]. The gateways play the role of border router precisely known as 6BR (6LoWPAN border router). 6BR having the capability to convert IPv4 address to IPv6 formats and vice versa. This capability of 6BR makes the interoperability of constraint devices with the internet possible. The raspberry pi module will act as a gateway for our proposed system.

Cloud and Big Data Support: Cloud is a very vital module of the system. BSN generates huge amounts of data. 6BR routes this data towards the cloud using a routing protocol. 6BR is a resource constraint device having less memory, computation and processing power. Hence, the proposed application uses the cloud to store the sensor data. Thereafter, data analytics and processing stakeholders will get the data with the help of mobile or web-based healthcare applications.

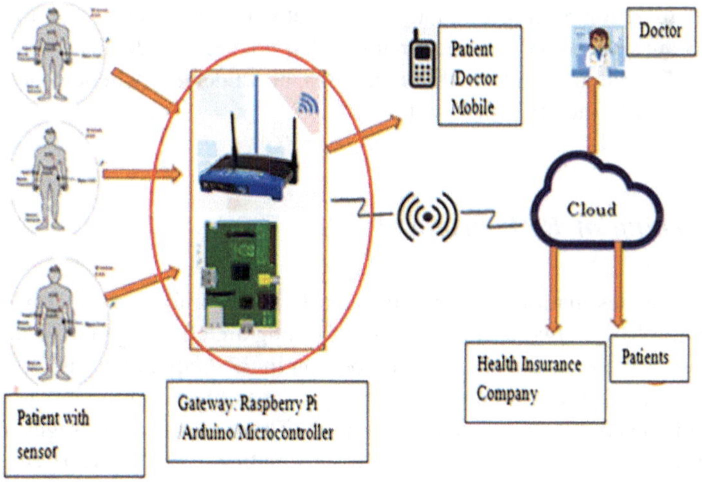

Fig. 3 The architecture of healthcare IoT

At the finer level, the working of healthcare architecture (Fig. 3) involves working of all the three modules precisely; data generated from sensors attached to the patient's body or room will be made available to authorized parties like family members and caregivers. It will give them the ability to check the subject's (patients) vital signs from anywhere at any time. Various sensors are connected to the patient body like (ECG, SPO2, and temperature, Blood pressure, etc.) to form a body sensor area network (BSN). This ubiquitous sensor captures the biomedical and context signals from the patient's room and body. BSN then transmit this data signals towards gateway (shown in Fig. 3 by Raspberry Pi module) using wireless protocols (Zigbee, wifi, LORA, 6LoWPAN, Bluetooth, etc.). Gateways support various protocols and generally act as a central server between sensor and cloud services. It performs protocol conversion of data before sending it to the private or public cloud. At last, cloud computing services are provided to the registered web client. Apart from data warehousing cloud supports various services like broadcasting and data analysis. The graphical user interface helps to visualize data in an organized and simple manner. Cloud also provides an efficient feedback mechanism that helps to improvise the system.

4 Attacks and Threats in Healthcare IoT

IoT is a revolutionary technology, it has the potential and surfeit of benefits to transform healthcare to become more reliable, accessible and fast. 6LoWPAN provides scalability as well as auto-configuration to the IoT networks which allow sensor nodes to move easily. Any sensor node can easily join or leave the 6LoWPAN networks. This allures the attacker to breach the security of the IoT network and agitate the normal routing. As mentioned in Sect. 3 the IETF's RPL protocol is used for routing the data in IoT networks. RPL falls prey to numerous attacks because of its constrained nature [24] to get the details about the RPL attacks, RPL structure and working need to be understood hence the following section elaborates the RPL protocol.

4.1 Working of RPL Protocol

RPL uses the destination vector routing (DVR) technique based on IPv6. The devices in the RPL network are connected forming a DODAG graph. The root node act as a data sink node. Based on node position in the graph, every node is having a rank value which is greater than its parent node. The objective function uses different metrics and constraints to decide the best path. The three main messages are used in the formation of DODAG (i) DODAG Information Object (DIO) (ii) DODAG Information Solicitation (DIS), and (iii) Destination Advertisement Object (DAO). Figure 4 shows the building of the DODAG graph.

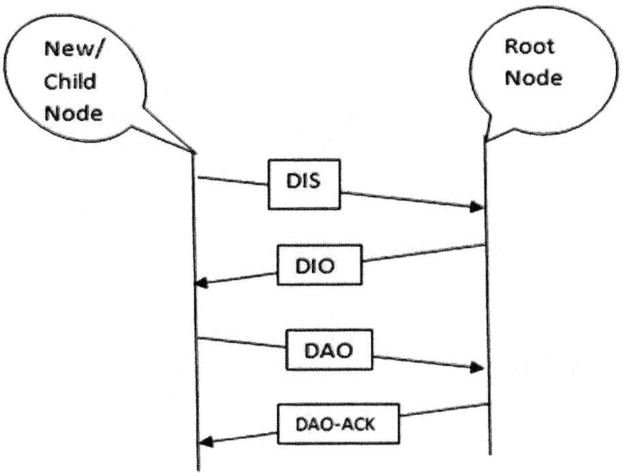

Fig. 4 DODAG formation

1. The root node initially built the DIO message (DODAG information object) with various parameters [25] like rank value, cycle time, delay root value (average delay of DAG root), Expected Transmission Count (ETX), wake up phase and sends it to the child node.
2. After receiving the DIO message, the child node determines its rank value with the help of objective function and includes the DIO sender node in its parent list.
3. Child node then forwards the updated DIO message which contains the updated state of the graph to its neighbours. This DIO message needs to be exchanged periodically by all the neighbours of DODAG.
4. All the nodes must route the data to the root node in an upward direction only. This is the default route which consists of all the preferred parent nodes.
5. If any new node wants to join the DODAG, it can forward the DIS message to the parent node and if the parent agrees it will send DIO message.
6. DAO messages are used by the child node to form the downwards route, DAO will be followed by DAO-ACK (DAO acknowledgment) message.

In healthcare, IoT system sensor nodes form different instances of DODAG and RPL protocol to routes the data. The compromising the security of a single sensor node may lead to the death threat of a patient because the compromised sensor node might be reading the crucial data like heart rate. Unfortunately, this RPL is prone to passive eavesdropping attacks as well as active modification attacks [15]. Mainly attacks are categorized into 4 types DoS, U2R (user to root), R2L (remote to local) and probe. The DoS attack will occur when an attacker will flood the network with the excessive number of packets so that the gateway will not provide access to the authenticated node. U2R is a user to root attack, which launched by an attacker to get root privileges illegally. In R2L attacks attackers will try to capture the entire network remotely. Probe attack launched by an attacker to get the information about the networks like IP addresses, security parameters incorporated in the networks,

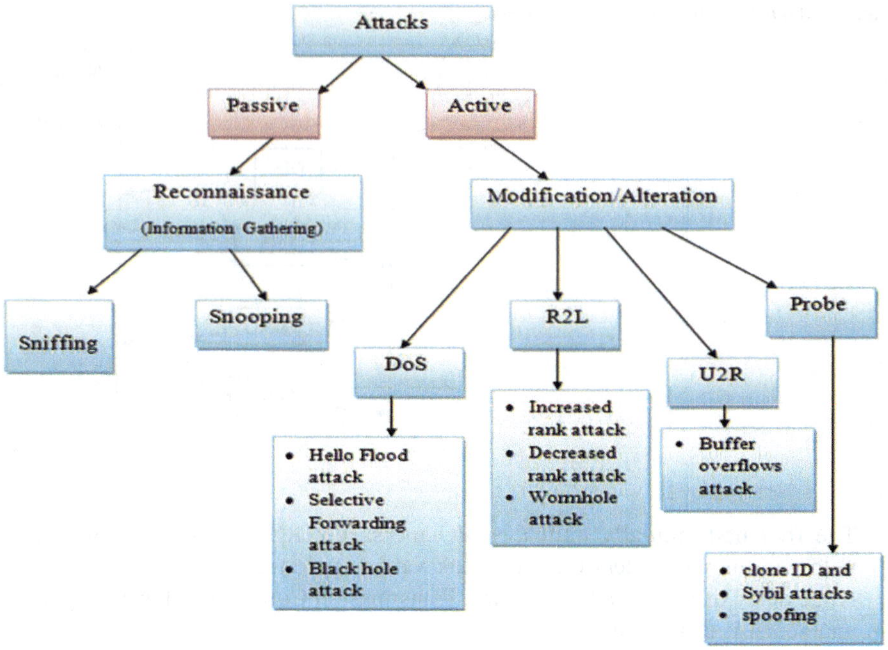

Fig. 5 RPL attack classification

open ports, to execute the attack. Figure 5 give the classification of different attacks followed by the detailed descriptions and analysis.

Hello Flood Attack: Initially for the formation of DODAG, DIO and DIS messages sent by root and child node respectively. Root provides DODAG information to the new node transferring the DIO message. These messages are acting like "hello" or the first message of the 6LoWPAN network. If attackers will ménages to send a large number of DIO or DIS messages in the network then the network will suffer from a hello flood attack. It is a kind of DoS attack that causes serve damage in the network. The causes of this attack were put forth by the authors [26].

Selective Forwarding Attack: This attack will disturb the routing path of the RPL protocol. A selective forwarding attack will take many forms. Sometimes they called a "black hole". In a black hole attack, the sensor node will not forwards or refuses to forward the data traffic. This will be possible when attacker updates the routing path and forwards only control messages hence rest of the other packets get dropped. A gray-hole attack is another form of selective forwarding attack where only a specific part of network traffic is discarded by a malicious node. This also is used to perform a DoS attack.

Wormhole Attack: The sensor nodes in the network are connected with specific connection configuration parameters. If an attacker connects malicious node in the network with out of band connections, then the attacker will forwards the massages at a very faster rate than the normal data transfer. This results in the formation of a wormhole in the network. It is a research challenge to counter this attack [27].

Clone ID and Sybil Attack: When an attacker succeeds in capturing the valid node and copying the clone of valid node at many places in the network, results in clone ID attack. If the network will have a large number of clones then the authenticated or valid node will be unreachable. Sybil attack is similar to a clone ID attack where a single node will take multiple logical identities at multiple places at the same time. Individually these attacks are not very dangerous but if they combine with other attacks network may incur heavy loss.

Snooping and Sniffing: The attacker can snoop into the RPL network to launch the various attacks or to breach the confidentiality of the RPL network. The messages exchanged for the formation of the network (like DIO, DIS) can be used to get the details about the network. Gathering information is the passive attack, which is the first step to perform an active attack. Sniffing is a technique, which used by an attacker to perform reconnaissance.

Buffer Overflow Attack: RPL specially designed for low constraint network where data traffic is low as compared to the conventional wireless network. However, in some emergency services applications of IoT, suddenly at the time of emergency heavy and dynamic traffic occurs in the network. When during such emergency the child sensor node transfers a huge amount of data towards the root node, the root node will experience a buffer overflow attack. The attacker adheres to the same technique and forwards the heavy traffic towards the root node causing buffer overflow attack. This attack will cause serve damage and loss of data. Root node redeems from buffer overflow by forwarding DIO message containing congestion information towards the child node and the child node will be acted upon it by changing its parent node. The congestion aware routing mechanism [28] proposed to reduce congestion in low constraint emergency applications.

Increased Rank Attack: In DODAG, the rank value gives the relative position of the node in the 6LoWPAN network. DIO messages are used to exchange rank values. The rank of the root node is always less than that of the child node. If an attacker succeeds in inserting the node in the network with the higher rank value then the proffered best parent node will be change and the worst node (attacker node) might act as a parent node. The increased rank attack will create loops in the network.

Decreased Rank Attacks: In a similar line with the increased rank attack, the attacker inserts the malicious node with decreased rank value hence the routing path of the network will get completely disturbs. This attack also inserts a loop in the network by changing the root node. The loops in the network adversely affect the network performance by making authenticated nodes unreachable.

4.2 Threats in Healthcare IoT

Healthcare cyber-attacks could be lethal or serious if the patient's healthcare records are revealed or altered. Cyber security researchers have been anticipating that cyber attack will sabotage the healthcare sector on a large scale [6]. The data breach, data modification, data inaccessibility, unauthorized access are the major threats in the system which demolish the system performance shown in Fig. 6 and also adversely affect the financial outcome of the system. The threat will not only curtail the patient security but it will hamper business continuity, brand reputation, and revenue and cost also.

To mitigate these threats and attacks at IoT based healthcare system must adhere to the basic security goals that are confidentiality, authentication, non-repudiation, integrity. The existing systems discussed in Sect. 2 surveyed critically and the attacks on those systems are summarized in Table 2. The table also gives the information about the security goals which are not satisfied by the systems.

The implementation of healthcare IoT also suffers from three main threats namely energy optimization, privacy and trust. Sensors are the main devices in healthcare IoT system. Sensors are low constraint battery powered devices which require battery charging periodically. Transferring the vital signs of patients with such energy inefficient devices will be dangerous. Hence, the reliability of the data will not be assured. Privacy is utmost important. Healthcare system carries very

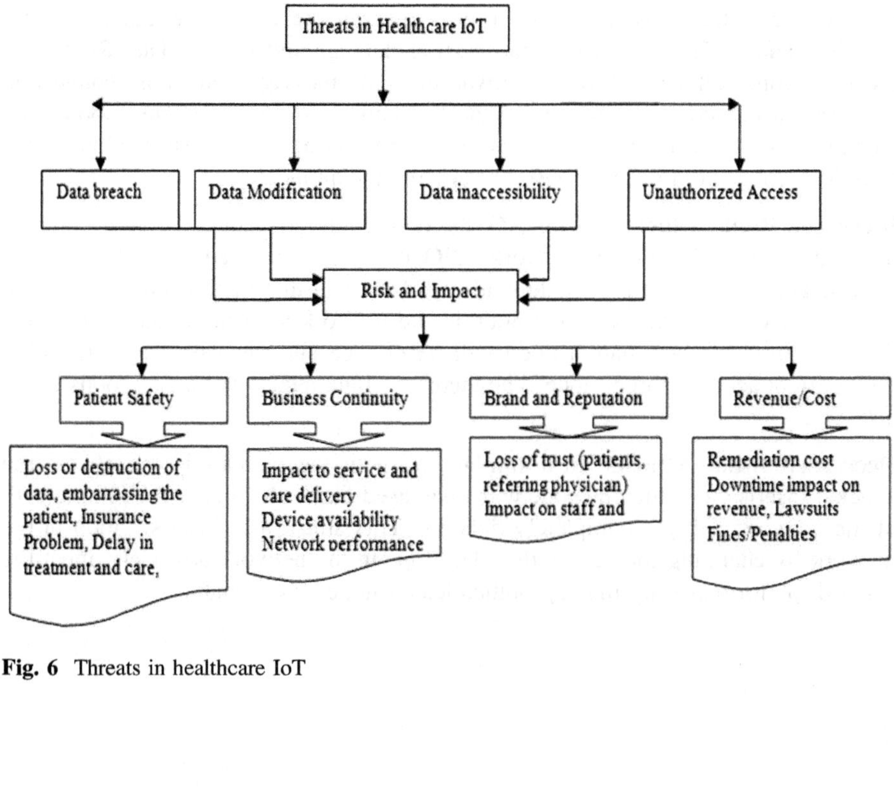

Fig. 6 Threats in healthcare IoT

Table 2 An analysis of lightweight security algorithms

Algorithm	Block size	Key size	Rounds	Analysis for healthcare system
AES	128	128	10	Consumes less power and fast as compared to other algorithms. This complex algorithm provides substantial security
PRESENT	64	128	31	Uses less memory (RAM), code size is also less
SIMON	32, 48, 64, 96, 128	Variable	32, 36, 42, 44, 52, 54	Throughput is high in FPGA and ASIC platforms, provides an easy, simple and optimal implementation
HEIGHT	64	128	32	Better cycle per block ratio. Hardware implementation used less number of GE (Gate Equivalent)
RECTANGLE	NA	80	NA	Bit slicing technique [29] is used low-cost hardware and improves the security of the algorithm

important data with the point of view of hospital and patients, trusting on any third party application and providing the personal as well as healthcare data may hamper the privacy of patient. Trust is aligned with the privacy. Automation of healthcare system requires collecting and transferring of huge amount data (big data) over internet and cloud. Attack on this data will give incorrect analysis and physician may provide inappropriate diagnosis which may lead to death threat of a patient.

Existing security proposals (Table 1) were struggling with various threats and attacks. Although the existing proposals put forth the security measures (Table 1) implies that these security measures are not adequate to alleviate the attacks. Hence, this research work proposes recommendations (Sect. 6) to improve the security of healthcare IoT systems.

5 Open Research Issues and Challenges

Patient Safety and Quality Improvement Act (PSQIA) states that healthcare privacy is the right of every individual. Healthcare institutions must provide security, privacy, availability to patient sensitive data.

The security of IoT is a vast area of research. Although the contributions of the researchers are significant still certain areas are not explored. In this section, we aim to highlight these research issues, some of which related to the different areas of security of healthcare IoT.

Confidentiality is the most preferred way to provide security. IoT networks are operated in a constraint environment; therefore, conventional confidentiality algorithms will not be suitable hence, there is an urge to design lightweight confidentiality algorithms. A lightweight algorithm means the algorithms that are suitable for low constraint devices. Although there are many lightweight security algorithms like AES [30], PRESENT [31], SIMON [32], HEIGHT [33], RECTANGLE [34]. The details of these algorithms are given in Table 2. All are analyzed based on various parameters and the survey reveals that AES is the most secure lightweight cryptographic algorithm [35]. Although AES depicted as a most secure algorithm, not every application requires the same high level of security that AES is promising to provide. When an application uses low constraint resources, it does not always make sense to lush them on an algorithm providing 128 (or 192 or 256) bits of security when 96 might adequate. So there is a research challenge to device a new lightweight security algorithm which will provide a substantial level of security.

Implementation of these symmetric lightweight algorithms needs the generation and distribution of secret key. The inimical network of resource constraint devices make the key generation task more challenging. Dynamic topology of the IoT network will promote the attack intrusion and further key generation and distribution become more difficult [36]. Recently, RFID is a core technology for IoT and most popular in healthcare systems too. The existing healthcare systems use RFID to build the healthcare system but they are not providing significant security. Authentication can be achieved with ease using the RFID technique. A research challenge is to save the chip area as an RFID authentication protocol may ask that 64-bit to be encrypted, and if 128 bits of state demanded when only 64 are necessary, a remarkable waste of chip area will be there. The RFID technology has the caliber to improve patient management in the healthcare system, but it also suffers a huge risk of healthcare data leakage. Therefore, while improving medical workflow management; there is a need to enhance the security of the healthcare system that uses RFID technology. In most of the existing research, the RFID reader is authenticating to the RFID tag but the tag is not authenticating the reader hence the existing mechanisms cannot be used for mutual authentication [17]. Bridges that gap and proposes the mutual authentication mechanism [17]. Mechanism but lacks in providing tag anonymity and suffer from a man in the middle attack. Furthermore, the existing algorithms are not specifying the secure way to store the message exchanged in the authentication mechanism. Moreover, the conventional authentication mechanism like MD4, MD5, SHA (secure hash algorithm) is not suitable for constraint IoT devices. As all these conventional algorithms require at least 7K logic gates [37]. However, widely used RFID card will provide only 2K–5K logic gates for security purpose [38]. Hence, there is a need to provide a lightweight authentication mechanism using RFID and required to propose a secure way to exchange the message.

Kerberos authentication method suggested [39] can also be use in sensor network.

Another area that can be explored to improve the security of healthcare IoT is intrusion detection and prevention (IDS/IPS) system. Implementing IDS/IPS is a challenging task as it requires devising a machine learning algorithms for constraint environments. A classifier will be trained using a high computational, complex machine algorithm so that it can detect the intruder from normal traffic. Signature-based, anomaly-based and hybrid are the three types of IDS that can be deployed either centrally or it can be placed on different nodes in a distributed deployment. A stored signature of an attack is used to detect intrusion in signature-based IDS. In an anomaly detection system, the normal behavior of the system is analyzed and it can be used to detect abnormalities. Hybrid IDS is the combination of a signature-based and anomaly-based IDS system. The signature-based system is not suitable for healthcare IoT systems as attacks signature is changing frequently. Hence research needs to proceed to implement anomaly or hybrid IDS.

Lastly, the end to end security at 6LoWPAN layer is provided by the IPSec protocol [40, 41] which is very complex and also creates a burden on low constraint devices with its extra header format. The research challenge is to develop lightweight end to end security protocol for IoT devices. The following section suggested some recommendations to overcome the research issues and challenges.

6 Recommendations

The healthcare IoT system mainly required to take care of the security of the system by implementing the research challenges discussed in Sect. 5 as nihility of security measure may ruin the system because of sensitive data loss.

The analysis given in (Table 1) of the security of different healthcare security system blazons the fact of lack of protection in the healthcare IoT system. Pre-dominantly, confidentiality and authentication are the two security goals which are failed to achieve by the system (Table 1). To improve the confidentiality mechanism lightweight confidentiality algorithms are there (Table 2) but they will not provide an adequate level of security with low cost and low power consumption. Hence researchers need to develop lightweight cryptographic algorithms that will consume less power and provide an appreciable amount of security. Confidentiality can be achieved using symmetric or asymmetric means with their pros and cons. The implementation of symmetric lightweight cryptographic algorithms requires the sharing of the key between the sender and receiver. This Key distribution is a basic problem while implementing symmetric lightweight algorithms. Key distribution triggers the concept of authentication since the same key must be shared with the intimate receiver. Even though RFC 6568 [42] specifies the Internet key management solution still researchers need to specify the mathematically based key management technique for IoT based applications. Public key cryptography (algorithm like ECC [43]) can also be used to achieve confidentiality. But public-key cryptography uses complicated mathematical calculations and larger key values that require more processing power and symmetric algorithms having the problem of key distribution. Therefore this proposal recommends researchers to develop the hybrid algorithms which combine the benefit of symmetric as well as asymmetric algorithms that will be best suited for IoT applications. Implementation of network layer security protocol like IPSEC is another solution for achieving confidentiality but the researcher must develop the compressed headers for serving the purpose of authentication header (AH) and encapsulation security payload (ESP) and which will be more appropriate for low constraint IoT environment.

Along with confidentiality authentication is another major security goal. Anonymous intrusions in the system will cause serve damage to the privacy of the data. The existing system (Table 1) uses only basic techniques to incorporate authentication. The survey finds out that sensors having an RFID tag can provide strong RFID based mutual authentication. Researcher [17] proposes a very good authentication mechanism using RFID but an anonymous intruder can get in the system and cause the man in the middle attack. Hence the recommendation is to devise the mutual authentication system where the researcher needs to find some other methods using RFID sensors to identify the authenticated gateway.

This research work strongly recommends the implementation of the IDS/IPS system as it is a promising way to mitigate the attacks. Sensors are tiny devices and transfer a small packet of information, encrypting every packet with complex encryption algorithms will pose a computational burden on low constraint devices. Hence IDS is a system that will detect the attack and prevent the system. Implementing IDS judiciously is a need of an hour as it could restrict every possible attack on the healthcare system. Machine learning, deep learning, pattern matching, artificial intelligence, genetic algorithms are some way to implement IDS.

If future research finds the best way to implement IDS as well as algorithms to achieve confidentiality and authentication then it will help to strengthen security and improves the security, performance, and utility of the healthcare system.

7 Conclusion

In the existing system, the architecture of healthcare IoT system suffers from various attacks, threats, and vulnerabilities. Security is the key consideration since a single security breach incident can wipe-out any cost savings or revenue provided by healthcare IoT application. Hence, it is of utmost important to provide substantial security to this patient's data. So, through comprehensive research of this paper, it is conveyed that RFID is encouraging technology to provide robust authentication and confidentiality. The paper proposed detailed architectural model for implementation of healthcare IoT system which provides the direction to implement the healthcare system. Existing lightweight security algorithms not provide either the substantial security. However, robust security will be provided at the cost of the huge computational power. Hence a lightweight security algorithm needs to be devised which provides the required amount of security and will be more suitable for resource constraint devices. Further implementation of IDS is a promising way to alleviate attacks. Paper provides a recommendation for improving security measures. The research and challenges put forth in the paper will provide the further research direction. The future scope of this paper is the implementation of the lightweight protocol for improving the security of the healthcare system.

References

1. Medagliani, P., Leguay, J., Duda, A., Rousseau, F., Duquennoy, S., Raza, S., Ferrari, G., Gonizzi, P., Cirani, S., Veltri, L., Montón, M., Domingo Prieto, M., Dohler, M., Villajosana, I., Dupont, O.: Internet of Things Application from Research and Innovation to Market Deployment (2014)
2. Khambete, N.D., Murray, A.: National efforts to improve healthcare technology management and medical device safety in India. In: 7th International Conference on Appropriate Healthcare Technologies for Developing Countries, pp. 1–5. IET (2012)
3. Dey, N., Ashour, A. (eds.): Classification and Clustering in Biomedical Signal Processing. IGI Global, Hershey (2016)
4. Abouelmehdi, K., Beni-Hssane, A., Khaloufi, H., Saadi, M.: Big data security and privacy in healthcare: a review. In: The 8th International Conference on Emerging Ubiquitous Systems and Pervasive Networks (EUSPN 2017) (2017)
5. Ellen, N.: Security Firm Finds Link between China and Anthem Hack. The Washington Post, 27 Feb 2015. Web, Feb 2016
6. Kwon, J., Johnson, M.: Protecting patient data—the economic perspective of healthcare security. IEEE Secur. Priv. (2015)
7. Dey, N., Ashour, A.S., Shi, F., Fong, S.J., Tavares, J.M.R.: Medical cyber–physical systems: a survey. J. Med. Syst. **42**(4), 74 (2018)
8. Alessandrelli, D., Mainetti, L., Patrono, L., Pellerano, G., Petracca, M., Stefanizzi, M.L.: Implementation and validation of an energy-efficient MAC scheduler for WSNs by a test bed approach. In: 2012 International Conference on Software, Telecommunications and Computer Networks (SoftCOM 2012), art. no. 6347615 (2012)
9. Catarinucci, L., et al.: An IoT-aware architecture for smart healthcare systems. IEEE Internet Things J. **2**(6), 515–526 (2015). https://doi.org/10.1109/JIOT.2015.2417684
10. Malan, D., Jones, T.F., Welsh, M., Moulton, S.: CodeBlue: an Ad-Hoc sensor network infrastructure for emergency medical care. In: Proceedings of the MobiSys 2004 Workshop on Applications of Mobile Embedded Systems (WAMES 2004), Boston, MA, USA, 6–9 June 2004
11. Wood, A.V., Doan, G., Cao, T., Selavo, Q., Wu, L., Fang, Y., He, L., Lin, Z., Stankovic, S.: ALARM-NET: wireless sensor networks for assisted-living and residential monitoring; Technical Report CS-2006-01; Department of Computer Science, University of Virginia, Charlottesville, VA, USA (2006)
12. Ko, J., Lim, J.H., Chen, Y., Musaloiu-E. R., Terzis, A., Masson, G.M.: MEDiSN: medical emergency detection in sensor networks. ACM Trans. Embed. Comput. Syst. **10**, 1–29 (2010)
13. Tung, H.C., Tsang, K.F., Lam, K.L., Tung, H.Y., Li, B.Y.S., Yeung, L.F., Ko, K.T., Lau, W. H., Rakocevic, V.: A mobility-enabled inpatient monitoring system using a ZigBee. Sensors (Basel) **14**(2), 2397–2416 (2014). https://doi.org/10.3390/s140202397
14. Dey, N., Ashour, A.S., Shi, F., Fong, S.J., Sherratt, R.S.: Developing residential wireless sensor networks for ECG healthcare monitoring. IEEE Trans. Consum. Electron. **63**(4), 442–449 (2017)
15. Winter, T., et al.: RPL: IPv6 routing protocol for low power and lossy networks. Internet Engineering Task Force (IETF), Fremont, CA, USA, Request for Comments: 6550 (2012)
16. Chen, M., Fang, Y.: Lightweight anonymous authentication protocol for RFID. IEEE Trans. Netw. **25**(3) (2017)

17. Fan, K., Member, IEEE, Jiang, W., Li, H., Member, IEEE, Yang, Y., Member, IEEE: Lightweight RFID protocol for medical privacy protection in IoT. IEEE Trans. Ind. Inf. **14**(4) (2018)
18. Tapia, D.I., Corchado, J.M.: An ambient intelligence based multi-agent system for alzheimer health care. Int. J. Ambient Comput. Intell. (IJACI) **1**(1), 15–26 (2009)
19. Archip, A., Botezatu, N., Şerban, E., Herghelegiu, P., Zală, A.: An IoT based system for remote patient monitoring. In: 2016 17th International Carpathian Control Conference (ICCC), Tatranska Lomnica, pp. 1–6 (2016). https://doi.org/10.1109/CarpathianCC.2016.7501056
20. Swaroop, K.N., Chandu, K., Gorrepotu, R, Deb, S.: A health monitoring system for vital signs using IoT. Internet Things **5**, 116–129 (2019). ISSN: 2542-6605. https://doi.org/10.1016/j.iot.2019.01.004. https://www.sciencedirect.com/science/article/pii/S2542660518300349
21. Pai, S., et al.: Confidentiality in sensor networks: transactional information. IEEE Secur. Priv. Mag. **6**(4), 28–35 (2008)
22. Hasan, M., Milon Islam, Md., Ishrak Islam Zarif, Md., Hashem, M.M.A.: Attack and anomaly detection in IoT sensors in IoT sites using machine learning approaches. Internet Things **7**, 100059 (2019). ISSN: 2542-6605. https://doi.org/10.1016/j.iot.2019.100059. https://www.sciencedirect.com/science/article/pii/S2542660519300241
23. Fedor, S., Collier, M.: On the problem of energy efficiency of multi-hop vs one-hop routing in wireless sensor networks. In: 21st International Conference on Advanced Information Networking and Applications Workshops (AINAW'07), Niagara Falls, ON, pp. 380–385 (2007)
24. Tsao, T., Alexander, R., Dohler, M., Daza, V., Lozano, A., Richardson, M.: A Security Threat Analysis for Routing Protocol for Low-power and Lossy Networks (RPLs), RFC 7416, Internet Engineering Task Force (2015)
25. Tang, W., Ma, X., Huang, J., Wei, J.: Toward improved RPL: a congestion avoidance multipath routing protocol with time factor for wireless sensor networks. J. Sens. **2016**, 1–11 (2016). https://doi.org/10.1155/2016/8128651
26. Le, A., Loo, J., Luo, Y., Lasebae, A.: The impacts of internal threats towards routing protocol for low power and lossy network performance. In: IEEE Symposium on Computers and Communications (ISCC'13), pp. 789–794 (2013)
27. Wallgren, L., Raza, S., Voigt, T.: Routing attacks and countermeasures in the RPL-based internet of things. Int. J. Distrib. Sens. Netw. (2013). https://doi.org/10.1155/2013/794326
28. Bhandari, K.S., Hosen, A.S.M., Cho, G.H.: CoAR: Congestion-Aware Routing Protocol for Low Power and Lossy Networks for IoT Applications Sensors (2018)
29. Zhang, W., Bao, Z., Lin, D., Rijmen, V., Yang, B., Verbauwhede, I.: RECTANGLE: a bit-slice lightweight block cipher suitable for multiple platforms. Sci. China Inf. Sci. **58**, 122103(15) (2015). https://doi.org/10.1007/s11432-015-5459-7
30. Sridevi, S., Priya, S., Karthigai Kumar, P., SivaMangai, N.M., Rejula, V.: FPGA implementation of efficient AES encryption. In: ICIIECS'15 (n.d.)
31. Bansod, G., Raval, N., Pisharoty, N.: Implementation of a new lightweight encryption design for embedded security. IEEE Trans. (2015)
32. Feizi, S., Ahmadi, A.: A hardware implementation of Simon cryptography algorithm. In: ICCKE-2014
33. Nascimento, F.M., dos Santos, F.M., Moreno, E.D.: A VHDL implementation of the Lightweight Cryptographic Algorithm HIGHT, Sept 2015
34. Zhang, W., Bao, Z., Lin, D., Rijmen, V., Yang, B., Verbauwhede, I.: RECTANGLE: A Bit-slice Ultra-Lightweight Block Cipher, Suitable for Multiple Platforms (2014)
35. Granjal, J., Monteiro, E., Sá Silva, J.: Security for the internet of things: a survey of existing protocols and open research issues. IEEE Commun. Surv. Tutor. **17**(3), 1294–1312 (2015) (third quarter). https://doi.org/10.1109/COMST.2015.2388550
36. Kimbahune, V.V., Deshpande, A.V., Mahalle, P.N.: Lightweight key management for adaptive addressing in next generation internet. Int. J. Ambient Comput. Intell. (IJACI) **8**(1), 50–69 (2017)

37. Bogdanov, A., et al.: Hash functions and RFID tags: mind the gap. In: Proceedings of the CHES, pp. 283–299 (2008)
38. Ranasinghe, D.C., Cole, P.H.: An evaluation framework, Chapter 8. In: Networked RFID Systems and Lightweight Cryptography. Springer, Berlin, Germany, Nov 2008
39. Pathan, S.K., Deshmukh, S.N., Deshmukh, R.R.: Kerberos authentication system-a public key extension. Int. J. Recent Trends Eng. **1**(2), 15 (2009)
40. Raza, S., Duquennoy, S., Voigt, T.: Securing communication in 6LoWPAN with compressed IPsec. In: Proceedings of the International Conference on DCOSS Workshops, pp. 1–8 (2011)
41. Raza, S., Duquennoy, S., Hoglund, J., Roedig, U., Voigt, T.: Secure communication for the Internet of Things—a comparison of link-layer security and IPsec for 6LoWPAN. Secur. Commun. Netw. **7**(12), 2654–2668 (2014)
42. Kim, E., Kaspar, D., Vasseur, J.P.: Design, and Application Spaces for IPv6 over Low-Power Wireless Personal Area Networks (6LoWPANs), Internet Engineering Task Force (IETF), Request for Comments: 6568, Category: Informational, ISSN: 2070-1721
43. Yaoa, X., Chena, Z., Tian, Y.: A Lightweight Attribute-Based Encryption Scheme for the Internet of Things. Elsevier (2015)

Lightweight Secure Technology Future of Internet of Things

Aruna Gawade and Narendra Shekokar

Abstract The Internet of Things (IoT) is a term which treats all objects around us as a whole, interconnected network. The 'thing' in IoT can be a person with a pulse monitor or smart meter to calculate electricity. These 'things' must have a unique IP address and able to communicate with other objects over internet without manual intervention. Though these new technologies are overwhelming, they also expose severe IoT security challenges. IoT applications suffer from different types attacks such as eavesdropping attack, spoofing and false data injection attack, man-in-the-middle attack, replay attack, denial of Service attack, jamming attack, flooding attacks etc. During the last decade, many cases of attacking IoT devices were identified. It is a direct threat to our privacy, our conversations everything that sums up our life. While the world today, is moving towards a more secure system, we must be aware of the security lacks and should provide a strong mechanism to defend ourselves against the same. Because IoT is an amalgamation of both Hardware and Software, the issues posed by IoT are also double fold. As hardware concern, objects are lightweight in terms of memory, energy, computing power etc. and as a software concern current security algorithms are designed for high end systems so will not suffice in IoT environment. This leads to the need for light-weight security mechanisms in IoT network. Currently protocols like PRESENT, Clefia, TEA etc. are designed for lightweight encryption but still suffers from Differential Attack, Related key attack etc. Lightweight integrity is also achieved with SHA2, SHA 3, Light-Weight One-way Cryptographic Hash Algorithm (LOCHA), Timestamp Defined Hash Algorithm (TDHA). So there is a need for a whole concrete lightweight system which will remove the shortcomings in the current lightweight encryption and integrity protocols and provides better security.

Keywords IoT security · Lightweight · Confidentiality · Integrity

A. Gawade (✉) · N. Shekokar
Department of Computer Engineering, D. J. Sanghvi College of Engineering, Mumbai University, Mumbai, India
e-mail: aruna.gawade@djsce.ac.in

N. Shekokar
e-mail: narendra.shekokar@djsce.ac.in

N. Dey et al. (eds.), *Internet of Things, Smart Computing and Technology: A Roadmap Ahead*, Studies in Systems, Decision and Control 266,
https://doi.org/10.1007/978-3-030-39047-1_14

1 Introduction

Internet plays a very important role in our daily life. Automation and ease of accessibility leads to the increasing demand of internet. If we can have automation of our daily needs using internet, then it will be a huge success. This term leads to the Internet of Things which states that Things like fan, fridge etc. can be accessed over the internet.

The Internet of Things (IoT) is a terminology that indicates how these everyday objects can be accessed over the internet. The everyday objects can be anything such as sensors, actuators etc. IoT Technologies are becoming more important as the need to better understand our environments and make them smart increases. Currently many organizations are entering into today's market by developing IoT solutions.

Some important key issues in IoT are [1]:

- Large no. of constrained devices
- Designed for low cost and power consumption
- Constraints w.r.t. Energy, Memory, Processing power
- Requirement of node sleep time for energy saving
- Different traffic patterns, dynamic topology.

The basic building block for IoT is a Wireless Sensor Network (WSN) as it offers many advantages by offering mobility and its ability to work with limited source of energy under the harsh conditions. As a result WSN intelligent devices and networks, are connected and integrated into IP based Internet applications [2]. The advancements in the WSN makes the IoT to come into existence. Though both technologies are similar there is always a difference between WSN and IoT as shown in Figs. 1 and 2.

In Fig. 1 different WSNs are connected to internet via gateway which has sensor collected data. These gateways are vendor specific which takes care of translation of protocols to and from the internet. On getting request the gateways refer the data from the database and respond accordingly.

This system does not provide information about the origin of data i.e. if it is old stored value or real time.

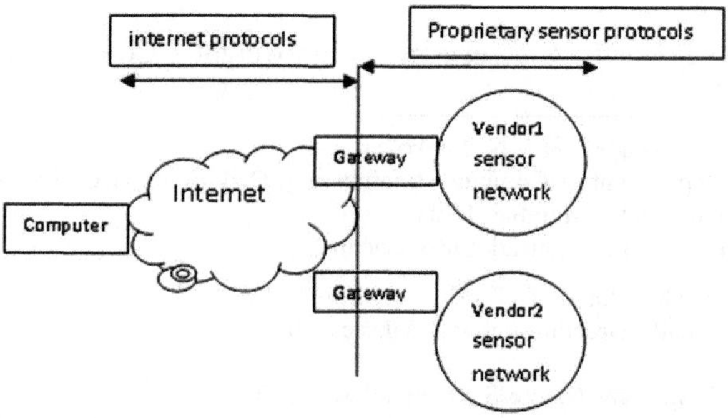

Fig. 1 WSN proprietary protocols

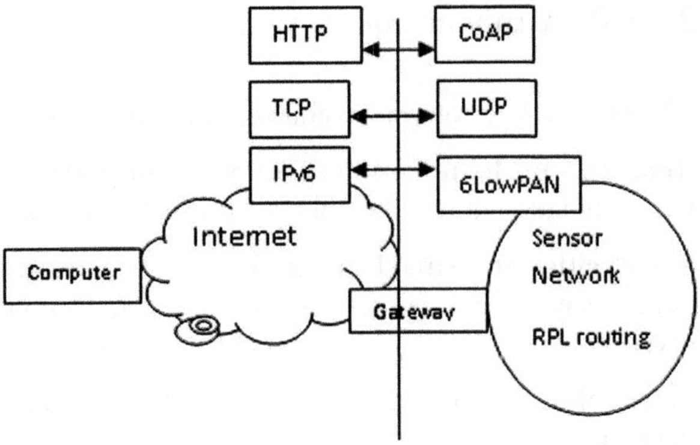

Fig. 2 WSN standard protocols

But as we know IoT implementation with all its key issues will not suffice with vendor specific constraints, some standardized protocols for network communication with constrained devices are needed.

As shown in Fig. 2 we can observe that, gateways are still used but now between the internet protocols like IPV6 and sensor network protocols like 6LowPAN. As these are standardized protocols, they allow to gather data from different sensors and also supports use of gateways from different vendors.

As shown in Fig. 3 various protocols such as CoAP (Constrained Application Protocol), DTLS(Datagram Transport Layer Security), UDP(User Datagram Protocol) and 6LowPAN (IPV6 for Low power Wireless Personal Area Network) are used in the stack.

CoAP: Constrained RESTful Environments (CoRE), works for making standard framework for resource constrained applications so that existing web services and IoT web services should work in similar manner. So the Constrained Application Protocol (CoAP), protocol is designed which can be used in IoT networks.

6LowPAN: Limitations of IPv6 protocol makes it unsuitable in IoT network. IPV6 has 40-byte header which is not possible to accommodate in 802.15.4 networks MAC payload because of the space limitation. To deal with this problem, the 6LoWPAN (IPv6 for Low-power Wireless Personal Area Network) was formed. This takes care of IPv6 protocol extensions required for IoT networks which uses 802.15.4 standard [3].

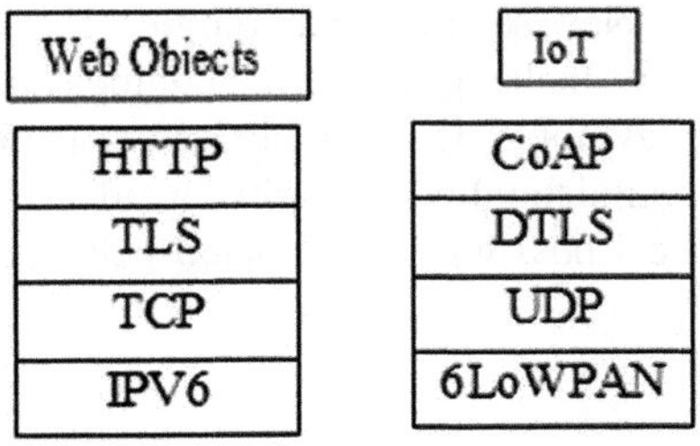

Fig. 3 Protocol stack for IoT

2 IoT Applications

IOT plays a key role in several real life applications which are listed below:

Building and home automation: Governing buildings and smart homes by providing functionalities like energy optimization, predictive maintenance etc.

Smart cities and smart meter system: To make cities smart, smart street lights, traffic management, surveillance cameras etc. can be done. This reduces cost and saves resources.

Wearables: Smart watches and wearables can be used for fitness and location tracking.

Health care: Smart patient remote monitoring system with drug tracking and hospital management To improve the quality of health care solutions can be used.

Smart manufacturing: Applications in smart manufacturing require inventory management, maintenance, safety of employees, flow optimization etc. For this a proper hardware and software combination is required which saves the design time.

Smart Farming: Automation in soil irrigation for smart farming can be done using various sensors such as humidity, moisture etc. Also the crops can be monitored for various diseases.

3 IoT Architecture Problems

Architecture of the current IoT is based on a centralized Client Server model. Which has many problems as:

IoT devices: Inadequate guidance for management of IoT devices.

Cost: High infrastructure cost associated with IoT network as it requires clouds, servers, and networking devices. Millions of devices are communicating with each other which has tremendous amount of information [4, 5] and maintenance and analysis of this information [6] incurs the cost.

Security and privacy: maintaining IoT network's security and privacy [7] is a complex task.

New security technologies will be required to protect IoT network and address new challenges. The IoT "things" use simple processors and operating systems that may not support sophisticated security approaches. Main objectives of security are confidentiality, integrity, availability, Non-repudiation but these can be violated. Few possibilities are listed below:

- **Confidentiality**: Device's memory can be accessed or altered, Actual payload can be spoofed, messages can be replayed.
- **Integrity**: Integrity can be hampered by tampering or replacing genuine data sent from the sensors to the data concentrator unit. This can be done either locally, or remotely through forging/injection/deletion of messages.
- **Availability**: Switching off the device, jamming communication channel, Denial-of-Service, and spoofing to shut down an IoT device.
- **Non-Repudiation**: Any compromised device can claim that whatever data transmitted was not sent by it.

IoT applications can suffer from different types attacks such as eavesdropping attack, spoofing and false data injection attack, denial of Service attack etc.

For example IoT application Smart Grid can suffer from following attacks as shown in Fig. 4.

Mainly these attacks are happened because of sending a large number of packets during routing. Out of these DoS and DDoS attack is very destructive as its main objective is to bring down a large portion or even the whole targeted network. These attacks range from simple jamming to application layer [8] where an adversary sends many more commands than expected to the gateways or on the other end to the servers. This will saturate the system to an extent that it is no more able to respond to the legitimate requests. It will essentially shut down the network or part of it for essential services. Such attacks can mostly be launched through the WAN.

To prevent security attacks is a big challenge in IoT application network as not a single routing protocol is fulfilling all these needs of any IoT network.

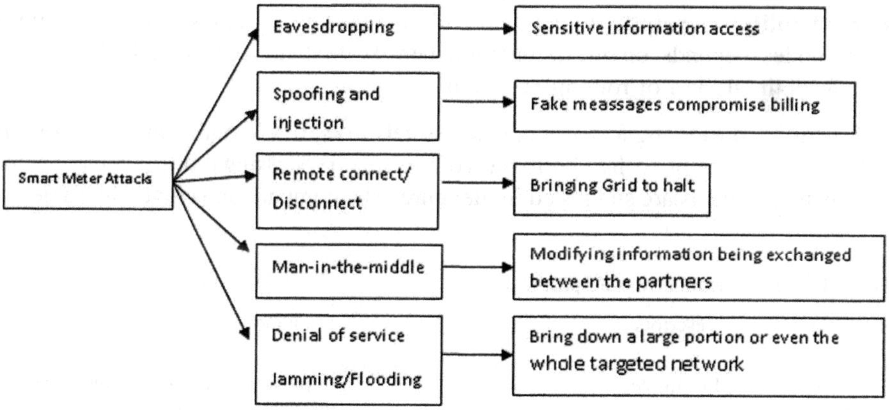

Fig. 4 Attacks in smart grid

4 IoT Routing Protocol Implementation Challenges

Factors that affect the routing protocol design for IoT discussed here [9]:

- **Node Heterogeneity**: The nodes involved in IoT network vary from each other w.r.t. CPU power, memory, battery etc. These nodes generate vast amount of data which again differs in terms of routing and service requirements. Some nodes may have multiple interfaces, and provides alternative routes.
- **Interoperability**: Interoperability is very crucial in IoT as the nodes and IoT networks may have different owners and makes.
- **Node placement**: As the network has fixed locations of nodes and are deployed initially there may be different possible network topologies. The decision of Placement of nodes as a data collector or leaf nodes affects the routing decisions. As all nodes will be forwarding data to collector there will be high number of collisions. So node placement should be done carefully.
- **Network dynamics**: Network is highly dynamic in nature as the topology changes may happen due to link failure or node mobility. This dynamics incurs new challenges in design of the routing protocol.
- **Security and privacy**: IoT nodes are kept in open and insecure environment where the physical attacks are possible. This leads to various kinds of attacks. For e.g. routing table alteration.
- **Quality of Service (QoS)**: QoS [10] i.e. reliability and availability should be given high consideration as information should be delivered in time and it should not be fake.
- **Bandwidth and QoS**: IoT applications may differ w.r.t. delay, throughput etc. [10] so accordingly data has to be generated and transferred at different pace and size. This impacts the performance of routing protocol.
- **Scalability**: The ability to cope up with increasing network size as the number of nodes depends on the population density in that area is scalability. This affects the design of routing protocols.

Primarily considering Routing in wireless networks, IPV6 routing architecture and different IoT concerns such as limitations of memory, processing power, battery power etc., many protocols are suggested in literature. The comparison is given in Table 1.

Table 1 IoT application routing protocols

Protocol name	Description	Limitation
Flooding based	Its simple	Implosion of messages, resource blindness can happen
Cluster based	Resource efficient, gives scalability and security	Selection of cluster head
Geographic based	Based on geographically close neighbor	Not considering link quality Distance calculation in advance
Self organizing	Based on communication distance Adapts well to link quality	Complicated

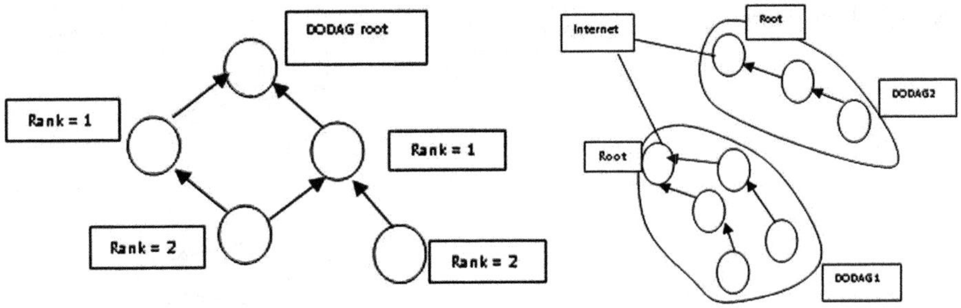

Fig. 5 DODAG in RPL

Out of these self organizing is the best category under which RPL protocol is best suited in constrained network and covers all the routing requirements of the IoT application domain.

5 RPL [11]

RPL (Routing Protocol for Low-Power and Lossy Networks) is working at 6LowPAN and supports various applications of LLNs such as home automation, building automation, smart grid etc.

It is a "route-over" architecture implemented at network layer which supports different types of network such as point-to-point, multipoint-to-point etc. It has a root or sink node and entire traffic moves to or from root.

RPL is a distance vector routing protocol which works on the basis of objective functions and constraints while building the topology (Fig. 5).

RPL supports two types of nodes i.e. non-storing and storing. Non storing nodes have very limited or no memory so only root will maintain the full route to all nodes and sends full route in every packet. It increases packet overhead. Whereas in storing mode all nodes stores the address of their children. They use routing table.

In a RPL network, nodes with more constraints behave only as a leaf node. For any kind of failure or link quality related issue, RPL proposes different repair mechanisms.

6 Working of RPL

The main principal on which RPL works is Directed Acyclic Graph (DAG). All DAGs are having sink or destination which forms Destination Oriented DAGs (DODAGs). For every node DAG rank has to be calculated which is done by the sink node by broadcasting DIO (DODAG Information Object) message. This

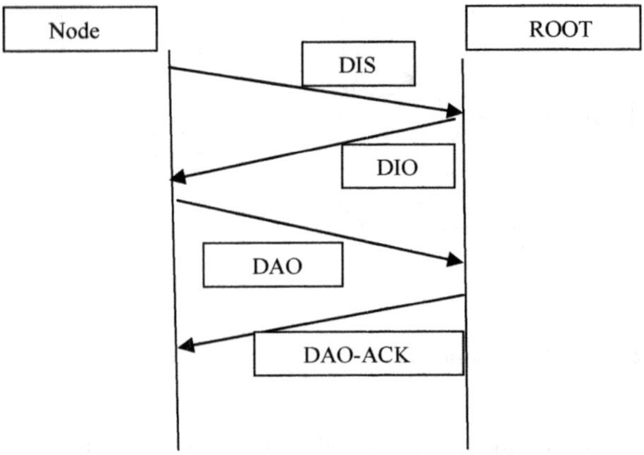

Fig. 6 RPL messages

message has information such as ID, rank information of the broadcasting node, and the objective functions for computing DAG rank.

Once this information is received the node calculates its rank and does necessary changes and broadcasts it.

To build an outgoing path DAO (DODAG Advertisement Object) is sent and DAO-ACK is sent by ROOT in return of that. A new node can send DIS (DODAG Information Solicitation) to join the network. RPL uses 4 values to identify and maintain topology: RPL instanceID, DODAGID, DODAG version number and Rank. The sequence of steps is shown in the Fig. 6.

7 Limitations of RPL

- RPL is a route-over routing protocol used in 6lowpan network. This adaptation layer requires fragmentation and header compression of the data packets. After the fragmentation, reassembly of packets at every intermediate node is required to know about the next hop. At each node all fragments are stored which requires extra memory space. But as we know memory is limited in constrained networks, this protocol will not be an optimal solution for IoT networks.
- As the sensor-to-sensor communication is always through the root or preferred parent. This can lead to congestion on the paths around the root.
- Link quality for a point-to-point route is not monitored in the RPL
- Mobile nodes usually switch from one parent to another which has the impact on reliability and energy consumption. So a parent can be selected which will live for a longer time.
- The existing specification, does not give any information about authentication, key recovery and secure joining using asymmetric mechanisms.

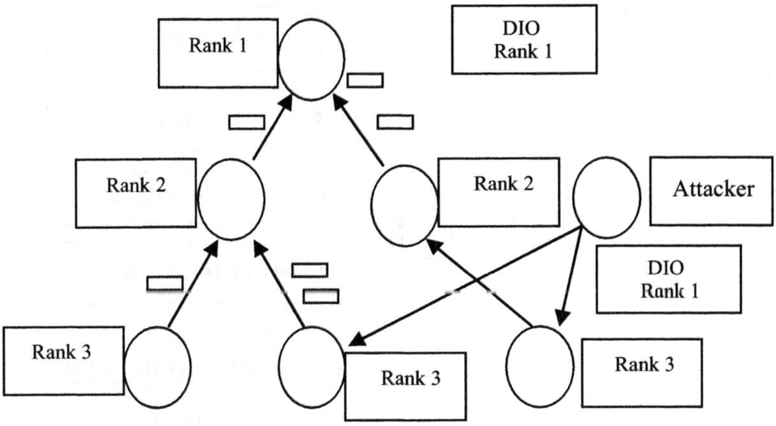

Fig. 7 RPL attack

- Rank plays important role in parent selection. Node with lowest rank is preferred as a parent. In case of multiple parents with same rank, the first node has given priority so that energy will not be wasted for determining link quality (Fig. 7).
- An internal attacker may be in possession of a node and required security keys, may selectively inject routing messages with malicious purposes. RPL provides mechanisms to secure routing communications against external attacks, but no provision against internal attacks. Research efforts are required to protect RPL communications and operations from internal attackers. For e.g. By sniffing the traffic, important information such as DODAG ID, Version Number etc. can be accessed easily and attacker can find out necessary information which can lead to different types of attacks. Analyzing traffic patterns can also lead to attacks which gives access to the topological information.

Many attacks are happened because of messages between root and other nodes while forming the topology. Few examples are shown below (Fig. 8).

From the diagram, it is very clear that in IoT, RPL though works fine in 6LowPAN network has many flaws that has to be taken care of.

8 Security Solutions

RPL has three security modes as specified in RFC [11]:

- **Unsecured**: It does not have any Security sections and assumes availability of link layer security.
- **Preinstalled**: A key is preinstalled and node can be host or router while joining network.

Fig. 8 DIO DIS DAO attacks

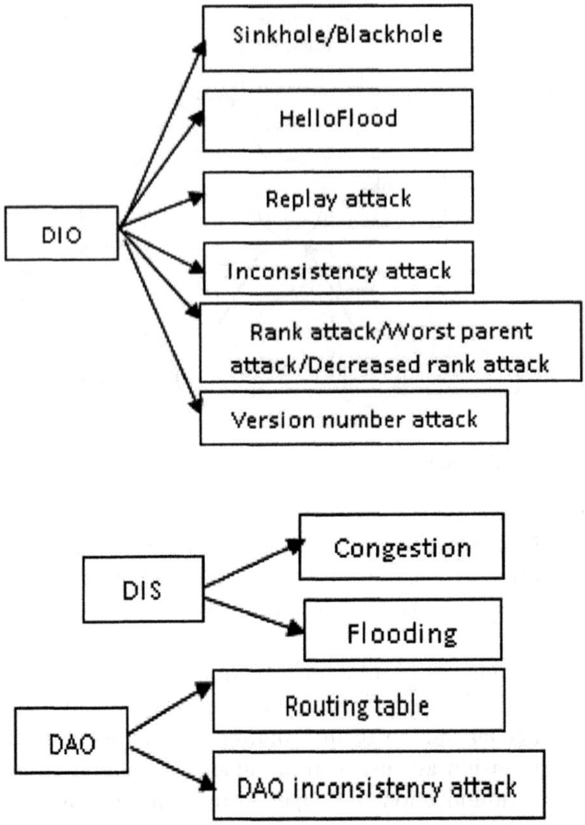

- **Authenticated**: Same as Preinstalled mode for joining as a host. For joining as a router, a second key is issued by a key authority after authenticating the node.

Currently concentrating mainly on confidentiality and integrity of data, RPL supports AES/CCMP as a standard encryption and hashing algorithm. This algorithm has a huge footprint area which slows down the system and in turn reduces throughput. It also requires a lot of computation.

In embedded IoT devices, which are severely resource and memory constrained, the use of AES can take up as much as thirty percent of RAM and makes inefficient use of the ROM available. AES/CCMP is designed for non-constrained environment and IoT is a network of lightweight constrained devices so there is a need for lightweight cryptography in IoT network. Lightweight Cryptography can be achieved by using lightweight private key generation, key exchange mechanism, encryption and hashing mechanism.

9 IoT Application Scenario

Consider a scenario where User A is a patient and has pulse sensor attached to his body. The pulse data to be sent to doctor periodically so that if anything goes wrong immediate help can be provided. At the same time we also want that the data should be accessed securely i.e. no third person other than doctor should be able to view the data. To achieve this we have to encrypt the data and to maintain integrity we have to apply hashing mechanism on the data.

Encryption and Hashing requires two steps:

1. Private key generation
2. Key exchange mechanism
3. Hashing mechanism.

10 Private Key Generation

In cryptography keys play an important role for any secure communication. Random numbers are the basic building blocks for generation of private and public keys, challenges and nonce. In literature various random number generators like True Random and Pseudorandom Number Generators (TRNG and PRNG, respectively) are studied. These current versions of TRNG and PRNG are designed for high end systems and require heavy computation.

As we know in IoT scenario there are hardware and software limitations. Existing random number generators for generating the private key on resource constrained device is a memory as well as power consuming task because of the sheer number of rounds a random number generator takes to achieve an expected level of security.

So we have to design a lightweight and cryptographically secure random number generator for generating private keys.

10.1 Lightweight Random Number Generation

Considering the same scenario User A i.e. patient sending data to User B i.e. Doctor. To send data securely we need to encrypt the data. For this they should agree on some keys. Both users will have their own private keys. let nA and nB be private keys of User A and User B respectively. Normally, these private keys are generated randomly using PRNG or TRNG, but to make it lightweight Fibonacci PRNG can be used which generates private keys that are random but computationally less heavy.

Table 2 Comparison of PRNGs

PRNG	Key size (bits)	Hardware complexity
LAMED	32	1585
Warbler	32	1238
ALFG	128	**1402**

Several work has been done in the lightweight category for random number generation. The Warbler PRNG [12] for low cost smart devices was proposed that was based on non linear feedback shift register. However, the entire family of this PRNG is susceptible to cryptanalysis and not secure enough. A PRNG called LAMED was discussed in [13] for RFID application in wireless sensor devices. Although Lpez [14] proposed the use of the Additive Lagged Fibonacci Random Number Generator (ALFG) in his paper due to its inherent advantage against the PRNG's discussed below. The paper implement all the above discussed PRNGs and compares their respective hardware complexities to then choose ALFG has the most obvious choice for resource constrained IoT device. The findings can be summarized using the Table 2.

From the table it is clear that ALFG works on more number of bits and gives moderate hardware complexity which is the basic need for any IoT application.

Using the same concept of ALFG which depicts Fibonacci sequence i.e.

$$\mathrm{Sn} = \mathrm{Sn} - 1 + \mathrm{Sn} - 2$$

We can generalize the above relation to produce a random number as

$$\mathrm{Sn} = \mathrm{Sn} - \mathrm{j} * \mathrm{Sn} - \mathrm{k} \bmod \mathrm{m},\ 0 < \mathrm{j} < \mathrm{k}$$

The operator denotes any binary operation. The initialization vector of the above proposed PRNG could be counter based on the internal clock of the IoT node. Once the random numbers are generated *nA* and *nB* will work as private keys that needs to be exchanged between two parties.

11 Lightweight Key Exchange Algorithm

Once the lighter key is generated, it should be shared between communicating parties so a key exchange mechanism is required [15].

Elliptic Curve Cryptography (ECC) is a public key cryptography developed independently by Victor Miller and Neal Koblitz in the year1985. Two obvious choices for key exchange are Modular Diffie Hellman (DH) [16] and Elliptic Curve based Diffie Hellman (ECDH) [17]. The ECDH algorithm offers an inherent computational advantage over DH as it does not make use of modular arithmetic. It favoured for IoT devices due to its robustness, added security and low power consumption.

Kalra et al. [18] put forth a method for multi-server authentication that was based on ECC. The use of attribute based ECC that uses one or more attributes of the IoT node for public key generation and subsequent encryption was also proposed. However, this method generally slows down the overall encryption process and also not provide enough flexibility.

A number of signature generation and authentication algorithms over ECC have been proposed and proved [19] through their findings that ECC fits well with real time IoT devices with constrained environments.

Elliptic Curve Cryptography Diffie Hellman (ECCDH) is also suggested since ECC algorithm has much stronger bit security when compared with RSA and it is easy to be realized on hardware. Goyal [20] proves beyond doubt the superiority of ECCDH over exponent based DH and RSA algorithms. Also, the generation of a shared secret key lowers down the key generation costs for subsequent communications between the two nodes. It also uses scalar multiplication on elliptic curve and can meet the lightweight requirement and is suitable for IoT [21].

From the Table 3 it is obvious that ECDH proves better when compared with DH as it took less power and latency. So the use of ECCDH can be used for key generation for communication between two IoT nodes.

Once the key is exchanged public keys kA and kB are generated using the formula given below:

$$\mathrm{k\,A = nA \cdot G\,k\,B = nB \cdot G}$$

These public keys are then shared between User A and User B. The security of ECC is based upon ECDLP, which indicates that it is hard to find k such that $Q = k \cdot P$ for a given elliptic curve and points P and Q on the curve.

Thus both set of private keys and public keys are thus determined.

User A computes point $(x_k, y_k) = nA \cdot kB$

User B computes point $(x_k, y_k) = nA \cdot kB$

It turns out the point calculated by both is the same and is the shared secret key.

By this User A and User B has a shared secret key which is generated by lightweight random number generation method and exchanged by lightweight ECCDH key exchange mechanism.

Table 3 Comparison of DH and ECDH

Algorithm	Power consumption (mW)	Latency (ms)
ECDH	0.570	22.4
DH	0.832	49.6

12 Lightweight Encryption Mechanism

Once both parties have a shared secret key the next step is encryption of data using encryption mechanism. Currently AES is the algorithm which is used in many IoT applications. But as we discussed for IoT we need lightweight encryption algorithm.

The survey of existing lightweight block ciphers is given below:

PRESENT [22]: It is an ultra-lightweight encryption technique which uses Substitution–Permutation network. It supports 80/128 bits keys and supports 64 bits block size. It has total 32 round keys each of length 64 bits.

CLEFIA [23]: It is lightweight encryption mechanism which supports key length of 128/192/256 bits and the block size is 128 bits. Instead of Substitution-Permutation network it supports Feistel structure and has four data lines of 32 bits. It has different number of rounds for different key sizes i.e. for 128-bit keys 18 rounds are taken, for 192-bit 22 rounds are taken and for 256-bit keys 26 rounds are taken.

Tiny Encryption Algorithm (TEA) [24] This algorithm supports 128-bit key and works on two 32-bit data. It also uses a Feistel structure and the number of rounds are 64 rounds. Its key schedule is very simple.

From the survey it is found that lightweight block cipher PRESENT proves better when compared with other techniques.

PRESENT works on Substitution–Permutation (SP) network. In SP network, P-box (permutation box) works at a bit level and allows permutation of bits. There are many bit permutation instructions such as PPERM3R, GRP and OMFLIP [25] which are given in literature.

GRP stands for group operations and its P box has very few instructions when compared with other techniques. OMFLIP (Omega and Flip) works on the concept of butterfly network and is complex when compared with GRP. GRP can be more suitable as it takes less memory requirement which makes it obvious choice in IoT cryptographic environment.

The tweak can be used along with key as additional input as they are easy to design and its cost is very low. So to provide a secure IoT protocol we can combine all techniques i.e. lightweight encryption with permutation using GRP and tweak can be used.

The proposed encryption algorithm will give a lightweight crypto system w.r.t. time and memory constraints.

13 Lightweight Hashing Mechanism

Once the data is encrypted integrity of the data can be maintained using hashing mechanisms. Existing hash functions are not suitable for IoT constrained scenario, as they require high power consumption. This lead to the design of lightweight hash

Table 4 Comparison Hashing Algorithms

Scheme	Computation overhead (clock cycles)	Communication overhead (bits)	Storage overhead (bits)
MD5	36360	128	12 registers of 32 bits
SHA-1	84272	160	12 registers of 32 bits
LOCHA	2952	96	4 registers of 16 bits and 18 registers of 8 bits
TDHA	159	288	3 registers of 8 bits

functions. The main difference between the conventional and lightweight hash functions is w.r.t. message sizes, output sizes etc.

SHA-1 being similar to MD5 was widely held for its time efficiency and robustness. But after a 2010 attack by created by Marc Stevens, it is no longer used. SHA-2 was also formulated by the NSA [26]. It has two hash functions SHA-256 (uses 32-bit words) and SHA-512 (uses 64-bit words). SHA-2 is more difficult to crack as compared to SHA-1, but it isn't as time efficient as the SHA-1 algorithm. SHA-3 is the latest addition to the SHA series, yet unlike the previous two versions SHA-3 varies internally. SHA-3 uses the sponge construction [27].

The Light-Weight One-way Cryptographic Hash Algorithm (LOCHA), is a light-weight hashing scheme, that was proposed in [28]. The hashing scheme describes and creates a relatively short-length, fixed hash digest from an input message of arbitrary length.

The Timestamp Defined Hash Algorithm (TDHA) [29], is a light-weight alternative to all the aforementioned algorithms. Unlike LOCHA, TDHA does not require large prime numbers. Also, since a timestamp value is considered for padding there is no need to store or compute this key beforehand. This reduces the computational overhead as well. Each message contains a digital signature of the sender to ensure that the message is circulated between authentic clients only.

Performance of MD5, SHA-1, LOCHA and TDHA can be compared as follows:

From the Table 4 it is clearly observed that TDHA takes less computation overhead i.e. 159 clock cycles whereas other algorithms are taking in thousands. TDHA also takes less memory space i.e. 3 registers of 8 bits. Which in turn gives lightweight solution for IoT constrained environment.

14 Conclusion

IoT being basic Wireless Sensor Network(WSN) inherits all problems related to WSN. IoT technology incorporates things which are lightweight in terms of processing power, memory, energy etc. These devices exchange real time data over the internet which can incur different types of security attacks. Existing security solutions of confidentiality, integrity, authentication etc. will not suffice in IoT environment as these algorithms are designed for high end systems.

To take care of security solutions in IoT, we have suggested a lightweight hybrid crypto system which starts with key generation techniques and ends with maintain integrity of data i.e. hashing techniques. For key generation many techniques are studied along with lightweight random number generators and ALFG is used as a lightweight pseudo random number generator which is given as a feed to generate private keys. These keys once generated are exchanged using ECCDH technique as it is proven to be better when compared with existing techniques. Once both parties have the keys the messages can be encrypted using lightweight encryption technique along with GRP and tweak. For maintaining integrity of data different existing mechanisms are surveyed and TDHA proves better as it is lighter, require lesser computational costs and improve processing capabilities. The resulting hybrid crypto system will be computationally efficient and better in resource constrained environments.

References

1. Christin, D., Reinhardt, A., Mogre, P.S., Steinmetz, R.: Wireless sensor networks and the internet of things: selected challenges
2. Heer, T., Garcia-Morchon, O., Hummen, R., Keoh, S.L., Kumar, S.S., Wehrle, K.: Security challenges in the IP-based internet of things (2011)
3. Lamaazi, H., Benamar,N., Jara, A.J., Ladid, L., El Ouadghiri, D.: Challenges of the internet of things: IPv6 and network management. In: 2014 Eighth International Conference on Innovative Mobile and Internet Services in Ubiquitous Computing (2014)
4. Dey, N., Hassanien, A.E., Bhatt, C., Ashour, A.S., Satapathy, S.C. (eds.): Internet of Things and Big Data Analytics Toward Next-Generation Intelligence. Springer, Berlin (2018)
5. Elhayatmy, G., Dey, N., Ashour, A.S.: Internet of things based wireless body area network in healthcare. In: Internet of Things and Big Data Analytics Toward Next-Generation Intelligence, pp. 3–20. Springer, Cham (2018)
6. Dey, N., Wagh, S., Mahalle, P., Pathan, M. (eds.): Applied Machine Learning for Smart Data Analysis. CRC Press, Boca Raton (2019). https://doi.org/10.1201/9780429440953
7. Tamane, S., Solanki, V.K., Dey, N. (eds.): Privacy and Security Policies in Big Data. IGI Global (2017)
8. Prabhakaran, K., Claudio, P.: Denial-of-service detection in 6LowPAN based internet of things (2013) (IEEE)
9. Saputro, N., Akkaya, K., Uludag, S.: A survey of routing protocols for smart grid communications (2012) (Elsevier)
10. Mahamure, S., Railkar, P.N., Mahalle, P.N.: Mathematical representation of quality of service (QoS) parameters for internet of things (IoT). Int. J. Rough Sets Data Anal. (IJRSDA) **4**(3), 96–107 (2017)
11. RFC 6550 RPL: IPv6 Routing Protocol for Low-Power and Lossy Networks
12. Mandal, K., Fan, X., Gong, G.: Design and implementation of Warbler family of lightweight pseudorandom number generators for smart devices. ACM Trans. Embed. Comput. Syst. **15** (1), 1–28 (2016)
13. Peris-Lopez, P., Hernndez-Castro, J.C., Estvez-Tapiador, J.M., Ribagorda, A.: LAMEDA PRNG for EPC class-1 generation-2 RFID specification. Comput. Stand. Interfaces **31**(1), 88–97 (2009)

14. Ore Lpez, A.B., Hernndez Encinas, L., Martn Muoz, A., Montoya Vitini, F.: A lightweight pseudorandom number generator for securing the internet of things. IEEE Access **5**, 27800–27806 (2017). https://doi.org/10.1109/access.2017.2774105
15. Kimbahune, V.V., Deshpande, A.V., Mahalle, P.N.: Lightweight key management for adaptive addressing in next generation internet. Int. J. Ambient Comput. Intell. (IJACI) **8**(1), 50–69 (2017)
16. Kallam, S., Diffie-Hellman: Key Exchange and Public Key Cryptosystems. Math and Computer Science Department, Indiana State University Terre Haute, IN, USA, 30 Sept 2015
17. Ahirwal, R.R., Ahke, M.: Elliptic curve Diffie-Hellman key exchange algorithm for securing hypertext information on wide area network. Int. J. Comput. Sci. Inf. Technol. **4**(2) (2013)
18. Kalra, S., Sood, S.: Advanced remote user authentication protocol for multi-server architecture based on ECC. J. Inf. Secur. Appl. **18**(2–3), 98107 (2013)
19. Dhillon, P.K., Kalra, S.: Elliptic curve cryptography for real time embedded systems in IoT networks. In: 2016 5th International Conference on Wireless Networks and Embedded Systems (WECON), Rajpura, pp. 1–6 (2016)
20. Goyal, T.K., Sahula, V.: Lightweight security algorithm for low power IoT devices. In: 2016 International Conference on Advances in Computing, Communication s and Informatics (ICACCI), Jaipur, pp. 1725–1729 (2016). https://doi.org/10.1109/icacci.2016.7732296
21. Asri, S., Pranggono, B.: Impact of distributed denial-of-service attack on advanced metering infrastructure. Wirel. Pers. Commun. (2015) (Springer)
22. Bogdanov, A., Knudsen, L.R., Leander, G., Paar, C., Poschmann, A., Robshaw, M.J.B., Seurin, Y., Vikkelsoe, C.: PRESENT: an ultra-lightweight block cipher. In: CHES. LNCS, vol. 4727, pp. 450–466. Springer, Berlin, Heidelberg (2007)
23. Shirai, T., Shibutani, K., Akishita, T., Moriai, S., Iwata, T.: The 128-bit blockcipher CLEFIA (extended abstract) (2007) (Springer)
24. https://www.tayloredge.com/reference/Mathematics/TEA-XTEA.pdf
25. Shi, Z., Lee, R.B.: bit permutation instructions for accelerating software cryptography (2000) (IEEE)
26. FIPS 180-1, Secure Hash Standard (SHS), National Institute of Standards and Technology, US Department of Commerce, Washington D. C. (1995)
27. Bertoni, G., Daemen, J., Peeters, M., Van Assche, G.: Sponge functions. In: ECRYPT Hash Workshop (2007)
28. Roy Chowdhury, A., Chatterjee, T., DasBit, S.: LOCHA: a lightweight one-way cryptographic hash algorithm for wireless sensor network Procedia Comput. Sci. 497–504 (2014) (Elsevier Science)
29. Mondal, A., Mitra, S.: TDHA: timestamp defined hash algorithm for secure data dissemination in VANET. In: International Conference on Computational Modeling and Security (CMS 2016) (2016)

Secure Data Transmission in Underwater Sensor Network: Survey and Discussion

Pooja A. Shelar, Parikshit N. Mahalle and Gitangali Shinde

Abstract The security of information has become a crucial issue in today's digital world. The reason being an ever-increasing number of web-based applications and the use of proficient hacking techniques by hackers to gain unauthorized access to confidential information. Digital data can be hacked from two positions, viz. static or dynamic. The dynamic position of digital data is synonymous to data which is in a transmission phase whereas static data is a standalone data in servers, sensors or memory chips. According to physics, developing security protocols for data that is in a dynamic position is always a great challenge as compared to static one. For instance, Alice transmits data packets to Bob with the motive of not just sharing the information but perhaps Alice aims to share the information by providing the highest possible security to the data so that the data is only accessed by authorized personnel (Bob). Data transmission can be divided into two types, namely Terrestrial and Underwater Data Transmission. To date, lots of research has been conducted for securing terrestrial data transmission, but very less research has been done to provide security to data which is transferred underwater. The state of art and its parametric analysis done in chapter proves that underwater data transmission in itself is a great challenge due to various limitations of underwater communication mediums like lower bandwidth, propagation delay, noise, multipath effect, doppler spread, path loss and so on. These challenges make the underwater networks as one of the most vulnerable networks for many different security attacks occurring in different network layer like a wormhole, sybil, homing, sinkhole and many more are presented in the chapter under the heading of, "Attacks in UWSN". According, to the complete survey and discussion done in this chapter, it can be stated that "Underwater Data Security is the heart of UWSN". Thus, underwater data security is a necessity as UWSN is prone to various security threats and malicious attacks in addition to problems of dynamic underwater environment and communication mediums. Earth is a water planet and around 71% of the earth's surface is covered by water. In spite of this water to land ratio of the world the "secure underwater data transmission" is a technologically lesser-explored area. Therefore, this survey work

P. A. Shelar (✉) · P. N. Mahalle · G. Shinde
Department of Computer Engineering, Smt. Kashibai Navale College of Engineering (Savitribai Phule Pune University), Pune, India

N. Dey et al. (eds.), *Internet of Things, Smart Computing and Technology: A Roadmap Ahead*, Studies in Systems, Decision and Control 266,
https://doi.org/10.1007/978-3-030-39047-1_15

will intrigue many researchers to work on underwater sensor networks and to find innovative solutions for transferring digital data securely in the underwater environment. The proposed future work is to develop an innovative architectural model named "ACOOP UWSN" with a strong security protocol by using both optical and acoustic waves as a medium of data communication.

Keywords Underwater sensor network (UWSN) · Underwater acoustic sensor network (UASN) · Underwater data transmission · Autonomous underwater vehicle (AUV) · Remotely operated vehicle (ROV) security protocols · Acoustic communication

1 Introduction

Data transmission in Underwater Sensor Network is a need of the day as the future of many underwater applications is depended on secure data communication in an underwater environment. The origin of underwater communication date in the year of 1490 when Leonardo da Vinci stated that, "If one head of a long tube is placed in water and the other end is placed near human ear, then it's possible to hear sound generated by ships which are a long distance away". Then after a significant amount of time, in 1687 Sir Isaac Newton published the first mathematical theory of how sound travels in water. Later in 1918, the very first underwater data transmission took place which was a submarine communication. This communication system was developed by the United States at the end of World War II. Research has since advanced in underwater acoustic communications. The underwater data transmission has achieved great attention and importance due to development in various underwater applications like Tsunami reporting systems, Study of marine life, Anti-submarine missions, Oceanographic data collection, Pollution Monitoring, Anthropogenic climate impacts, Oil Leakage Detection, Intrusion detection and Surveillance, Mine Reconnaissance, Water quAlity moniToring sEnsors netwoRk (WATER) and many more newly emerging applications. Also, many more innovative applications can be developed by combining UWSN with IoT [1, 2]. Therefore, rapid development in various underwater applications suggests a great urgency for developing a strong security protocol in an underwater environment. Post-2000, every aspect of the underwater sensor network came under research studies which included MAC protocols for UWSN, Routing protocols for UWSN, UWSN architecture, the energy consumption of sensor nodes in an underwater environment, Localization of sensor node, Time synchronization in sensors and so on. But, the security aspects for underwater sensor networks are rarely focused. Wireless Underwater Communications is becoming a discretionary part of everyday life and could befall a threat if security is not considered before the deployment of an underwater sensor network. Thus, few findings on security to underwater data transmission are discussed in the below sections of this paper so that the readers will be motivated to find a solution in the field of security to underwater data transmission.

Motivation: The motivation for this research work is a Microsoft project called "Natick". This Microsoft Research Project has put Cloud in the Ocean for the first time. Natick is a project to manufacture and operate an underwater datacenter. The Natick mainly aims to reduce the cooling cost of datacenters by keeping them in a natural cooling environment like "Ocean Water". The success of Natick has made researchers forecast the day when there will be numerous underwater datacenters that will be wirelessly connected to the Internet. And if this is the vision for underwater networks, then it is time to develop a high-level security protocol for underwater data transmission.

To maintain the systematic view and proper flow of contents, the research paper is divided into sections which are as follows, Sect. 2 covers Basics of UWSN, Sect. 2.6 presents Attacks in an underwater sensor network and Sect. 3 discussing on what is the importance of security in an underwater environment. Section 4 tells about the applications of UWSN and Sect. 5 gives a brief discussion about related works in secure underwater data transmission. Section 6 shows the Analytical view whereas Sect. 7 describes the methodology used in short terms. In conclusion, Sect. 8 focuses on the future scope of this research study.

2 Basics of Underwater Sensor Networks

This section is to discuss different features of UWSN such as data transmission medium in UWSN, Architecture of UWSN and comparative study between TWSN and UWSN. These general studies about UWSN are to get a complete overview of what is UWSN and how it works.

2.1 Different Data Communication Mediums in UWSN

Underwater communications can be accomplished via three means radio, optical and acoustic waves. The selection of the propagation medium for an underwater network is directly proportional to the success rate of data transmission. To have the correct choice of communication medium, below is a brief discussion about different ways of underwater wireless communication and possible challenges faced by them in water.

1. **Radio Waves**:

The high-frequency radio waves range from 3 kHz to 300 GHz. These waves are a form of electromagnetic frequency, which makes them infeasible for underwater communications as the electromagnetic waveforms have a tendency of absorption and attenuation in an underwater environment. This is the main reason for renouncing radio waves as a medium for underwater communication.

The radio waves can travel in water but only at extremely low frequencies (3–300 Hz) and covering a distance from 1 mm to 100 km. The extremely low frequency of radio waves has drawbacks like limited bandwidth and extremely short propagation length making radio waves as the infeasible medium for communication.

2. **Optical Waves**:

Optical waves can transmit data to a maximum distance of 10 meters in an underwater environment. Optical waves can be a good channel for underwater communication for specific underwater applications which requires very high bandwidth and secure mode of data transmission. The only drawback of an optical wave is its scattering phenomena in water, which may lead to unsuccessful data transmission. And scattering indirectly leads to the energy loss of the original signal, huge absorption loss and also high attenuation.

3. **Acoustic Waves**:

Acoustic waves also known as sound waves are the most commonly used communication medium in an underwater environment. The reason behind the popularity of acoustic waves as an underwater communication medium was its low-frequency band which ranges from 20 Hz to 20 kHz. The speed of sound waves increases as the depth of water increases. The high-speed acoustic waves can hence transmit data up to a few kilometers. The only drawback of acoustic waves is its limited bandwidth.

2.2 Comparative Study of Different Underwater Communication Mediums

From the Table 1, it can be noted that each communication medium has its own positive and negative background. Thus, the selection of communication mediums for the underwater environment should be application-oriented.

2.3 Application of Underwater Communication Mediums

Radio Waves

- It is used as communication link between UWSN and TWSN/offshore station.
- It is also used to transmit sensed data among the sensor nodes and then to the base station.
- Mainly used in applications which require high-speed data transmission.

Table 1 Comparative table of different UWSN communication mediums

Sr. no.	Differentiating criteria's	Radio waves	Acoustic waves	Optical waves
1.	Rate of attenuation	Highest, as attenuation increase with increasing frequency (3.5–5 dB/m)	Moderate, here attenuation is distance-dependent (0.1–4 dB/km)	Lowest, as intensity of light reduces as distance between sender and receiver increases (0.39 dB/m)
2.	Bandwidth	Lowest	Moderate, as its distance dependent. In few kHz	Highest, in MHz's
3.	Type of water	Prefer in shallow water	Mostly used in deep water	Used in pure sea water
4.	Data rate	Moderate, in Mbps	Lowest, in Kbps	Highest, in Gbps
5.	Latency	Moderate	High	Low
6.	Propagation speed	3×10^8 m/s	1500 m/s	2.25×10^8 m/s (It is almost at the speed of light)
7.	Scattering phenomena	Absent	Absent	Present
8.	Cost	High	High	Comparatively low
9.	Energy consumption	High	High	Low
10.	Doppler effect	Negligible	Considerable	Negligible

Acoustic Waves

- It is the most commonly used underwater communication medium.
- It can be used as a single communication medium for the entire underwater sensor network.
- Acoustic waves are mostly used in applications which requires message exchanges at long distance sensor nodes.

Optical Waves

- It is used in applications which generate messages that have a large number of bits.
- It can also be used by applications which require a frequent exchange of messages at short distance.
- Optical waves can also be used in applications which requires security in data transmission.

2.4 Comparative Study of Terrestrial and Underwater Sensor Network [TWSN vs. UWSN]

If we compare terrestrial and underwater wireless sensor networks, it can be concluded that there exist many difficulties in communication within UWSN which can be seen with the help of Table 2. Therefore, developing a secure communication protocol for UWSN is a great challenge for researchers as compare to TWSN. In summary, UWSN uses acoustic waves by acoustic modems, to transfer digital information in underwater which poses grand challenges for designing efficient security protocol.

2.5 Challenges of Underwater Environment

There is below the list of challenges given for UWSN which is as follows,

1. **Low Propagation Delay**: The time taken by a wave to transmit from source to its destination is known as propagation delay. If propagation delay is low, means data transmission is at the fastest speed vice versa if propagation delay is high then data transmission is at a slow rate. But due to different characteristics like noise, the multipath effect of underwater networks the propagation delay is always higher. Thus, designing of security protocol for UWSN becomes more difficult.
2. **Limited Bandwidth**: The underwater communication channel mostly has limited bandwidths because bandwidth is directly proportional to range and frequency among nodes, it can be stated as,

Table 2 Comparative table between TWSN and UWSN

Parameters	TWSN	UWSN
Minimum propagation speed	300,000 km/s	1500 m/s
Communication medium	Electromagnetic waves	Optical waves, radio waves and acoustic waves
Bandwidth	890–960 MHz	1–10 kHz
Maximum transmission delay	10 μs	50–1000 ms
Signal transmission rate	1200–9600 bps	Tens of bps
Cost	Inexpensive	Expensive
Nodes mobility	Stable	Stable or moving sensor nodes
Latency	Low	High
Error probability	Comparatively low	High

(i) The longer-range is equal to smaller communication frequency which is in few kilohertz vice versa
(ii) The shorter range is equal to larger communication frequency which is 100 kHz.

For instance, if the range between two sensor nodes is less than 0.1 km then the bandwidth will be more than 100 kHz and if the range is 1000 km then bandwidth will be less then 1 kHz. As a result, longer the distance, narrower is available bandwidth. And lower bandwidth makes a simple adaption of traditional security solutions (e.g. Digital Signatures) practically infeasible.

3. **Path Loss**: The Path loss also known a Path attenuation means a reduction in the signal strength as the distance among the nodes increases. The main reason for path loss in an underwater environment is the absorption nature of acoustic and radio signals in water. The absorption of acoustic signals in water leads to the conversion of sound energy into heat, thus reducing signal strength. The path loss always increases with distance. There are various other reasons for path loss like scattering, dispersion, and refraction of the acoustic signal in water.
4. **Noise**: Signal to Noise Ratio is defined as the strength of acoustic signal carrying information to that of unwanted interference. Signal to noise ratio for deep and shallow water is more at a lower frequency and is less at a higher frequency. The SNR ratio in underwater communication is expressed in terms of transmission loss, ambient noise, directivity index, and source level. The source of noises in underwater are sounds produced by submarines, radars, and ships.
5. **Multipath Effect**: Acoustic waves take multiple paths to pass through a transmitter to a receiver which is the main reason for the occurrence of the multipath effect. Multiple travel paths arrive as acoustic waves get reflected at the surface and bottom boundaries of the acoustic channel. Even with the multipath effect sound signal can travel thousands of kilometers as there is no loss occurring due to reflections. The degradation of the acoustic communication signal mainly occurs due to multipath propagation since it generates Inter-Symbol Interference (ISI).
6. **Doppler Spread**: The high data transmission rate becomes a challenging task in an underwater environment due to Doppler spreads. It leads to degradation of performance in underwater communications. The spreading is dependent on the distance a signal travels and its frequency. Thus, shorter the range higher is the Doppler spread whereas longer the range lower is Doppler spread.
7. **Properties of Water**: The Chemical and Physical properties of water like the high polarity, the high heat of vaporization, density, salinity, color, hardness, conductivity causes the underwater communication mediums to be spacial and temporally variable.
8. **Energy Constraints**: In underwater communication, short-range Multihop transmissions become more energy efficient than Long-Range One Hop ones. All nodes of UASN are battery operated and also acoustic signal is the type of mechanical waves that undergoe alternating compression and rarefactions in the

water medium, thus requiring high power for data transmissions. Battery replacement is also difficult in UWSN. Thus, it is a crucial requirement to avoid message losses occurring due to security attacks which ultimately reduces sensors energy.

The above list of challenges in UWSN suggests that it is difficult to apply security algorithms to underwater data transmissions. And the most conflicting challenge is, “Increasing security performance of the underwater network by saving the resources of underwater nodes”. Supposedly if there are some extra algorithms to be run for security then it will surely surge the power usage of nodes. This causes a dilemma when designing a secure communication channel. Cryptographic Protocols for underwater networks must always be communication efficient in terms of size and number of messages. Hence, applying traditional techniques such as ciphers, digital signatures in an underwater environment requires many modifications to be done.

2.6 Architecture of UWSN

Network architecture is a blueprint that shows different network functionalities. For example, a communication model of the architecture is used for designing and implementing an application related task. An underwater network has to face many challenges like Energy constrained, Multipath effect, Path loss, Limited bandwidth and many more. Thus, it becomes a necessary job to predesign underwater sensor network architectures for each different underwater application while considering all different aspects involved. Underwater Sensor Networks (UWSN) is a type of ad hoc networks deployed in an underwater environment. The building blocks of UWSN are underwater sensors (nodes), Autonomous Underwater Vehicle (AUV), Remotely Operative underwater vehicles (ROVs), surface buoys(sink) and an offshore station which are interconnected by wired or wireless underwater communication mediums. The underwater networks are said to be the most unsecured networks due to different characteristics of the underwater environment. Thus, constructing underwater network architecture along with high-level security protocols is the main motive of this research work.

1D-UWSN Architecture: As shown in Fig. 1, the 1D type architectures for the underwater network consist of floating buoy, AUVs and sensor nodes. In this type of architecture, each sensor node is an individual network. In summary, the network with only one single node in an underwater sensor network is called One-Dimensional architecture (1D-UWSN). And these networks are capable of processing sensed data and then finally transmitting the information to offshore stations.

2D-UWSN Architecture: As shown in Fig. 1 the 2D type architectures for the underwater network consist of many fixed sensor nodes forming different groups

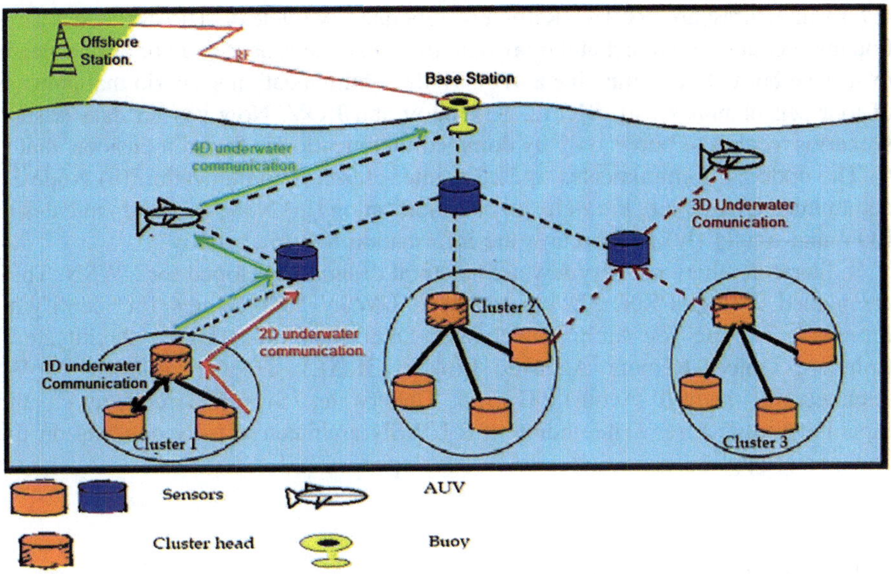

Fig. 1 Underwater sensor network architecture

which are named as clusters. The sensor nodes are fixed using anchors at sea bottom. The minimum capacity of each cluster is at least two sensor nodes, where anyone node is declared as the cluster head also known as an anchor node. In this type the sensed data reaches to buoys by doing communication in two dimensions, that is (i) each cluster member sends its sensed data to the cluster head using horizontal link while (ii) each cluster head sends processed sensed data to the surface buoy using vertical links. Due to such a two-dimensional nature of communication, this type of underwater architectures is named as 2D-UWSN.

3D-UWSN Architecture: As depicted in Fig. 1 the 3D type architectures for the underwater network consists of many floating sensor nodes forming clusters. The communication among sensors become three-dimensional due to floating sensor nodes present on variable heights. Hence this type is named 3D-UWSN.

Let us imagine a 3D-UWSN communication scenario that consists of four sensor nodes (S0, S1, S2, S3) and a buoy. Assume that S0 and S1 belongs to cluster A while S2 and S3 belong to Cluster B and also S0 is head of the cluster A while S3 is head of cluster B. Now sensed data in such scenarios reaches to buoys only by doing communication in three dimensions, that is (i) The node S1 communicates to S2 i.e. inter-cluster communication (ii) Node S2 communicates to S3 i.e. intra-cluster communication and finally (iii) Node S3 communicates to surface buoy.

4D-UWSN Architecture: As shown in Fig. 1 the 4D type architectures for the underwater network consist of some floating and fixed nodes. The floating nodes in

4D architectures are ROVs (Remotely Operated Vehicles). The ROV act as a floating mediator to collect and relay data from a single sensor node or cluster heads to surface buoy. Let us imagine a 4D-UWSN communication scenario that consists of four sensor nodes (S0, S1, S2, S3), buoy and ROV. Now sensed data is such scenarios reaches to buoys only by doing communication in four dimensions, that is (i) The node S1 communicates to S2 i.e. inter-cluster communication (ii) Node S2 communicates to S3 i.e. intra-cluster communication (iii) Node S3 communicates to ROV and finally (iv) ROV relays the information to surface buoy.

In literature, there are very few architectural designs developed for UWSN. They are named as Aqua-Net, 3D cluster-based UWSN architectures for identifying Intruders, Tic-Tac-Toe-Arch, Underwater Direct-Diffusion (UDD) Architecture, Common Object Remote Acoustic Link (CORAL), Trees of Wheels (ToW) Architecture, Hybrid Acoustic-Optical Underwater Swarm Architecture, and Pipeline Architectures. The choice of a UWSN architecture truly depends on the underwater application which is to be developed.

2.7 *Attacks in UWSN*

There are many underwater applications among which 90% of applications are implemented for national security, like Submarine Communication Systems, Tsunami Detection Systems, Water Monitoring System and so on. All these applications and their data transmission systems are of prime importance to each nation of the world. These applications are the main origin of national confidential digital information from underwater networks and hence are very much prone to different attacks. The Fig. 2 shows all types of attacks within each layer of the underwater sensor network. The nomenclature of attacks for UWSN is almost as same as the terrestrial sensor networks. But the large difference occurs in the attacking environment and the strategy used. The study of underwater attacks is one of the most important prerequisites for any researcher who is working with underwater networks.

Jamming Attack

The name jamming originates from the concept of "jammer", where jammer is nothing but a device/corrupted sensor node that intentionally disrupts the communication among the underwater sensor nodes. A jammer can disrupt any communication in two different ways. Firstly, by sending the excess amounts of data packets to drain out the senor batteries. And secondly by injecting malicious packets into the network. Jamming attacks have thrown serious threats to underwater sensor networks. Thus, for reliable and secure monitoring services it is necessary to address jamming attacks, at least for applications like water monitoring systems that carry sensitive information about lakes, rivers or ocean water. The characteristics of acoustic waves such as severe attenuations, large delay, severe multipath and low frequency of underwater acoustic channels, and the energy

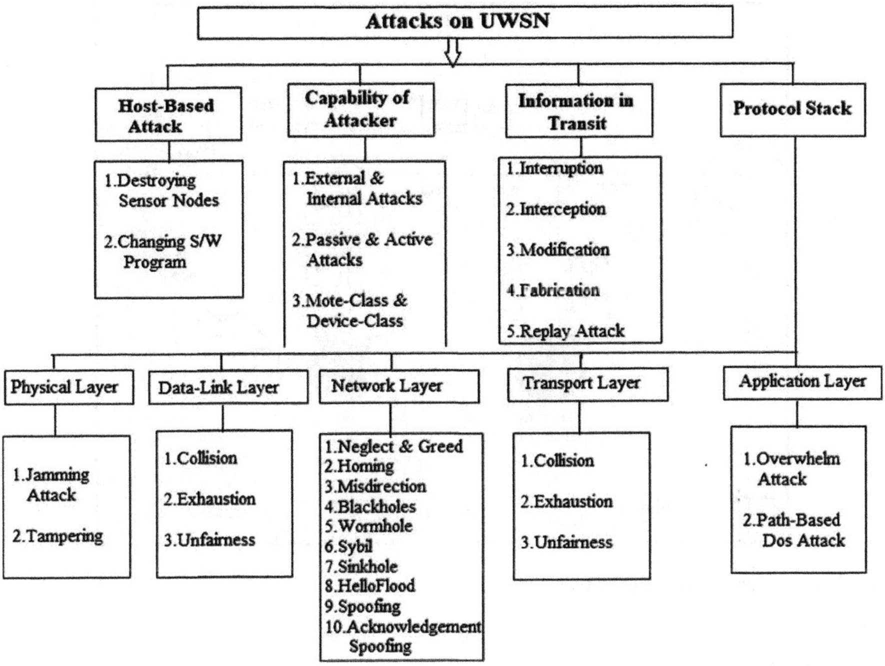

Fig. 2 Taxonomy of attacks

constraints of underwater sensors are some of the main reasons that underwater networks are vulnerable to jamming attacks.

As shown in Fig. 3, the sender is named Alice whereas the receiver is named Bob. There are also two types of Jammer/Eve,

(i) Dummy Jammer:
It is also known as signal jammer and its job is to generate noise signals to jam the communication channels.
(ii) Smart Jammer:
It is also known as a deceptive jammer. It is named smart as this jammer has some pre-information about the network protocols being used. Hence the smart jammer uses corrupted data packets to jam the network.

Jamming Model:

Step 1: Eve will jam the signal such that the message from Alice to Bob cannot go through.
Step 2: After some time, Eve replay the message or steal the information and replace it with some false message.
Step 3: Finally, Bob will receive the message from Eve and believe it is sent by Alice.

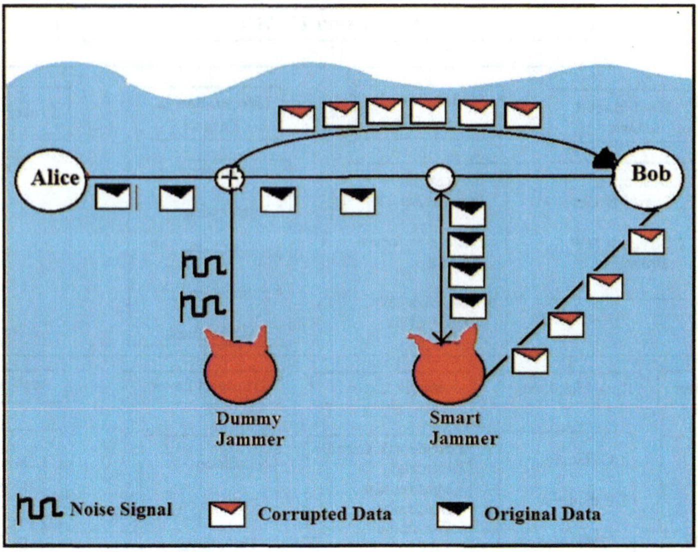

Fig. 3 Overview of jamming attack

Tampering
In underwater surveillance application, there may be hundreds of nodes spread at the bottom of the ocean. In such cases, it's difficult to ensure the security of all sensor nodes because there is a high possibility of tampering of nodes deployed in enemy territory. Tampering is a type of attack where the attacker can damage or replace the sensor node. He can also decode sensitive information such as cryptographic keys to gain unauthorized access to confidential underwater data transmission. Tampering of node also means capturing and modifying the nodes internal memory. Hiding of nodes can be used as one of the defensive measures from a tampering attack.

Neglect and Greed Attack
The word neglect is used as a corrupted node/attacker sends the acknowledgment signals to source node but they neglect the original data packets which are supposed to be sent to the destination node. Moreover, the attacker becomes greedy as he gives higher priority to his messages. The countermeasure to such attacks will be reducing the number of routing hops or to use multiple paths for routing data.

Homing
Most of the underwater sensor network architectures are cluster-based. As many underwater applications require the deployment of hundreds of sensors and to manage this many numbers of nodes a cluster-based architecture is the best-fitted option. But the main drawback of this architecture is sink node and cluster-head which is the single node having special responsibilities like key management, data aggregation, localization and so on. Therefore, cluster-heads become a center of

attraction for attackers. Hence compromising the cluster-head or sink node to break down the whole network is known as a homing attack. The countermeasure for the homing attack is to hide the cluster-heads by using Encryption and to make each hop of data transmission trustable by using Trust-based algorithms.

Misdirection

As the name suggests it attacks where the sender is misguided to forward messages along the wrong path. Also, the attacker provides the wrong routing table information to misdirect the route. The countermeasure for this type of attack can be the "Identity-based Authentication Scheme", where only the authorized nodes can exchange the routing information.

Selective Forwarding and Black Hole Attack

It is the type of attack in which the corrupted node becomes the relay node for several neighboring nodes. Then this corrupted relay node decides selectively which data packet should be forwarded or dropped. It is the most common type of attack in UWSN as underwater networks are multi-hop networks. If the corrupted sensor node, selectively forwards the packets, then it is said as "Selective Forwarding" and if the node drops all the packets passing through it then it is said as "Black Hole Attack". The countermeasure to this type of attack can be the use of braided paths, multi-path routing, implicit acknowledgments and many more.

Wormhole Attack

The term wormhole means, a speculative structure linking two points in spacetime. The same concept is used by the attacker in a wormhole attack, where a tunnel between two malicious nodes is made as shown in Fig. 4. This tunnel is then used by an attacker to launch different types of attack like modifying packets, breaking

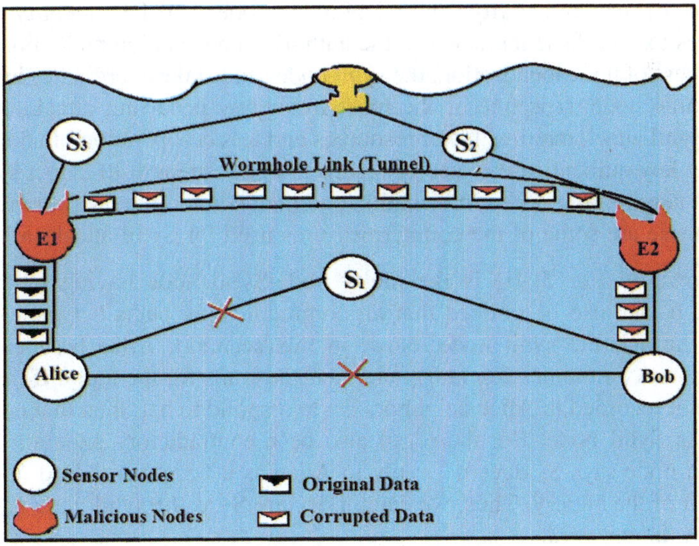

Fig. 4 Overview of the wormhole attack

the order of packets, selectively dropping of packets, destroying routing protocols, changing original network topology, man-in-middle attack, denial-of-service attack, and network hijacking.

Steps in Wormhole Attack:

Step 1: Tunnel Creation: The wormhole attack requires two attackers which create a wormhole link called a tunnel, to create the illusion of the shortest routing path in the network.
Step 2: Packet Pre-processing: The packet when enters the tunnel, it is pre-processed, such as the head of the packet is extracted and the direction of a packet is identified.
Step 3: Tunneling Recorded Messages: Attackers then route the messages recorded at one end of tunnel to the other. Hence normal nodes are tricked into believing that there exists a shortest path between them, which is the wormhole link between two compromised nodes.

As shown in Fig. 4, the sensors are using a shared acoustic channel to send information in underwater environments, the malicious nodes E1 and E2 can eavesdrop on the packets, tunnel them to another location in the network, and retransmit them. Thus, generating an illusion that the original sender (Alice) is in the neighborhood of the remote location (Bob).

Sybil Attack
The attack is named Sybil after the famous multiple personality disorder patient "Sybil" (Shirley Ardell Mason.). It was named so as the single underwater sensor node can also show multiple identities. Such nodes are called the faulty nodes or sybil nodes. There are two ways to obtain an identity for a sybil node, (i) create a new identity or (ii) steal an identity from a legitimate node. This fake identity is used by sybil nodes to send false messages to the authorized destination node. For example, in the process of node localization, the sybil node sends fake coordinates information to the source node. It is one of the most frequently occurring attacks in UWSN because Sybil attack requires fewer resources and less co-operation of another node. The secret key authentication, use of received signal strength to detect sybil node, collecting neighboring nodes information and then counting the appearance of each identity, these are some of the countermeasures used for Sybil attack in UWSN.

As shown in Fig. 5, E_1 is the faulty node/Sybil node having multiple fake identities. It can also be noticed that all communication paths between Alice and Bob are through the sybil node. Now, in this scenario, Alice broadcasts query packets in her communication range and then she waits for the reply. If the node E_1 which is not recorded in Alice neighbors list to respond to her, then that node can be declared as Sybil node. But there can also be a contradictory situation where an authorized node says S_3 does not reply to Alice as it becomes a dead node due to exhaustion of the battery. Thus, we cannot declare S_3 as the sybil node and remove them from the network.

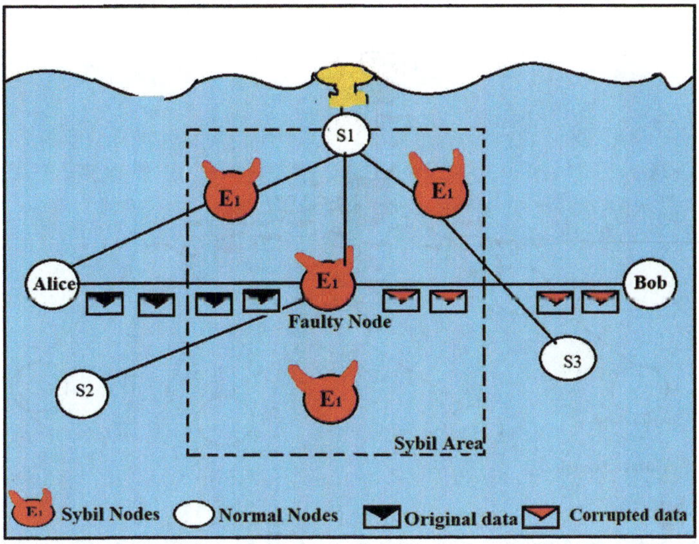

Fig. 5 Overview of the sybil attack

Sinkhole Attack
It is an attack implemented in the network layer. The attacking node becomes like a sinkhole where the attacker decides whether to drop all or selective packets. As shown in Fig. 6, the sinkhole attack aims to prevent the buoy present at the surface of the water, from receiving the complete and correct sensed data by underwater sensor nodes. To overcome such attacks in UWSN multi-level trust and reputation-based solutions have been proposed. The countermeasure for sinkhole attack is routing of data packets from sensor nodes to multiple buoys present at the sea's surface or to do authentication of nodes exchanging routing messages.

Modeling of Sinkhole Attack

Step 1: Sinkhole Node Creation
The attacker compromises the normal node and makes it corrupted to create a sinkhole node. These malicious nodes are then used to launch a sinkhole attack. The sinkhole node uses its routing metric to lie his neighboring nodes to launch the attack.

Step 2: Sending Fake Information
Then the compromised node sends fake information to neighboring nodes about its link quality. As shown in Fig. 6, the red-colored data packets denote the fake routing information send by the sinkhole node to its neighboring nodes.

Step 3: Passing of original data through Sinkhole node
Then all the packets from his neighbors will pass through a sinkhole before reaching to the base station. Thus, preventing buoy/base station from acquiring a complete and correct sensing data from nodes.

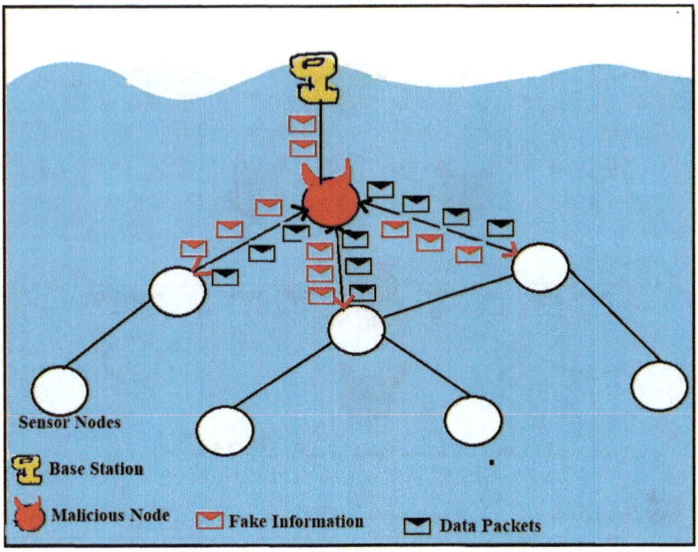

Fig. 6 Overview of the sinkhole attack

Hello Flood Attack

The attacker floods the network by using the hello packets, so named "Hello Flood Attack". The hello packets are generally used to announce an authorized node presence to all its neighboring nodes. In this attack, an intruder sends these hello packets to all the sensor nodes in the network so that they consider this malicious node as their neighboring node. Hence, resulting in large number of nodes sending data packets to their fake neighbor's. Therefore, to overcome such attacks authentication of the neighboring nodes is one of the solutions.

Spoofing Attack

Spoofing is an art of imitating someone. According to terminology in this attack, the malicious node (spoofer) imitates like an authorized sensor node to the network. Thus, in the above figure, the spoofer is shown in two colors the white portion indicates that the node is imitating like a normal node but the red half indicates that it is a malicious node. Spoofing in sensor networks is done is by creating a fake MAC address and by falsifying the data. Due to which the spoofer node has all the access rights the same as the normal node. As shown in Fig. 7, Alice is sending all data packets to the spoofer node assuming that it's an authorized neighbor. But the spoofer then alters and replay all the routing information to interrupting traffic and create congestion. The countermeasures for such attacks are the use of reinforcement learning techniques or appending each message with an authentication code.

Acknowledgement Spoofing

The adversary misguiding the neighboring nodes by generating false acknowledgment packets to broadcast false information about network links is called

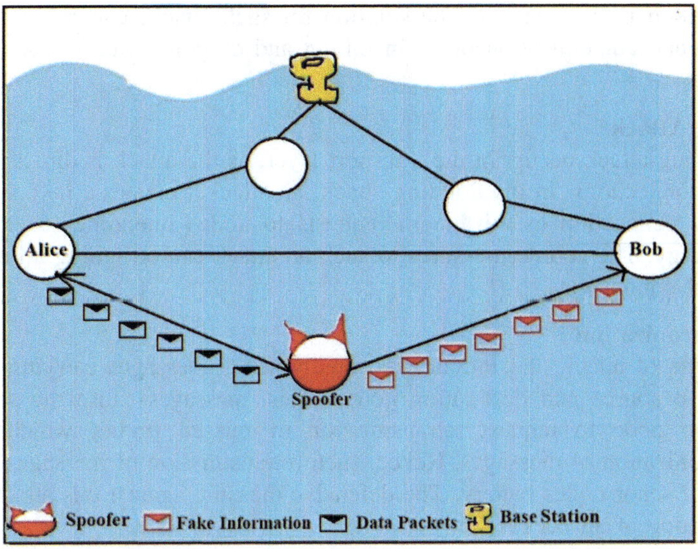

Fig. 7 Overview of spoofing attack

"Acknowledgement Spoofing". In the scenario shown in Fig. 8, Alice is sending data packets in reply to which Bob should send back acknowledgments data packets to Alice. But spoofer in middle tricks Alice that he is Bob and sends her false

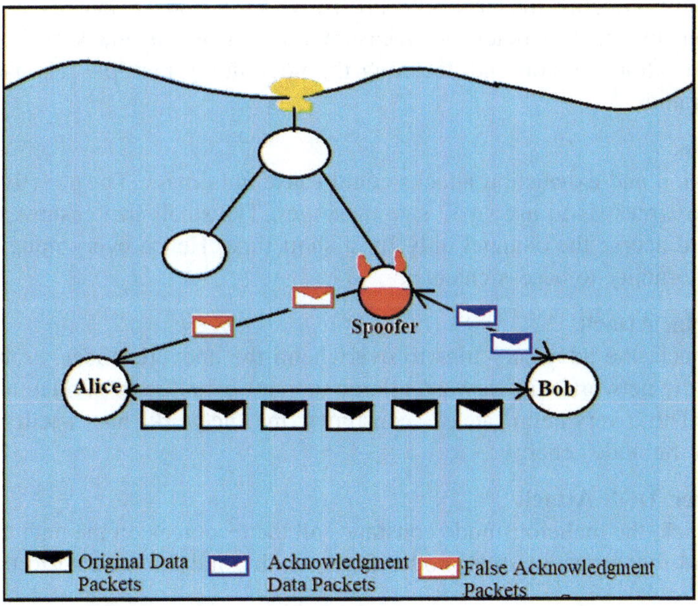

Fig. 8 Overview of acknowledgement spoofing attack

acknowledgment data packet. The solution for such attacks can be encryption of messages and authentication of all incoming and outgoing data packets from the node.

Flooding Attack
The flooding attack occurs in the transport layer, as the attack is directly done on network connections. In the flooding attack, an attacker creates a fake scenario of too many connection establishment requests to nodes intending to exhaust the nodes energy. The countermeasures to such an attack can be setting a limit on the number of connections.

De-synchronization
In this type of attack, the intruder repeatedly sends messages carrying sequence numbers to source and destination nodes. This, messages cause the source and destination node to request retransmission of missed frames which leads to desynchronization of messages. Hence, such retransmission of messages results in wastage of senor nodes battery. The defensive measure against this attack is to do authentication of all packets.

Collision
Collision attack occurs mainly due to a strong multipath effect, in UWSN. The reason behind multipath is the challenging underwater environment where the underwater network topology changes rapidly and also the tendency of the acoustic waves to travel from two or more paths. The collision attack indirectly leads to the reduction of communication quality and the high error rate.

Exhaustion
Intentionally reducing the networks life by wasting node resources is called an exhaustion attack. The defensive measure for exhaustion attack is limiting the number of retransmission and declaring the node doing multiple retransmission a malicious node.

Unfairness
The collision and exhaustion attacks could cause unfairness. The effective defense against unfairness is to use small size messages. The small size ensures that every node could access the channel only for a short time. Hence every node will have equal opportunity to access channel.

Overwhelm Attack
In this attack, the adversary tries to overwhelm the network nodes so that all the nodes in the network will forward a large amount of traffic to surface buoy (base station). The overwhelm attack consumes the network bandwidth and also decreases the nodes energy.

Path-Based DOS Attack
In this attack, the malicious node consumes all the resources on the path to the base station and thus prevents other nodes from sending the sensed data to the surface

buoy. The adversary implements this attack by injecting false data packets into the network which causes starvation of the network traffic.

External Versus Internal (Node Compromise) Attacks
These attacks are also known as Outsider and Insider attack.

External or Outsider Attacks: This attack is done by nodes which are not part of the underwater network so-called an outsider attack. The external attacker secretly listens to data transmissions and then inject spurious data packets into the network into consume network resources and launch denial of service (DoS) attack.

Internal or Insider Attacks: This attack is done by nodes which is part of the underwater network so called an insider attack. The insider attacker forces the authorized sensor nodes of the underwater network to start with abnormal behavior. Internal attacks are more harmful as compared to the external attacks as the insider attacks know important and secret information of the network.

Passive Versus Active Attacks

Passive Attacks: In this attack, the attacker is said passive as he listens to all unencrypted messages to collect all sensitive information from the network so that it can be used to launch some sever attack. Examples of Passive attacks are traffic analysis, decrypting weakly encrypted traffic, capturing authentication information and monitoring communications.

Active Attacks: In this type of attack, the attacker is said active as he listens and then takes major action on transmitted data. The action taken by an attacker are modifications of the data packets and making of a fake data packet. Examples of active attacks are jamming, DoS, fabrication, modification of data, wormhole, black hole, sinkhole, replay, spoofing, man-in-middle attack, flooding, overwhelm, Hello flood, selective forwarding, node subversion, lack of cooperation, etc.

Mote-Class Versus Device-Class Attacks
In mote-class attacks, an attacker uses two or more motes (sensor nodes) having the same capabilities to attack a UWSN whereas, in device class attacks, an attacker uses some powerful devices like ROVs and AUVs to attack a UWSN.

Interruption
In a sensor network, sensors are used to detect and respond to the behavior of the underwater environment to buoy or sink node. While sending these sensed data, it may be possible that the information in transit is attacked by the attackers and fake data is then sent to the base stations or sinks. Therefore, sending of false information or breaking of the communication link to make network services unavailable is said as an interruption attack.

Interception
In interception, an adversary captures a sensor node and then gains unauthorized access on the network nodes or on the data. This attack is generally launched in the application layer and it mostly affects data confidentiality.

Modification
In the modification attack, the malicious node not only accesses the data but also modifies it. This kind of attack is mainly launched in the network and application layer to affect the integrity of the original message.

Fabrication
An attacker inserting false data stream into the network with a motive to lose the authenticity of messages is known as fabrication attack.

Replaying Existing Messages
This is also a type of attack where the attacker forces the captured node to send old messages again and again with the intention of reducing nodes energy and message freshness. A novel key management scheme can be used as a countermeasure for relay attacks [3].

Host-Based Versus Network-Based Attack

Host-Based Attacks
The attack done on a specific machine(host/node) is called a host-based attack. Each machine has three major components that are user, software and hardware. Therefore, depending on these components host-based attack is also further divided into three attacks which are as follows,

User compromise: In this attack, the users of UWSN which is an offshore station situated on land is compromised. Users are cheated to reveal sensitive information of the network e.g. passwords or keys about the sensor nodes.
Hardware compromise: This attack is directly done on the hardware that is the sensor node. The attack is done with the intention of hacking program code, data, and keys stored within a sensor node's hardware.
Software compromise: This attack breaks the software which is running on the sensor nodes. Buffer overflow is one of the examples of software attacks.

Network-Based Attacks
The attack done on the network by devices with the intention of network usage is called a network-based attack. The network-based attacks mainly deviate the network protocol from its pre-planned functioning.

3 Importance of Security in UWSN

The underwater networks come across many security threats due to the energy constraints and poor communication capabilities of sensor nodes. As shown in Table 3, there are many threats in underwater sensor networks that suggest the importance of implementing secure data transmission in an underwater sensor network. In underwater sensor networks, denial of service attack and wormhole attacks are most common due to the unique characteristics of underwater acoustic

Table 3 Mapping of security threats with attacks

Threats	Attacks
Unavailability of network	DoS attack Exhaustion attacks and host-based attack
Shorter battery life	Power exhaustion attack Flooding attack Overwhelm attack
Modification of original message	Tampering message
Unreliable data transfer	Jamming, sinkhole, spoofing, blackhole, wormhole attacks
Hacking of cluster-heads	Jamming cluster head nodes or homing
Delays in end-to-end data transmission	External attacks and wormhole, misdirection, sybil attack
Compromised neighbor discovery phase	Blackhole attack
Loss of network integrity	Spoofing-based attack

channels. The countermeasures proposed to stop wormhole attack in terrestrial wireless networks cannot be used for underwater wireless networks, because all the schemes for TWSN only ensure that the transmitter and receiver are at a distance of one-hop. But these proposed schemes cannot be used to stop underwater wormholes shorter than one hop distance. Thus, there is a need to develop new techniques for protecting underwater networks from wormhole attacks.

There are also many other possible security threats in underwater networks. For example, underwater sensor nodes can be hacked by an attacker and malicious information can be injected into it. And if this is the scenario then all underwater applications, especially military application are in danger zone. Therefore, to protect UWSN from DOS, wormhole and any other kind of attacks there's a need to develop a strong security protocol.

4 Applications of UWSN

The underwater sensor networks have proven to be a boon for many different underwater applications. UWSN had made it possible for humans to develop a new technological world below the water surface. The term new technological world is emphasized due to the rapid development of many different underwater applications like Underwater Datacenters, Tsunami Monitoring Systems, Aquatic Monitoring, Underwater Oil Field Discovery, Surveilling Water Territory of a Country via UWSN, Submarine Target Localization and many more. The development of different underwater applications is indirectly related to the development of a nation. Therefore, before starting any project with underwater sensor networks, it's a prerequisite to have complete knowledge about UWSN and its applications developed so far.

Aquatic Monitoring

In aquatic monitoring, an underwater acoustic sensor network is constructed which consists of distributed autonomous sensor nodes with limited transmission range. As shown in Fig. 9, sensor nodes are randomly deployed in the underwater environment and each node can only communicate with the neighbor nodes in its transmission range. For example, if the source node (say node 1), is monitoring the aquatic conditions. Now if the source node detects any interesting activity about aquatic life, then it tries to send that critical piece information to a sink node as soon as possible. Since sink nodes are outside of the source node transmission range, the data packet will be relayed by node 2, node 3 and node 4 until they reach the destination. Therefore, these types of applications require the network to be enabled of handling, multi-hope data communications, random network topology and the end to end delays.

Water Quality Monitoring Systems

"Water", one of the basic needs of mankind, and it's essential for survival. There are more than one billion people all over the world which do not have access to a safe water supply. In India, ground-water is being used as raw water for 85% of public water supply. So, the people in rural areas are always prone to loss of their lives due to different water-borne diseases. Therefore, it's the prime responsibility of governance to maintain the quality of drinking water to maintain the health of its citizens. To achieve this goal, in 2004, the concept of an underwater sensor network has been proposed to implement water quality monitoring systems. Such systems

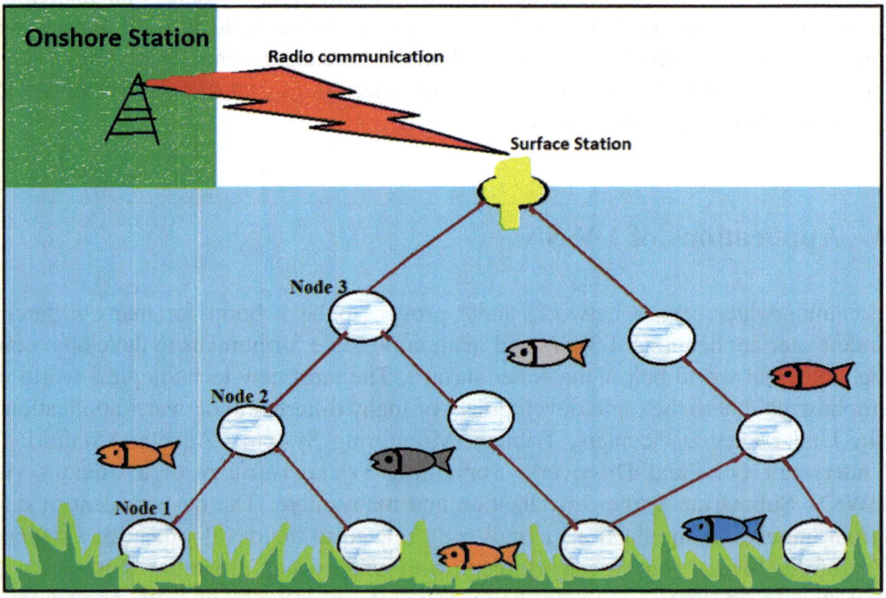

Fig. 9 Aquatic monitoring using UWSN

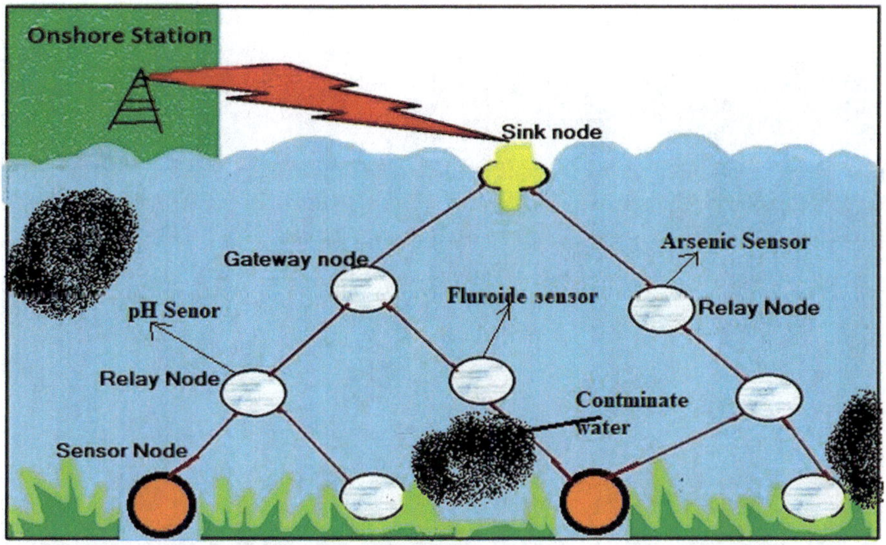

Fig. 10 WATER monitoring system

were named as Water quAlity moniToring sEnsors netwoRk (WATER). As shown in Fig. 10, WATER constitutes low costs water sensors for measuring fluoride, arsenic, pH and many other factors of water. These sensors are deployed at different depths and connected by a multihop wireless transmission medium to relay the sensed data. Water gets contamination due to chemicals, heavy metals, nitrate, fluoride, arsenic making water unpotable. Therefore, these contaminants can be measured using the WATER protocol. The water quality monitoring networks are one of the attacking objects because of their high importance to public healthcare. Therefore, it is necessary to build a water monitoring system that is secure as well as energy efficient.

Underwater Oil Field Discovery
The use of UWSN for discovering the underwater oil fields is one of the set examples to showcase the importance of underwater applications. And considering the socio-economic value of oil field discovery, it's necessary to encrypt all the communications among sensors that are used for discovering oil fields. As shown in Fig. 11, the ROVs are used as prime nodes in the application of underwater oil field discovery. ROVs are responsible to make a detailed map of the seafloor so that the subsea infrastructure and pipelines for oil fields are built and installed in most cost-effective manner. Also, the simultaneous operation of multiple ROVs with acoustic and optical sensors are used to detect mine-like objects. The first oil field discovery project, to start production in the Barents Sea off Norway was named "Galiot". In Galiot project, data about the number of oil wells, the discovery of new

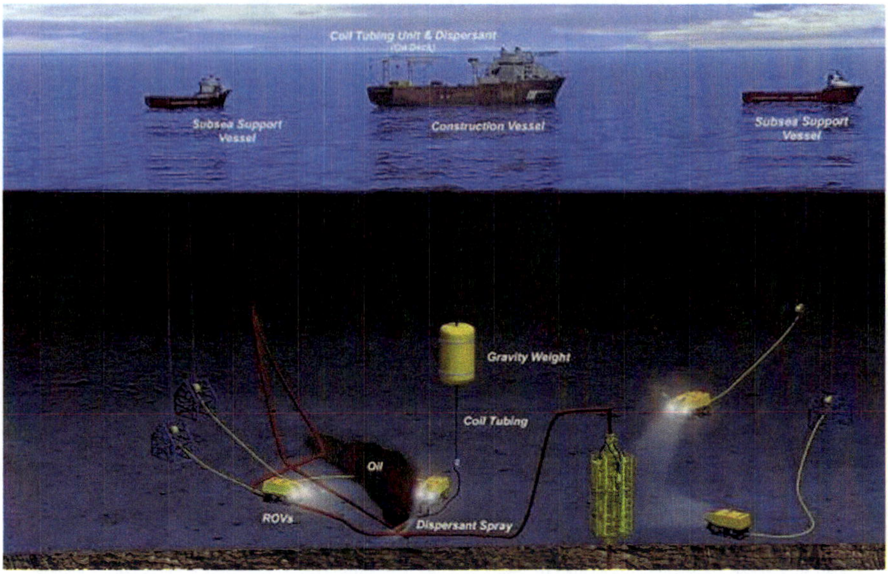

Fig. 11 Underwater oil field discovery [4]

oil mines, amount of oil storage, presence of any enemies to destroy mines can be sensed by sensors and then relayed to the offshore base station, therefore requiring secure data transmission in the underwater sensor network.

Microsoft Project Natick

The concept of "Cloud in Ocean", was the first time developed by Microsoft Research Project called Natick. The term "Cloud in Ocean" is used because as shown in Fig. 12, Natick had kept its datacenters in oceans. The main aim of the project was to reduce the cooling cost of data centers by keeping them in a natural cooling environment like "Ocean Water". Now, why data centers require a cooling environment. Let's understand it by taking the example of our laptops or CPUs of desktop computers, which starts generating heat after a while. The excess generation of heat leads to data loss. Now just imagine the amount of heat produced by thousands of hard disks kept in servers which process and stores the worlds digital data. Thus, companies like Google and Facebook have moved their data centers to cold countries than paying air conditioning bills. But Microsoft had found another option for data centers, which is underwater. As water has a natural cooling property at the bottom which is 60 °F and it is enough temperature for cooling of data centers. The success of the Natick project, makes researchers forecast the day when there will be many underwater data centers that will be wirelessly connected to the Internet. If these happen then there would be a great emergency of high-level security protocols for underwater data transmission.

Fig. 12 Conceptual illustration of underwater data centers [5]

Disaster Prevention
As shown in Fig. 13, the underwater sensors network is also used to measure seismic waves from remote locations. Thus, using UWSN we can pre-determine tsunami and can thus alert coastal areas. At the same time, disaster prevention systems can also keep volcano activities under surveillance. One of the examples of disaster prevention was in August 2011 where a 3-month-old volcanic eruption off the oxygen coast MBARTS seafloor was prevented. The frequent seismic monitoring is also of great importance in oil extraction industries.

Defense Application
As shown in Fig. 14, the underwater sensor networks are also used for some defense applications which include submarine communications, submarine detection, removal of sea mines, assisting navigation of battleships or locating submarines in enemy sea areas. The detection, classification, and tracking of underwater terrorist targets are the necessary components of underwater sensor networks used for defense. UWSN provides an advanced solution over traditional SONAR systems, which are impractical in some missions as they need to be mounted on ships. The UWSN detects the intruder using two steps (i) firstly, detect target on basis of radiation and mechanical sensors, (ii) then detected target is classified into one of the following target types based on the data coming from acoustic and magnetic microsensors whether it is a drive, a submarine or a mine.

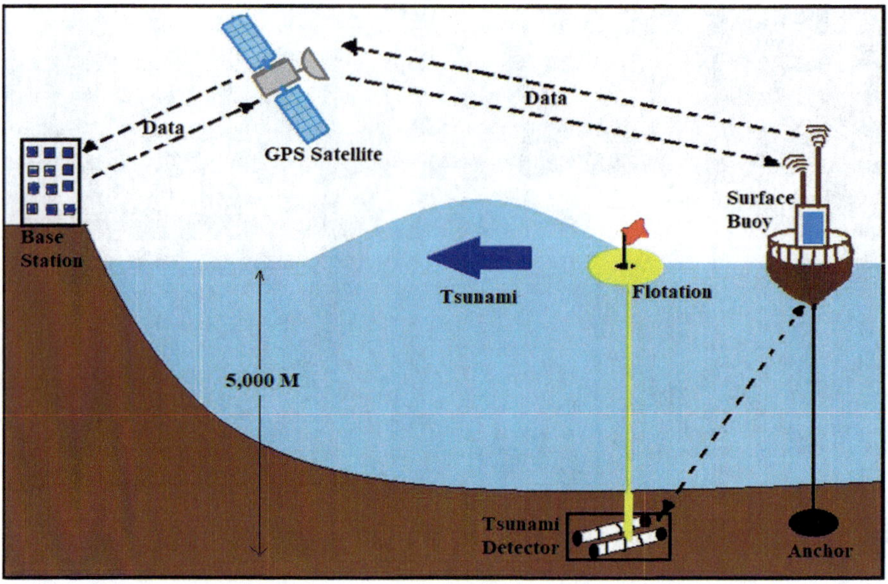

Fig. 13 Tsunami detection system

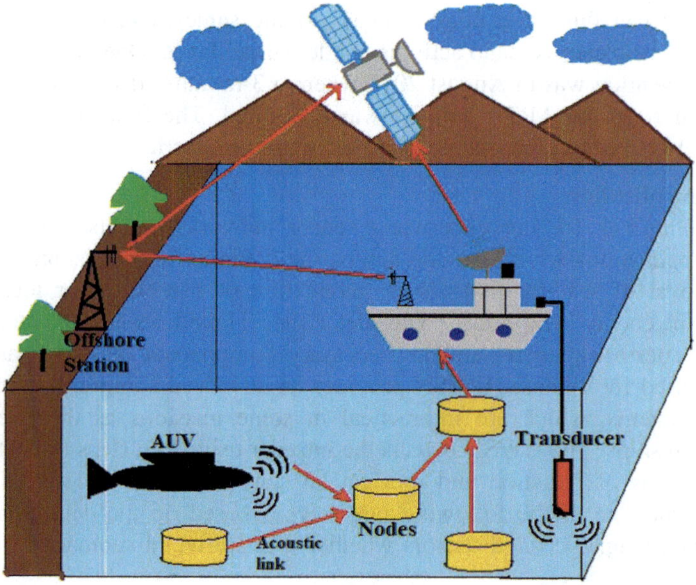

Fig. 14 Underwater surveillance network

For example, sea-web (Rice 2007) is a large-scale UWSN used for defense, which is being developed by the US Navy since the 1980s. AUVs and fixed underwater sensor nodes are collaboratively used for surveillance, reconnaissance, targeting and intrusion detection systems.

Submarine Target Localization
Submarine localization done using traditional methods had a major drawback of heavy computation. This drawback was overcome after the invention of UWSN based systems which was found to be an economical solution for submarine localization problems. The network with AUVs and some underwater sensor nodes were used to localize submarines. It was concluded that AUV provides the optimal solution to the localization of submarines with less manpower and less risk.

UWSN and IoT
The underwater networks can be connected to the outside Internet world using the "Internet of Underwater Things", (IoUT) [6]. The big data generated by underwater networks can also be managed using IoT [7]. This great combination of UWSN and IoT can create magic in different applications like Motorola projects in IoT [8] and to save the life of aquatic species by monitoring and preventing their health using the same concept of using IoT for human healthcare [9, 10].

5 Literature Survey

In spite of the unique characteristics of underwater sensor networks, the security aspect of underwater networks is less focused due to which UWSN is vulnerable to various malicious attacks. There are a few research works done to overcome the attacks occurring in underwater networks, out of which we are discussing some of the cream findings. The complete overview of all attacks to secure communication in an underwater sensor network is discussed by Han et al. [11], they have surveyed and then listed protocols that are used for secure communication in each layer for UASN. Table 4, shows a list of protocols that are proposed by Han for securing communication in each layer of underwater networks. It can be concluded from this work, that research on security for the underwater network is still in its infancy. This work has only focused on the interaction of nodes to detect the affected nodes. It doesn't consider message integrity and confidentiality.

Rafiqul Islam et al. [12] proposed a cluster-based USN architecture for application of surveillance of water territory of a country which consisted of many sensor nodes mounted at the bottom of sea forming clusters and among all nodes one node of each cluster is declared as cluster head, which is installed at the surface of the water. Each node of the cluster is equipped with multiple sensors to detect activities of an intruder entering into the territory of that node. Each member of the cluster reports the sensed data to the cluster head which then sends data to an offshore station via the wireless link. Thus, this research work intended to secure the communication between the cluster members and the cluster head or vice versa.

Table 4 Secure communication protocols in each layer [11]

Layers of UASN	Malicious attacks	Countermeasures
Physical layer	Jamming attack	Underwater Jamming Detection Protocol (UWJDP) [11], Jamming through Analog Network Coding (J-ANC) [11]
Data link layer	Power exhaustion attack, collision attack	Cluster-Based Secure Synchronization Scheme (CLUSS) [11]
Network layer	Flooding, sybil, wormhole, sinkhole and blackhole attacks	Distributed Visualization of Wormhole (Dis-Vow) [11], Wormhole-resilient Secure Neighbor Discovery (WSND) [11], Secure Routing Protocol and a set of Cryptographic Primitives (SRCP) [11]
Transport layer	Flooding attack and de-synchronization attack	Digital signature schemes for end-to-end authentication in UASNs and Key generation system (KGS) [11]

The proposed suite had two parts Authentication Key Exchange and Data Exchange. The scheme was resistible to many attacks but had a drawback of increasing network overload due to sending and receiving multiple messages for authentication and data exchange thus making the scheme insufficient for UWSN.

According to the work of Jiang et al. [13], their framework offers flexible and configurable features to provide security in underwater sensor networks. The framework serves authentication and confidentiality to messages by using both asymmetric and symmetric-based cryptographic techniques. It has built up a strong security model by implementing the concept of cross-layer protection from the link-layer to the application layer. The Galois Counter Mode was used as a mode of operation for symmetric-key cryptography while the asymmetric-key based authentication was done with short digital signatures which indirectly reduces the energy consumption in acoustic transmissions. The short digital signatures schemes which the author proposed where BLS, ZSS and Quartz. Simulation results showed that this framework is capable of meeting the security requirements of underwater networks at an affordable cost. But at the same time, the framework fails to deal with many other known attacks in the network layer.

Lal et al. [14] proposed two techniques which are Multi-Metric Reputation and Adaptive Monitoring systems. The Multi-Metric Reputation systems were able to achieve security by assigning a reputation value to each sensor node, where the reputation value was calculated by using the history of past transmissions like the total number of succeeded data transmissions by a node. The other technique named Adaptive Monitoring was used to detect the unknown underwater networks by the use of covert channels. But the proposed future solutions in the research work of Lal are remaining an issue that is widely open for future investigations.

Dini and Duca [15] proposed SeFLOOD (Secure FLOOD) protocol to protect the network against attacks that are performed during the neighbor discovery phase. Researchers have first briefly studied the previous network discovery protocol called FLOOD and then concluded it an unsecured FLOOD protocol. They have then developed a protocol named SeFLOOD (Secure FLOOD) which is an extended version of FLOOD as it can protect underwater networks against spoofing-based attacks. There is the four-step procedure in the SeFLOOD protocol (i) discovering neighbors, (ii) authenticating the neighbor, (iii) organizing cluster and finally (iv) distributing the cluster key. The SeFLOOD protocol is generally used in multi-hop network scenarios, where each node maintains a routing table. But this protocol is still experiencing spoofing attacks as the broadcasting of secret keys is not secure.

There are also some data transmission protocols in UWSN, which have used identity-based cryptography for protecting the transmitted data. Identity-based cryptographic schemes mean transmitting vehicles (AUV/ROVs/submarines) original identities along with secret messages, which makes data transmission among the vehicles vulnerable to attacks. Adversaries can hack real identities of vehicles to collect important information about them, like tracking the location of underwater vehicles. Thus Wan et al. [16] proposed two non-interactive identity-based data transmission protocols which are named as (i) Bilinear Map-Based Protocol and (ii) Algebraic Signature Based Protocol. Both protocols ensure that underwater vehicles should not transmit any surplus messages for key establishment and node authentication. The proposed scheme is secured to some extent but at the same time, it is also cost in-efficient in terms of energy consumption.

Souza et al. [17], research work is all about different types of digital signature schemes, to get an appropriate cryptographic technique for underwater networks. After a great state of the art, the authors have concluded that the short and aggregated digital signatures are the best solution for underwaters as they will be cost-efficient. Further, they have discussed different types of digital signature schemes which are (i) Elliptic Curve Digital Signature Algorithm (ECDSA), (ii) Boneh-Lynn Shacham Algorithm (BLS) [17] and (iii) Zhang-Safavi-Naini-Susilo (ZSS) [17]. The only drawback of this work was that the researchers have evaluated the cost of signature generations and have not considered the energy consumption of sensor nodes.

Luo et al. [18] proposed a key generation technique based on received signal strength for underwater secret communications. According, to research fellows the received signal strength- based key generation is a feasible solution as it eliminates the need to deploy an additional key distribution center, which was the major requirement of UANs. The RSS-Key Generation technique usually consists of 4-stages which are Signal Pre-processing, Key Extraction, Information reconciliation, and Privacy amplification. But like other symmetric key generators, an RSS-based scheme has no mechanism for device authentication, thus it requires the establishment of an initial secure link between communication entities.

Yuan et al. [19] work focused on matrix addition and multiplication for implementing the security protocols for underwater networks. The proposed scheme has used Vandermonde vectors in base station nodes for secret key generation and distribution. The use of this vector had decreased memory costs and eventually increased computational efficiency. The energy cost is also greatly reduced by simply replacing the matrix multiplications by matrix addition. A zero-knowledge proof protocol is also used to verify the paired nodes in the authentication and to also prevent the confidentiality of secret keys. The analysis showed that this scheme is computationally efficient and a less complex security solution for UWSNs. But still, the memory space and the processing capacity of underwater sensors give a major limitation on matrix computation.

The secret keys which are to be used for encryption and decryption, are always established before sending any messages. Such pre-generated secret keys are not secured if they are compromised. Hence, an alternative solution to generate dynamic secret keys is by using correlated channels. Thus Huang et al. [20] explored the concept of correlated channel measurements in underwater channels. They proposed a protocol that is capable of dynamically generating secret keys based on the Channel Frequency Response (CFR) in Orthogonal Frequency-Division Multiplexing (OFDM) systems.

Kulhandjian et al. [21] came with a new security approach where a sender and receiver can exchange the data in the presence of an eavesdropper. This magic is possible due to the concept of friendly jammers that used analog network coding. There is generally five-step procedure to send a message in presence of eve which is, (i) Allow simultaneous transmission of two signals, one from sender and another one from eve, (ii) Allow both signals to intentionally interfere with each other, (iii) Now, receiver first listens to interfered signal (i.e. corrupted signal) before doing any retransmissions, (iv) Then the corrupted signal is destroyed by receiver, (v) After signal suppression, receiver is now ready to decrypt the original signal. The evaluation results have shown that a higher bit error rate is achieved at the attacker node while only minimal error rate is achieved at the sender. Hence jamming through analog network coding is quite beneficial. But this approach does not work for all types of attacks.

An authentication protocol based on ticket generation was proposed by Yun et al. [22]. This protocol was able to reduce the size of data to be transferred and also the total number of data transmissions required. Such reductions make the protocol energy saver and thus friendly to an underwater environment. The researchers have used a gateway and base station to authenticate and re-authenticate the sensor nodes. Other than a gateway, a message sending the ticket and a reauthentication ticket are the two types of tickets used in this protocol. The message sending ticket is issued after normal authentication is done. And it's used to send the sensed data along with sensor ID whereas the reauthentication's tickets are sent by gateways or cluster heads to confirm the previous authentications. Hence according to studies, the ticket-based protocols can reduce the work of base stations and increase the data security by doing reauthentication.

The authors Martin and Rajasekaran et al. [23] have then used an architecture that is designed by keeping data at its paramount. Such data-centered architectures are only for analyzing and reducing security threats in underwater networks. This type of architecture was named Information-Centric Network (ICN) architectures which are specially designed to handle sensor node mobility due to water currents.

The job of ICN architectures is to reduce the injection of false data packets into underwater networks by making use of Named Data Networking (NDN) [23]. The NDN is all about a tree-structured naming convention used to handle all types of data. The use of NDN has proven to be a beneficial way to route data with minimum cost. The NDN and ICN techniques have also been succeeded in greatly reducing DoS attacks but have failed to overcome the advanced versions of malicious attacks. Till the date, lots of research work is still going on to protect underwater networks from malicious attacks.

Ahmed et al. [24] proposed a novel algorithm based on the theory of Dempster and Shafer, which worked on the concept of evidence. The DST [24] thus combines different evidence to identify a malicious node in underwater networks. This algorithm receives all the neighboring nodes parameters, to make a decision based on DST. Then the expected output of DST is the identification of malicious node with the help of abnormal parameters. The DST can find malicious nodes as their physical and transmission parameters are always different from the normal nodes. Simulation results produced different evidence and with the help of DST software, it was able to show whether a node is normal or abnormal.

Many UWSN applications like water monitoring systems, aquatic monitoring, oil pipeline, disaster prevention systems and many more are improving due to growing research in the domain of underwater networks. But all wireless sensor networks used in underwater applications have a drawback of the lower computational capacity of nodes. Now adding one more feather to advancement in research of UWSN, would be to make the design of underwater sensor nodes such that they are capable of communicating with each other in an underwater environment. Thus, Lima et al. [6] have achieved this motive of internal communication among underwater sensor nodes which is popularly known as ubiquitous scenario, similar to the Internet of Things (IoT). And they named such networks as "Internet of Underwater Things" (IoUT). The authors have tried to achieve intra-communication among sensor nodes same as the terrestrial internet networks. They have also proposed water ping, which is a header suppression method for ICMP and ICMPv6 protocol messages. These ICMP protocols are building blocks of IoUT. The proposed protocol also allows underwater sensor nodes to communicate by using echo request and echo reply messages, which is one of the most widely known internet applications. The results and analysis have shown that the proposed header suppression method is saving 93.75% of energy consumption and also underwater sensor networks could communicate with the internet.

The UWSN architecture consists of two types of nodes that are fixed and moving nodes. The fixed nodes are anchored to the bottom of water while moving nodes can move in a predefined space, such as ROVs and AUVs. Now, most of the research work done on UWSN has not considered nodes mobility during the development of protocols for underwater networks. The mobile nodes indicate that UWSN has dynamic topology. And one of the major drawbacks of dynamic topologies is poor communication conditions where data packets are not rapidly delivered to the sink node present at the water surface (i.e. buoy). Thus, dynamic topology is not a good house for various real-time underwater applications, which requires the time-sensitive data to be forwarded as quickly as possible. Therefore, authors Chang et al. [25], have first formulated the main three problems for data forwarding in underwater networks, which are (i) timeliness (ii) energy consumption required and (iii) mobility of nodes. For the above-stated problem authors have proposed a solution named "Reinforcement Learning". The solution given is an important type of machine learning concept where an agent (i.e. sensor nodes) automatically learns how to behave in an environment by observing the results. Finally, the performance of the machine learning technique was evaluated through simulations. And simulation results showed that the proposed method is a valid method for data forwarding along with the considerations of timeliness, energy consumption, and passive nodes mobility.

In today's date, many underwater applications are developed starting from underwater monitoring systems to underwater datacenters. All these applications are improving as the year passes and they are also generating a huge amount of complex data which includes pictures, videos, data from sonar and data from sensors. These, complex data generated at the bottom of the ocean are required to be transmitted to its destination which is at the water surface known as a buoy or offshore station. Therefore, for reliable transmission of data to surface, the authors Basagni et al. [26] have proposed a multi-modal communication system. This system allows two sensor nodes to communicate with each other via multiple devices at the same time. For example, if the multiple nodes embedded with acoustic and optical modems allow a way to combine short-range and high bit rate for data transfer along with long-range and low bit rate for data transfer. Thus, reliable data transfer can be achieved using multi-modal communication. The authors have also proposed a reinforcement learning technique due to which the nodes are trained to find the best possible route to reach the destination. Sensor nodes acquire and use the knowledge about link-quality to its neighboring nodes, reliability of link and multiple possible routes available to reach the destination. Therefore, the proposed multi-modal reinforcement learning based routing technique is along with the quality of service. It is abbreviated as MARLIN-Q. The SUNSET [27] simulation tool was used by the authors to evaluate the performance of proposed schemes and the result showed that MARLIN-Q can achieve a higher packet delivery ratio in the multi-modal scenario.

Now, various security protocols are developing for UWSN, so there is a need for a good simulator tool to test the project before its real-time deployment in water. Therefore, Petrioli et al. [27] proposed a Simulation and Emulation real-life Testing i.e. SUNSET [27] framework for underwater where implementation and testing of protocols for underwater networks can be done. It has enabled to test the underwater architectures and protocols in its pre-deployment phase. There are mainly two disadvantages of simulations, first that simulations cannot be the same as real-time environments and generally they also do not consider hardware features of the network. SUNSET is an open-source software that is based on the ns-2 network simulator. Thus, using SUNSET researchers and developers can perform quality work with cost-efficiency.

6 Analysis of Literature

Analysis of literature work is required so that the other researcher will get motivated to fill the gap between security protocols that are already developed for securing underwater data transmissions but are still not able to provide a complete security solution to underwater data transmission. This analysis will provide an answer to the question of "Why Underwater data transmission is not completely secured from attacks, instead of a large number of applications developing in underwater…??".

The below mentioned is a list of reasons why UWSN is not secured,

1. **Developing Protocols Based on Only Acoustic Communication**: Most of the protocols are developed based on acoustic waves, but other communication mediums like optical waves, can provide more bandwidth and security than the acoustic waves. The Table 5, shows that almost all the previous schemes are based on the only acoustic communication medium.
2. **Lack of Cross-Layer Design**: Resources of the network will soon get exhausted if security algorithms are developed against attacks in each layer. Hence, the cross-layer security design is a necessity. One should design the security system architecture from the whole network point of view. The Table 5, shows that 99% of previous work lack in providing cross-layer design.
3. **Communication Overhead due to the use of ECC Based Protocols**: The use of elliptic curve cryptography in underwater networks reduces the computational overhead but simultaneously increases communication overhead. Hence, there is a need to develop a new security protocol for underwater which should be more efficient than ECC.
4. **Key Distribution Mechanism**: The key distribution mechanisms for underwater communications are developed based on the concept of non-interactive protocols. The non-interactive protocols are developed in such a way that the data is sent from the sender to the receiver without sending any acknowledgment messages. These schemes are generally based on bilinear maps which require about 384 bits to exchange secret keys. Therefore, due to a large number of memory requirements of previous schemes, there is a need to develop a newer key distribution mechanism.
5. **SeFLOOD Protocol**: The secure communication suite developed for UWSN, aims to gain integrity, confidentiality, and authenticity to data being exchanged. But these protocols are still experiencing spoofing attacks because the secret keys used for encryption and decryption are not exchanged securely.

Table 5 Parametric analysis of literature survey

Ref. no.	Communication medium used	Memory required	Cross-layer design (present/absent)	Cost-efficient (Y/N)	Energy efficient (Y/N)	Protection from attacks
[11]	Acoustic	Large space	Present	No	No	Collision, sybil, wormhole and flooding attacks
[12]	Acoustic	Large space	Absent	No	No	All known attacks
[13]	Acoustic	Small space	Absent	Yes	No	Data confidentiality, integrity and authentication
[14]	Acoustic	Large space	Present	No	No	External or internal attacks
[15]	Acoustic	Larger space to store routing tables	Absent	Yes	No	Spoofing attack
[16]	Acoustic	Small space	Absent	Yes	No	Integrity and confidentiality
[18]	Acoustic	Larger space	Absent	No	No	Data confidentiality, authenticity and replay attacks
[19]	Acoustic	Large space for keys and Vandermonde matrix addition	Absent	No	Yes	Authentication
[20]	Acoustic	Large space to store keys	Absent	No	No due to the dynamic generation of keys	Secure key transmission
[21]	Acoustic	Large space	Absent	No	Yes	Jamming attack
[22]	Acoustic	Large space to store tickets	Absent	No	No	Authentication
[24]	Acoustic	Large space to store the DST algorithm	Absent	No	No	Malicious node detection

7 Methodology

The research work aims to define, design, develop, analyze and simulate the solution of the research problem. The goal is to promote a systematic approach to achieve proposed work in a distributed manner and research objectives in the context of Acoustic-Optical underwater sensor network architecture for the various underwater applications. In particular, this work defines the term ACOOP USN, as the future of underwater networks thus differentiating it from other existing methods in the context of the wireless sensor network. In order, for simplicity we further break the problem into four submodules, (i) the Architectural module (ii) Connectivity module (iii) Secure Data Transmission module (iv) Data aggregation software module.

Architectural Module

The first module is an architectural module which is for the development of Acoustic-Optical Underwater Sensor Network architecture. In these modules, the three main objectives will be (i) to find the position of each sensor into the network, (ii) to estimate the total distance required between two sensors and (iii) cluster formation of optical sensors to achieve high propagation speed. As shown in Fig. 15, the network will also make use of acoustic sensors for long-distance data transmission in water. Acoustic sensors are used as mediators for connecting cluster heads to the surface base station. Below shown is the pictorial view of the proposed ACOOP UWSN architecture.

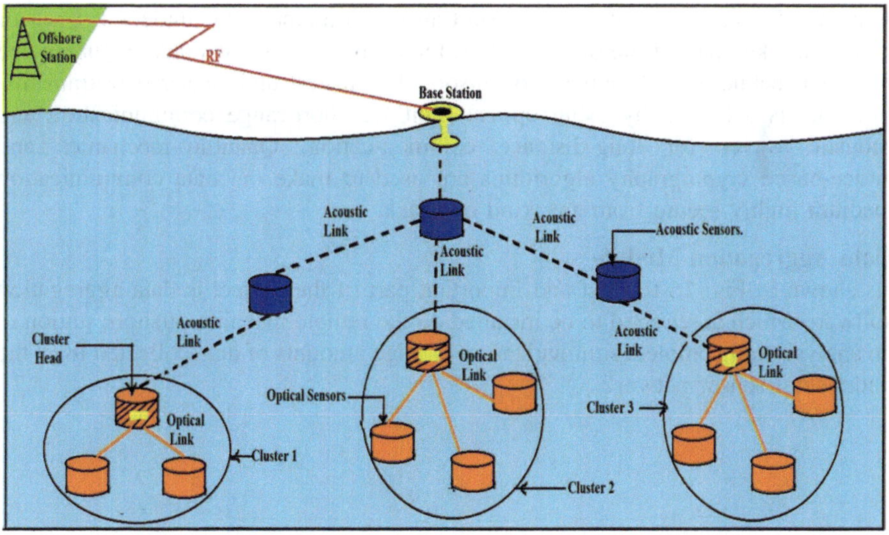

Fig. 15 Proposed ACOOP UWSN

Connectivity Module
The connectivity module is responsible for achieving the wireless connection between sensor nodes. As shown in Fig. 15, the connectivity module can be subdivided into three sub-modules (i) Optical Connectivity (ii) Acoustic Connectivity (iii) RF Connectivity modules.

1. Optical Connectivity Module:

The optical light beam is used for inter-cluster communication as optical sensors emit LED light which contains important sensed data. The propagation speed of the optical light below water is 10–20 m, therefore the maximum distance which we can keep between the two optical sensors is 20 m.

2. Acoustic Connectivity Module:

The acoustic waves are used for connecting the entire network to the base station. Acoustic waves are the connectivity bridge among all cluster heads and the surface base station. The acoustic wave propagation speed is 10–100 m, which is good for long-distance communication. Thus, the maximum distance which we can keep among the two acoustic sensors is 100 m.

3. RF Connectivity Module:

The last part of connectivity is an RF module which is required to connect the surface base station to an offshore station or the remote machine. This module can be implemented using WiFi or Zigbee technology.

Data Transmission Module
Accurate, Faster and highly secure data transmission among the underwater sensor node is the key part of the project. Thus, the accuracy of sensed data is guaranteed by using technologically advanced sensors. The fastest underwater data transmission rate is achieved by using optical light for short-range communication and acoustic waves for long-distance communication. Quantum-mechanics and lattice-based cryptography algorithms are used to make the data communication medium highly secure from any kind of attack.

Data Aggregation Module
As shown in Fig. 16 the last and important part of the project is data aggregation software which is required to be installed on the remote machine/offshore station to do analysis and complete summarization of huge amounts of data collected from the underwater sensor network.

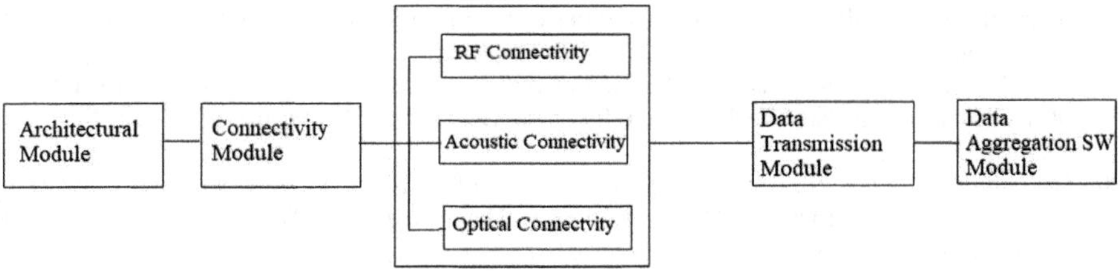

Fig. 16 Block diagram for proposed methodology

8 Conclusion and Future Work

The underwater sensor network has become a very wide area for research and it is fascinating many researchers to work on it. Every aspect within UWSN is in the study like UWSN Architectures, Communication mediums, Advanced Sensors for UWSN, Localization of sensor nodes, Synchronization among sensors and so on. But the security aspect of UWSN is still less focused. The rapid development in UWSN applications from all over the world suggests that security in underwater sensor networks should be given prime importance. The parametric analysis is done in the above section also showed that there is a need to develop a new high-grade security protocol which is resistible to all known attacks and is also a cost-efficient solution. In today's technological generation, it can be said as "Providing security to UWSN is like indirectly providing security to our country".

In our future research work, we aim to develop an innovative architectural model for the underwater environment which will run computationally efficient security algorithms by considering acoustic and optical waves as the mode of communication in UWSN.

References

1. Dey, N., Hassanien, A.E., Bhatt, C., Ashour, A.S., Satapathy, S.C. (eds.): Internet of Things and Big Data Analytics Toward Next-Generation Intelligence. Springer, Berlin (2018)
2. Bhatt, C., Dey, N., Ashour, A.S. (eds.): Internet of Things and Big Data Technologies for Next Generation Healthcare (2017)
3. Kimbahune, V.V., Deshpande, A.V., Mahalle, P.N.: Lightweight key management for adaptive addressing in next generation internet. Int. J. Ambient Comput. Intell. (IJACI) **8**(1), 50–69 (2017)
4. https://raymondpronk.files.wordpress.com/2010/05/oil_spill.jpg
5. http://theliquidgrid.com/2017/10/24/underwater-data-centers/
6. Lima, F.H.M.B., Vieira, L.F.M., Vieira, M.A.M.: Water ping: ICMP for the internet of underwater things. Elsevier 1389-1286(19)30024-6 (2019)
7. Sarowar, M.G., Kamal, M.S., Dey, N.: Internet of Things and its impacts in computing intelligence: a comprehensive review–IoT application for big data. In: Big Data Analytics for Smart and Connected Cities, pp. 103–136. IGI Global (2019)

8. Schaller, A., Mueller, K.: Motorola's experiences in designing the internet of things. Inte. J. Ambient Comput. Intell. (IJACI) **1**(1), 75–85 (2009)
9. Dey, N., Ashour, A.S., Bhatt, C.: Internet of things driven connected healthcare. In: Internet of Things and Big Data Technologies for Next Generation Healthcare, pp. 3–12. Springer, Cham (2017)
10. Elhayatmy, G., Dey, N., Ashour, A.S.: Internet of Things based wireless body area network in healthcare. In: Internet of Things and Big Data Analytics Toward Next Generation Intelligence, pp. 3–20. Springer, Cham (2018)
11. Han, G., Jiang, J., Sun, N., Shu, L.: Secure communication for underwater acoustic sensor networks. In: IEEE Security & Privacy in Emerging Networks in 2015
12. Rafiqul Islam, Md., Azad, S., Monzur Morshed, Md.: A secure communication suite for cluster-based underwater surveillance networks. IEEE Trans. Inf. Commun. Technol. (2014)
13. Jiang, J., Capossele, A., Gjanci, P., et al.: SecFUN: security framework for underwater acoustic sensor networks (2015). IEEE 978-1-4799-87368/15/
14. Lal, C., Petroccia, R., Conti, M., Alves, J.: Secure underwater acoustic networks: current and future research directions. In: IEEE NATO STO Centre for Maritime Research and Experimentation, La Spezia, Italy (2016)
15. Dini, G., Duca, A.L.: SeFLOOD: A secure network discovery protocol for underwater acoustic networks (2011). IEEE 978-1-4577-0681-3/11/
16. Wan, C., Phoha, V.V., Tang, Y., Hu, A.: Non-interactive identity-based underwater data transmission with anonymity and zero-knowledge. IEEE Trans. Veh. Technol. (2015)
17. Souza, E., Wong, H.C., et al.: End-to-end authentication in under water sensor networks (2013). IEEE 978-1-4799-3755-4/13
18. Luo, Y., Pu, L., Peng, Z., Shi, Z.: RSS-based secret key generation in underwater acoustic networks: advantages, challenges, and performance improvements. In: IEEE Underwater Wireless Communications and Networks: Theory and Application (2016)
19. Yuan, C., Chen, W., et al.: A low computational complexity authentication scheme in underwater wireless sensor network. In: IEEE in 11th International Conference on Mobile Ad-hoc and Sensor Networks (2015)
20. Huang, Y., Zhou, S., Shi, Z., Lai, L.: Channel frequency response based secret key generation in underwater acoustic systems. IEEE Trans. Wirel. Commun. (2016)
21. Kulhandjian, H., Melodia, T., Koutsonikolas, D.: Securing underwater acoustic communications through analog network coding (2014). IEEE 978-1-4799-4657-0/14
22. Yun, C.-W., Lee, J.-H., Yi, O., et al.: Ticket-based authentication protocol for underwater wireless sensor network. In: IEEE in Development of the Wide-Band Underwater Mobile Communication Systems (2016)
23. Martin, R., Rajasekaran, S.: Data centric approach analyzing security threats in underwater sensor networks (2016). IEEE 978-15090-1537-5/16
24. Ahmed, M.R., Aseeris, M., Shamim Kaisert, M., Zenia, N.Z., Chowdhury, Z.I.: A Novel Algorithm for Malicious Attack Detection in UWSN (2015). IEEE 978-1-4673-6676-2/15
25. Chang, H., Feng, J., Duan, C.: Reinforcement learning-based data forwarding in underwater wireless sensor networks with passive mobility. MDPI Sens. (2019). https://doi.org/10.3390/s19020256
26. Basagni, S., Di Valerio, V., Gjanci, P., Petrioli, C.: MARLIN-Q: multi-modal communications for reliable and low-latency data delivery. Elsevier 1570-8705(18)30555-9 (2018)
27. Petrioli, C., Petroccia, R., Potter, J.R., Spaccini, D.: The SUNSET framework for simulation, emulation and at-sea testing of underwater wireless sensor networks. Ad Hoc Netw. Spec. Issue Adv. Underw. Commun. Netw. (2014)

Rethinking Decentralised Identifiers and Verifiable Credentials for the Internet of Things

Parikshit N. Mahalle, Gitanjali Shinde and Pathan Mohd Shafi

Abstract In digital era of Internet of Things (IoT), internet-connected devices are becoming smart and autonomous. In this distributed world IoT device operate not only with the owner of the device but also need to operate with third parties, which leads to have identification for each IoT device. In the coming years, deploying IoT solutions at large scale will have to address complex issues of managing digital identity, security and identity ownership. Most of the current IoT solutions rely upon centralized client-server mechanism for providing connectivity and access to devices and services. However, when billions of devices will begin connecting simultaneously, then existing centralized solution will not suffice. Now days, we rely on federated identities however in this user need to depend on third party as users are not owner of their own identity, this leads to have Self-Sovereign Identities (SSI). In the sequel, there is a need of decentralized method for managing identities in near future. In this chapter, various identification methods are analyzed. Further, Decentralized Identifiers and Verifiable Credentials are discussed, the discussion focus on whether DID and VC are applicable to the IoT as IoT includes resource constrained devices at base level. It also presents smart home use case utilizing DID and VCs in order to evaluate its applicability. In the same line threat analysis is presented to analyze applicability of the DID and VC for the IoT applications.

P. N. Mahalle
Senior Member IEEE, Professor & Head of Department of Computer Engineering, Smt. Kashibai Navale College of Engineering, Pune, India
e-mail: aalborg.pnm@gmail.com

PostDoc Researcher, Center for Communication, Media and Information Technologies (CMI), Aalborg University, Copenhagen, Denmark

G. Shinde
Assistant Professor, Smt. Kashibai Navale College of Engineering, Pune, India
e-mail: gr83gita@gmail.com

P. M. Shafi (✉)
Professor, Department of Computer Engineering, Smt. Kashibai Navale College of Engineering, Pune, India
e-mail: shafipathan@gmail.com

N. Dey et al. (eds.), *Internet of Things, Smart Computing and Technology: A Roadmap Ahead*, Studies in Systems, Decision and Control 266,
https://doi.org/10.1007/978-3-030-39047-1_16

1 Introduction

In the era before IoT, the world wide web, internet, web 2.0 and social media made people's lives comfortable by providing web services and facility to access personal data irrespective of their location. Further, to save time and improve efficiency, there is a need for the machine to machine communication, automation and ubiquitous access to personal devices. This need gave birth to the phenomenon of Internet of Things (IoT). In the IoT, internet-connected devices are becoming smart and autonomous. IoT defined in many various ways in the literature.

As per CASAGRAS Eu project [1], IoT is a

> A global network infrastructure, linking physical and virtual objects through the exploitation of data capture and communication capabilities. This infrastructure includes existing and evolving Internet and network developments. It will offer specific object identification, sensor, and connection capability as the basis for the development of independent cooperative service and applications. These will be characterized by a high degree of autonomous data capture, event transfer, network connectivity and interoperability.

International Telecommunication Union (ITU) defined IoT as,

> A global infrastructure for the information society, enabling advanced services by interconnecting (physical and virtual) things based on existing and evolving interoperable information and communication technologies [2].

According to the Internet Engineering Task Force (IETF) [3], IoT is,

> The basic idea is that IoT will connect objects around us (electronic, electrical, non-electrical) to Provide seamless communication and contextual services provided by them. Development of RFID tags, sensors, actuators, mobile phones make it possible to materialize IoT which interact and cooperate each other to make the service better and accessible anytime, from anywhere.

In the literature, many authors have also defined IoT with different aspects. According to Dorsemaine et al. [4], IoT is a

> Group of infrastructures interconnecting connected objects and allowing their management, data mining and the access to the data they generate.

According to Botterman [5] IoT is,

> Things having identities and virtual personalities operating in smart spaces using intelligent interfaces to connect and communicate within social, environmental, and user contexts.

Authors Rayes and Salam [6] defined IoT as,

> The network of things, with clear element identification, embedded with software intelligence, sensors, and ubiquitous connectivity to the Internet.

Despite such different definitions of IoT, it consists of mainly three basic elements. First, things (devices), second, data (information), and third, communication technologies. These three main pillars of the IoT along with their uses are depicted in the Fig. 1.

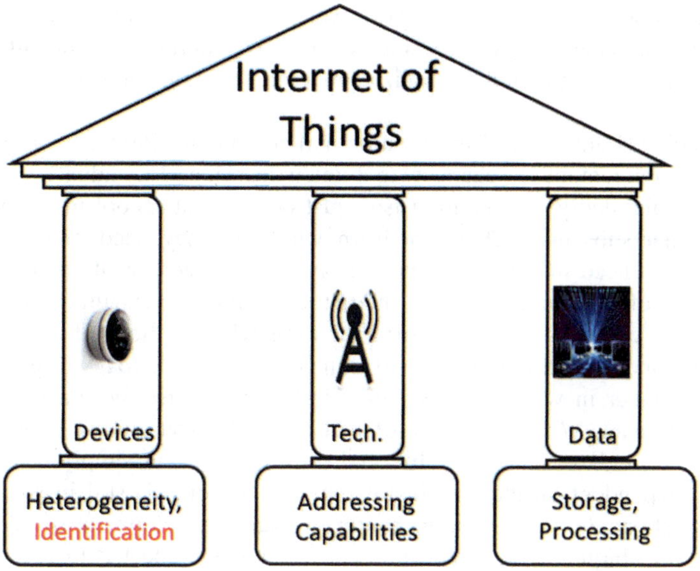

Fig. 1 Pillars of IoT

The first basic element-Things, it could be any physical/virtual elements around us, i.e. sensors, actuators, objects with RFID tags, objects in the home, office so on. The Second element—Information, is the parameters sensed by the IoT devices, which may be an environmental parameter or user-health parameter and any other depending on the domain. These parameters must be converted to appropriate services to provide its benefit to the user. The third element of IoT is the communication technologies which connect devices to the user.

1. Things: Devices/objects for collecting/sensing information. Devices are heterogeneous in size, computational capabilities, and energy source and having a unique identification, sensing and communication capabilities.
2. Data: Data is the information collected from IoT devices and provided for further process. On the basis of processed data IoT services are defined to make users life better.
3. Communication Technologies: Communication technologies are the backbone of communication in the IoT. They are the interface between the user and the device to obtain the advantages of IoT services. These are heterogenous in structure, energy demands, communication range and addressing capability.

IoT device operate not only with the owner of the device but also need to operate with third parties, then the identification becomes a major problem. In the coming years, deploying IoT solutions at scale will have to address complex issues of managing digital identity, security and identity ownership. Most of the current IoT solutions rely upon centralized client-server mechanism for providing connectivity and access to devices and services. However, when billions of devices will begin connecting simultaneously, then existing centralized solution will not suffice. In the sequel, there is a need of decentralized method for managing identities in near future.

The quick advancement of information and communication technologies (ICTs) has taken arrangement of changes to human learning. Individual's expectation adapting uninhibitedly, the learning isn't just occurred at school in the meantime, with a similar substance. They can learn whenever they need, think about any substance they need unbounded time and spot. In the course of recent years, the advancement of communication and internet accomplish every things to reality [7, 8]. The future of IoT is more intriguing than this where billions of things will talk one another and human mediation will turn out to be least. IoT will get full scale move the manner in which we live and work. AI changes the machine comprehension to human, With the ability of managing Big Data, those application are getting to be fundamental to our life and those application simply can make the prediction dependent on the past history, what's to come is still in some way or another questionable, that is the enchantment. Rapid development of IT promotes the advent of ubiquitous age. The Internet of Things, expected to become exponentially throughout the following half decade, will produce the basic information that AI Model need to automate industry. So the future acknowledgment of IoT's guarantee is reliant on AI to discover the patterns, correlations and anomalies that have the capability of empowering upgrades in pretty much every aspect of our day by day lives. This seminar survey the application related to Convergence of IOT, Machine and Deep learning [9–11]. IOT based Deep/machine learning model will conquer major limitation of IOT model. This Seminar also brief about machine learning algorithm like unsupervised and supervised algorithm also Deep learning algorithm like neural network. In this seminar we had survey some basic application related to IOT and deep learning related to healthcare, Utilities (energy, water and gas), Manufacturing, Consumer goods and retail, Transportation and Agriculture. We also survey about approaches to secure the communication between IOT devices. We study Machine to machine communication using Fuzzy approaches and how deep learning or machine learning technique used to manage the trust between machine to machine communications.

We survey applications based on 1. Application area such as to healthcare, Utilities (energy, water and gas), Manufacturing, Consumer goods and retail, Transportation and Agriculture, 2. Solution or services, 3. Sensor and devices, 4. Machine and deep learning technique, 5. Interface, 6. Outcome. We study technique used in google Assistant, Jarvis, Crotona, Google self-drive car and Alexa. After using IOT, deep and machine learning in healthcare application we get lower cost treatment/care, Improve patient outcome, Real-time diseases management, Detecting diseases in early stage of it. In Transport application we get improved service level, Lower cost and lower the carbon footprint. In Consumer goods and retail we can create a novel value-added application for consumer like alert on expiry date etc. And we can make application for only targeted advertisement. In Utilities application Consumers connected to smart networks have seen significant cost and resource savings [12, 13]. In industry like manufacturing when we use IOT machine and deep learning we can reduced field support costs, lower breakdowns, improved operational efficiency Optimal scheduling of production lines Anomaly detection and emission detection improved quality and lower energy costs. In agriculture application we can detect diseases in early stage and can find the solution of it also solution of diseases can be automated. Over all we can conclude that deep learning, machine learning and IOT can be used in various field or industry to improve the quality, Performance of previously made model and develop an application/system using technique mention above which will make life easy and world smarter [14].

The analytic technique/machine learning technique/deep learning technique which can be used in above application can be Patten recognition, anomaly detection, historical correlation, visualization, prediction technique, optimization technique, Decision support system, Classification Technique.

IOT model + Deep learning and machine learning technique = Smarter world. Internet Of Things and AI (Machine or deep learning) can solve many real life problem by just analyzing the data come from IOT devices and after analyzing that prediction/result can again be used by IOT devices to give outcome [15].

The main goal of identity systems is trust establishment and online identity enable secure and private interactions. In nutshell, there are three models of identity. The basic model is centralized identity where user has account at different organizations based on user id and passwords. This centralized identity model standards like HTTPS, SSL and TLS are used for security. The second type of model is federated identity model where service provider is there in the middle between user and organization and service provider federate user identity to different organizations like logins to social networks. Standards used in federated identity model includes SAML [16], OpenID Connect etc. [17]. Now days, we rely on federated identities where we use Internet giants like Facebook and Google as trusted third party and we can login to other service providers using identity on their site. In federated identities, online identity is owned by service provider and not by user.

Subsequently, for example if one of the service providers which you considered as trusted third party decides to terminate user's account, then users will lose access to and their identity all the service provider that user have consented to rely on trusted third party. This shows that users do not have ownership of their identity, and therefore the concept of third identity model i.e. self-sovereign identities (SSI) comes in [18]. Christopher Allen, Co-author of TLS Security Standard, defined SSI as:

> Self-sovereign identity is the next step beyond user-centric identity, and that means it begins at the same place - The user must be central to the administration of identity.
> —C. Allen [18]

The meaning of self-sovereign identity means individuals or organization can create, manage, assert and self-administer their identity without relying on a third-party identity provider. Every peer, you share SSI with, you have a connection and peer includes an app or connected device. As per the recent move of European Commission's move on General Data Protection Regulation (GDPR) which only protects individual human being [19, 20] states that user should have more control on their personal data. It also addresses the problem of current federated identity solutions and states the requirement that user should have privilege to approve what information is shared and have the ability to delete it.

With SSI, each user has control on his own identity and its related data. SSI is verifiable and enables users to make claim about them. These claims can be validated and signed by trusted provider. In SSI, user becomes self and central administrator of their identity and it requires user's identity to interoperable across multiple platforms and user to be autonomous. In SSI, users have complete ownership of their digital identities and control over sharing and using their personal data. In this case, users can also provide selective disclosure of the data enabling security and flexibility. Identities are persistence and transportable and users must agree to use their identity in SSI. In addition to this, users do not require an intermediate entity to present the claims. The bigger picture of SSI is depicted in the Fig. 2 along with the ten guiding principles as stated in [8].

Let's consider the example of OpenID connect [17] for which Google is one of the biggest client which helps you use your Google login to other service providers. This concludes that, Google also knows what service you use from other service

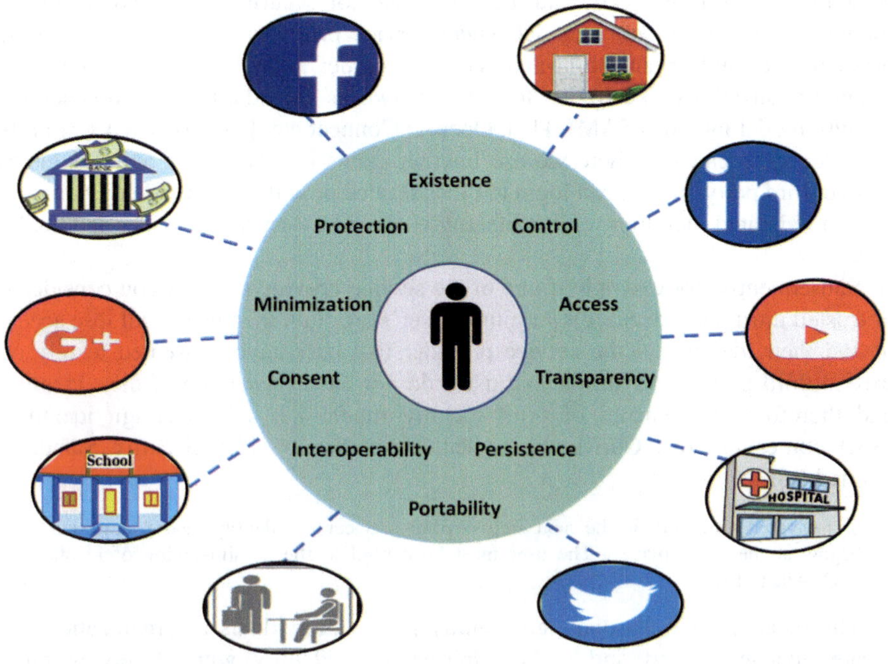

Fig. 2 Bigger picture of SSI

providers along with the Google services. Decentralized identifiers (DIDs) are fundamental building blocks of SSI. DIDs are globally unique and gives lifetime encrypted private channel with another person, organization or device. There is no central registration authority for DID. However, every DID is registered directly by user on public or private blockchain or distributed network. They are persistent, resolvable and cryptographically verifiable identifiers i.e. ownership can be proved using cryptography and decentralized in nature. They are used in applications benefiting from self-administered, cryptographically verifiable identifiers such as personal, organizations or Identifiers for IoT scenarios. Unlike A Universally Unique IDentifier (UUID) [11] and Uniform Resource Names (URNs) [12], DIDs have unique ability to cryptographically verify ownership of the identifier. A credential is certified container of attributes issued by issuer to a user and usually presented a proof of something about ourselves. It includes government ID, Passport, University degree etc. DID can be defined as:

DID is a portable new type of URL like identifier that is associated with an entity. DIDs are often used in credential and linked with recipients. It is globally unique, highly available and cryptographically verifiable.

The processing of DIDs is done by Decentralized Public Key Infrastructure (DPKI). In PKI, trusted third party is involved to issue certificate and it is used as identity proof. There is no trusted third party involved in DPKI and therefore it is decentralized in nature. DPKI uses DID instead of certificate as a proof of identity. There are manly two DID-based roles. First role is the DID subject which is an individual or device that a given DID identifies. The second role is DID owner or identity owner the individual or entity who holds and controls the private keys associated with that DID. DID infrastructure represent global key-value database in which the database is distributed networks, key is DID and value is DID document. The main goal of DID document is to describe the public keys, authentication protocols, and service endpoints necessary to bootstrap cryptographically-verifiable interactions with the identified entity. The format of a DID is presented in the Fig. 3. First part represents scheme and second part represent DID method which specifies how DID work on particular decentralized network. DID method specification define operations for particular target system which include create (DID entry), delete (DID entry), read (DID document) and update (DID document).

However, there are some problems associated with credentials like they can be easily forge, easy to impersonate the true owner, can be lost and cannot be scale

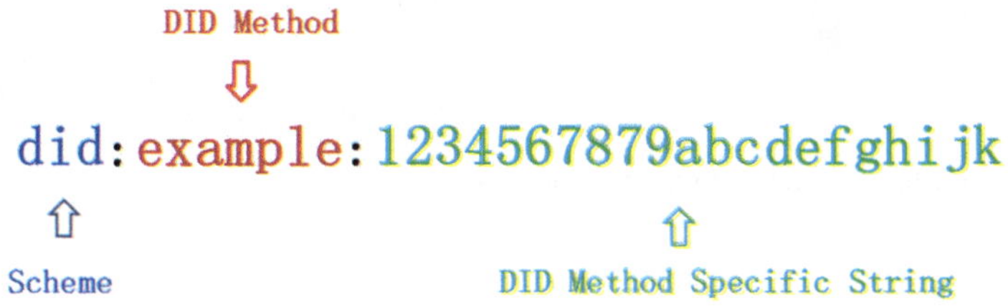

Fig. 3 Format of a DID

easily. It is also very difficult to verify the authenticity of these credentials online. To address these issues, we need verifiable credentials (VCs) [21] defined as a digital attestation of one identity owner to another. Digital VCs overcome all the problems of credential mentioned above. In addition to this, important and unique feature of DID and digital VC is zero knowledge proof (ZKP). ZKP is kind of like proving that something is true without revealing the thing we are proving. In simpler words it is a method by which one party (the prover) can prove to another party (the verifier) that they know a value "x", without conveying any information apart from the fact that they know the value "x". It is like proving you're over 21 without revealing your birthday or even your age.

VC uses decentralized public key infrastructure and has following three roles:

- Issuer: They create and issue credentials.
- Holder: They receive, hold and share credentials with verifiers.
- Verifier: They receive and verify proofs from holders.

The relationship between an entity which can be individual, organization or IoT device, DID and VC is presented in the Fig. 4. Entity can create, control, reveal and destroy any number of identifiers. Any identity owner can create verifiable claims. Creating verifiable claims means attaching attributes to any other identity. To achieve the vision of SSI, the DID-based solution consists of three architectural layers as shown in the Fig. 4.

Decentralized identifiers is the foundation layer which enables self-administered identifiers which any entity can create and control their own identifiers. Verifiable credentials is second layer which enable entity to store and control its digital credentials which includes passport, driving licenses etc. Third layer is decentralized identity management which enables entity control access to their verifiable profile using software applications and digital wallets.

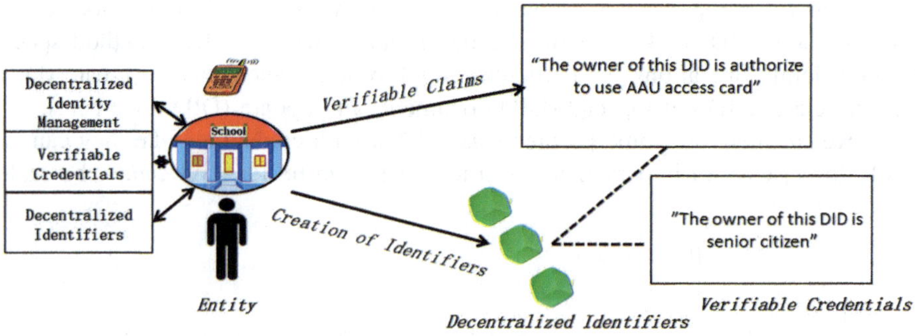

Fig. 4 Relationship between entity, DIDs, and VC

2 SSIs, DIDs and VCS for IoT

Looking at the identifier of IoT devices, the privacy related questions can be raised like can device be persistently identified by the identifier, does identifier reveal any information about user or device and whether the Identifiers used in different contexts provide correlation for the user's data across contexts. These privacy related issues will be exploited each device has only a single identifier that is used for all services. Looking at the threat analysis and cyberattacks in IoT [22, 23], IoT device identifier must be persistent, attestable and unique.

Today, IoT device identifiers are not unique and persistent and are never attestable. The device identity can be cryptographically verified using attestation which will help to reduce the risk of IoT device being impersonated. In IoT context, IoT devices may also have DIDs associated with them and they also may be DID subjects. However, the private keys for these DIDs need to be controlled by an individual or organization acting as DID owner or identity owner. These private keys will also be valid for the IoT devices associated with that DID. Like in the human notion, children or disabled people need caretaker or guardian relationship to someone who has self-sovereignty. In the similar way, IoT devices like bots, sensors or digital agents will need a guardian-type relationship who has self-sovereignty.

2.1 Smart Home—DID-Based Use Case

This use case presents mechanism to optimize home services. Smart home is equipped with multiple smart things which includes intelligent lighting system, wall, chair, floor etc. and they interact with each other at different levels. Different types of sensors and devices using multiple wireless communication technologies like ZigBee, Wi-Fi and Bluetooth are also deployed in the smart home for monitoring and notification purpose. The data generated from various sensory sources is aggregated at central place and used for notification and defining alarms for users. Location-aware services are also provided by the smart home. Monitoring service running at home gateway is continuously monitoring the groceries kept in the refrigerator using pressure sensors and ultrasonic sensors. Mother is a working woman and after her office work, she is visiting a super market which is on the way to home. The moment, she entered into the super market, she receives notification on her smart device that few items are under stock and recommended to buy. Here the location is considered as a context and mapped to the grocery requirement by monitoring and notification service running at the home gateway. Here, mother can bring in the concept of "Bring your own identity" i.e. DIDs while buying the products. Mother buys the grocery items and while checking out, she just provide her DID to online super market using the app or plugin installed on her smart device. This way, mother would be able to share payment information without even having to create an account with the super market's website. Mother can also link her customer loyalty card which we get in super market to avail small discounts on purchasing to the ID field of her DID. Smart home can also maintain distributed address book in order to stay connected with your close friends, relatives and businesses, vendors who provides customer care services to you.

In this case, DIDs enable lifelong connections with DID owners securely and will help you to share your personal address and other profile information with full control and data portability. Whenever, there is change in your profile information, DID will help you to notify this to all your connections automatically. In addition to this, ZKP property of DID helps to make verifiable claims without burdening the relying party and keeping your data secure. This can help IoT to use biometric, location or other contextual data to establish between user, organizations and devices.

3 Threat Analysis

However, IoT includes constrained devices and the practical limitation of applying DIDs with IoT devices is still a matter of discussion. For capable devices, there is no problem. However, the tiny devices lack with necessary processing power, energy and even the memory to apply DIDs. Delegation and access policy management in constrained IoT environment is presented in [24]. The printing use case is referred to demonstrate the functioning of proposed work. Authors claim that flexible authorization and management of access control policies can also be achieved using decentralized identifiers and verifiable credentials. IoT device with hardware identifier or any other permanent identifier interacting with third parties helps attackers to facilitate tracking of the devices resulting into information leak of device's owner and user. Application of DIDs to IoT devices to reduce the risk on privacy is presented and discussed by Kortesniemi et al. [25]. Authors claim that, DIDs are also suitable for constrained devices. However, an application of DIDs to IoT needs thorough threat analysis to conclude whether DIDs are right candidate for the IoT devices.

DID-based smart home use case described above uses capable device to create and process DIDs. Use of DIDs and VCs on constrained IoT devices requires sufficient processing power for cryptographic operations, enough energy to support these operations, permanent storage space for keys and code and sufficient entropy source to generate cryptographic keys. All devices and device-to-device communication in IoT mostly rely on the wireless communication technologies. These wireless mediums are also vulnerable to privacy and security attacks [26].

DIDs face few challenges like distributing public keys, keeping changes made to the keys and processing all the other information. To handle these challenges distributed ledger or blockchain for public DIDs are used. Blockchain technology is promising technology and the purpose is to build a secure application with trusted timestamping service that uses the decentralized blockchain to store anonymous, tamper-proof time stamps for any patient's data. This service allows patients to hash records, and subsequently store the created hashes in the blockchain as well as retrieve and verify time stamps that have been committed to the blockchain. Across the healthcare spectrum, there is a great deal of uncertainty and unfamiliarity when it comes to how blockchain technology could be applied. While most would agree that the industry can benefit from leveraging blockchain in various ways, one recent report points to how it's potential in precision medicine is gaining traction. The report, from New York City-based consultancy Frost & Sullivan, examines the global healthcare crisis, looking closely at chronic disease, aging populations, and rising costs, while exploring the challenges inherent in precision medicine—including the big data conundrum, interoperability, cyber-security, value-based care reimbursement, data ownership and healthcare consumerism, however, limitations to using conventional cryptographic primitives and access control models to address security and privacy concerns in an increasingly cloud-based environment. These DIDs are not public but private, and to get access of these for processing purpose specific channels are used in order to distribute the information. The current DID technologies includes Sovrin [27], verses One [28] and uPort [29]. All of these technologies use distributed ledgers or blockchain to process DIDs. However, as per the guidelines issued by EU GDPR [19, 20], it states that use of distributed ledger and blockchain (also known as nonmutable platform) for storing personal records might be very risky. Hence Sovrin and Verses One have taken all personal records off the storage. This course of action can be taken towards DIDs for IoT devices if they try to reveal owner's personal identification.

Distributed ledgers will allow IoT applications to create verifiable records for transaction. It provides a platform to build trust without centralized entity such as cloud which has several performance and security issues. Due to high number of transactions in IoT applications, the use of distributed ledgers would demand for high-throughput design, however the constrained IoT devices will be the major bottleneck. Constrained IoT devices are not capable of performing computational-intensive and storage-intensive tasks. Ubiquitous deployment of IoT devices makes them vulnerable to key leakage (if keys are stored on the device) or tampering unless devices utilizes solution like trusted platform module. To summarize, application of distributed ledgers to IoT application may require different design not only at application level but also at device level.

For any application of DID and blockchain to IoT, there are two main points to focus, first is storage capacity and second is scalability. As days are passing the numbers of IoT devices participating in operations are increasing at an alarming rate. These days will give more number of DIDs to be involved, which will increase the growth of chain at a rate of 1 MB per block every 10 min [30]. This chain will be there on every device in network participating in operations. In cases like constrained IoT devices the storage requirement will be the major bottleneck to be handled. Yet, the proxy-based is proposed by Kortesniemi et al. [25], where the capable proxy devices are used for storage instead of constrained devices in order to process of DIDs and distributed ledgers or blockchain. These proxy devices handle all complex operations (like public-key cryptography) for constrained devices.

However the evolvement of these proxy protocols and their analysis will remain the major issue to be resolved.

Even though, DIDs enable decentralized identity management in IoT context, the problem of anonymity and data privacy is serious problem. Consider the IoT-based healthcare scenario, where the device is linked with the person and anonymity needs to be guaranteed. When wearable devices send personal healthcare data, they should have ability to hide the identity of person who is owner of the data. During the use of DIDs, there is no direct relationship between entity and wallets, anonymity seems to be compromised.

Latency plays a crucial role while designing of any IoT application. In Block chain operations, after verifications are done the blocks are added to the chain. The time taken by these blocks will differ depending on the implementation in the context. If way refer to more specific example the time taken by Bitcoin verification is 10 min [31] where as in same context Ethereum block takes around 16 s [32] for the same. These transaction times should not be long for any IoT use-cases. Consider the following use-case, if we want to perform the verification of user before he/she gets in the car then ideally transaction should not exceed couple of seconds. If the transaction time exceeds those few seconds then the result will still be there but in terms of user experience it will lead to an unsatisfactory results.

Considering the threat analysis presented above, we may conclude that there is no full proof decentralized alternative solution which will replace traditional CA-based PKI that is suitable for use in IoT. To apply DPKI to IoT, there is a need to design alternative faster distributed ledger technology which will enable more efficient and faster transactions. Storage limitation and power consumption to process DIDs and VCs on IoT devices are the main challenges in IoT context. This gave impression that, DIDs for IoT needs rethinking as there is no strong base for services that users and companies would be able to trust enough for a PKI replacement.

4 Conclusions and Future Outlook

In era of Internet of Things (IoT), huge number of devices are connected to each other through wired/wireless medium. Devices are of heterogenous in nature in all respects i.e. size, operating capability, energy source etc. In such diversity, securing IoT application is crucial design parameter for any IoT application. Furthermore, nowadays deployment of IoT applications is increasing exponentially, and hence managing identity of devices is major challenge that IoT is facing. Identity management plays important role in securing interaction and traditional centralize approach doesn't fit in this distributed IoT world. Hence federated identity mechanisms are used, in this third party have complete control over user's identifier, which loses the user's ownership on their credentials. Hence SSI comes in the picture.

In the sequel, there is a need of decentralized method for managing identities in near future, Decentralized Identifiers and Verifiable Credentials are discussed in this chapter. To analyze their applicability to the IoT applications, smart home use case is presented. In the same line threat analysis is also done based on various parameters like processing power, energy, storage capacity and scalability, anonymity, data privacy and Latency etc. Considering this threat analysis, we may conclude that we may not apply DPKI as it is to the IoT, there is need to design alternative faster distributed ledger technology which will enable more efficient and faster transactions.

The future outlook of this work is to design distributed identity framework using DIDs and VCs that can be applicable to the constrained IoT devices. Performance analysis and evaluation of this mechanism against known attacks will be another interesting area of the research.

References

1. Project CASAGRAS: CASAGRAS Final Report: RFID and the Inclusive Model for the Internet of Things. Scientific American. https://docbox.etsi.org/zArchive/TISPAN/Open/IoT/low resolution/www.rfidglobal.eu CASAGRAS IoT Final Report low resolution.pdf (2009). Accessed 28 Mar 2017
2. Soldatos, J.: Overview of the Internet of Things. International Telecommunication Union (ITU). https://www.itu.int/rec/T-REC-Y.2060 (2004). Accessed 28 Mar 2017
3. L. Gyu Myoung et al., The Internet of Things Concept and Problem Statement draft lee IoT problem statement Status. https://tools.ietf.org/html/draft-lee-iot-problem-statement-00 (2011). Accessed 28 Mar 2017
4. Dorsemaine, B., Gaulier, J.P., Wary, J.P., Kheir, N., Urien, P.: Internet of things: a definition and taxonomy. In: Proceedings—NGMAST 2015: The 9th International Conference on Next Generation Mobile Applications, Services and Technologies, pp. 72–77 (2016)
5. Botterman, M.: Internet of Things: An Early Reality of the Future Internet. Report, European Commission Information Society and … http://cordis.europa.eu/pub/fp7/ict/docs/enet/iot-prague-workshop-report-vfinal-20090706_en.pdf (2009). Accessed 25 Mar 2017
6. Rayes, A., Salam, S.: Internet of things (IoT) overview. In: Internet of Things From Hype to Reality, pp. 1–34. Springer International Publishing, Cham (2017)
7. Park, H., JeeSook, E., Kim, S.-H.: Crop diseases diagnosing using image-based deep learning mechanism. In: Second International Conference on Computing and Network Communications (CoCoNet'18) (2018)
8. Xue, R., Wang, L., Chen, J.: Using the IOT to construct ubiquitous learning environment. In: 2011 Second International Conference on Mechanic Automation and Control Engineering (2011)
9. Fang, H., Hu, C.: Recognizing human activity in smart home using deep learning algorithms. In: Proceedings of the 33rd Chinese Control Conference, Nanjing, China, 28–30 July 2014
10. Ahamed, F., Farid, F.: Applying internet of things and machine-learning for personalize healthcare: issue and challenges. In: 2018 International Conference on Machine Learning and Data Engineering (iCMLDE) (2018)
11. Javed, A., Larijani, H., Wixted, A.: Improving Energy Consumption of a Commercial Building with IOT and Machine Learning. Published by the IEEE Computer Society 1520-9202/18/$33.00 ©2018 IEEE

12. Shanthamallu, U.S., Spanias, A., Tepedelenlioglu, C., Stanley, M.: A brief survey of machine learning methods and their sensor and IOT application. In: 2017 8th International Conference on Information, Intelligence, Systems & Applications (IISA) (2017)
13. Kanawaday, A., Sane, A.: Machine learning for predictive maintenance of Industrial machines using IOT sensor data. In: 2017 8th IEEE International Conference on Software Engineering and Service Science (ICSESS) (2017)
14. Mahalle, P.N., Thakre, P.A., Prasad, N.R., Prasad, R.: A fuzzy Approach to trust based access control in internet of things. In: Wireless VITAE 2013 (2013)
15. Crespo-Perez, G., Ojeda-Castro, A.: Convergence of cloud computing, internet of things, and machine learning: the future of decision support system. Int. J. Sci. Technol. Res. **6**(07) (2017)
16. Ordonez, L.D., Schweitzer, M.E., Galinsky, A.D., Bazerman, M.H.: Executive Overview SAML 101 Executive. Academy of Management Perspectives. http://marcomconsultant.com/samples/pi-saml.pdf (2009). Accessed 09 May 2017
17. OpenID Connect | OpenID. http://openid.net/connect/. Accessed 14 Feb 2015
18. Allen, C.: The path to self-sovereign identity. In: Life With Alacrity. http://www.lifewithalacrity.com, Apr 2016
19. EU GDPR Report: A new era for data protection in the EU What changes after May 2018. https://ec.europa.eu/commission/sites/beta-political/files/data-protection-factsheet-changes_en.pdf
20. Regulation (EU) 2016/679 of the European Parliament and of the Council of 27 April 2016 on the protection of natural persons with regard to the processing of personal data and on the free movement of such data, and repealing Directive 95/46/EC (General Data Protection Regulation)
21. Sporny, M., Burnett, D.C., Longley, D., Kellogg, G.: Verifiable Credentials Data Model 1.0—Expressing verifiable information on the Web. W3C Editor's Draft, 7 Dec 2018. https://w3c.github.io/vc-data-model. Accessed 7 Dec 2018
22. Cazorla, L., Alcaraz, C., Lopez, J.: Cyber stealth attacks in critical information infrastructures. IEEE Syst. J. **12**(2), 1778–1792 (2018)
23. Stellios, I., Kotzanikolaou, P., Psarakis, M., Alcaraz, C., Lopez, J.: A survey of IoT-enabled cyberattacks: assessing attack paths to critical infrastructures and services. IEEE Commun. Surv. Tutor. **20**(4), 3453–3495 (2018) (Fourth quarter)
24. Lagutin, D., Kortesniemi, Y., Fotiou, N.: Enabling decentralised identifiers and verifiable credentials for constrained internet-of-things devices using OAuth-based Delegation. In: Proceedings of Workshop on Decentralized IoT Systems and Security (DISS) 2019, San Diego, CA, USA, 24 Feb 2019. ISBN 1-891562-56-8
25. Kortesniemi, Y., Lagutin, D., Elo, T., Fotiou, N.: Improving the privacy of IoT with decentralised identifiers (DIDs). J. Comput. Netw. Commun **2019**(Article ID 8706760), 10 (2019)
26. Adina, P., Venkatnarayan, R.H., Shahzad, M.: Impacts & detection of network layer attacks on IoT networks. In: Mobile IoT SSP'18, Proceedings of the 1st ACM MobiHoc Workshop on Mobile IoT Sensing, Security, and Privacy Article No. 2 Los Angeles, CA, USA, 26 June 2018
27. Sovrin Foundation: Identity for All, Sovrin Foundation, Northampton, MA, USA (2018). https://sovrin.org/
28. VeresOne: VeresOne: a globally interoperable blockchain for identity. https://veres.one/
29. Leach, P., Mealling, M., Salz, R.: A Universally Unique IDentifier (UUID) URN Namespace, RFC 4122, July 2005
30. Reyna, A., Martín, C., Chen, J., Soler, E., Díaz, M.: On blockchain and its integration with IoT challenges and opportunities. Future Gener. Comput. Syst. **88**, 173–190 (2018)
31. Nakamoto, S.: Bitcoin: a peer-to-peer electronic cash system. Technical Report
32. Ethereum Foundation: White paper: Ethereum. Technical Report (2016)

The Invisible Eye—A Security Architecture to Protect Motorways

Neel Patel, Pratik Panchal, Yash Shah, Pankaj Sonawane and Ramchandra Mangrulkar

Abstract There have been proposals and even some successful deployments (by official bodies) of various road safety and traffic **control** systems. Emerging technologies like the Internet of Things (IoT) and Image Processing have found effective uses for developing the same. Although road security hasn't been considered a very pushing topic. Particular motorways hold severe significance when it comes to a nation; security concerns. And hence it is very crucial for these motorways to have security systems in place and sometimes even have officials deployed for checking all passing vehicles which seem vulnerable to them using barriers and other equipment. Two peculiar problems that come into the picture when considering road security are first, lack of availability of automated, intelligent and feasible systems on motorways that can detect vulnerabilities in passing automobiles and second, knowability and visibility of any security mechanism used for checking the vehicles is an easy task for anyone traveling by the road under surveillance. The prior makes it easy for malevolent individuals (one's carrying any illegal weapons, equipment, etc.) to escape and the latter makes them well aware about which security-checks they will have to tackle to surpass the checking zones without being caught red-handed. Hence need for a system which eliminates these two problems becomes vital to increase the security of motorways. This chapter introduces an intelligent cop assisting system TIE (The Invisible Eye). The system senses parameters that aren't direct knowable by anyone at their first glance on the vehicle like weight, etc. and make them available to the security personnel(s) deployed to check those vehicles. Also, the system would provide a vulnerability index and it can be deployed in an unnoticeable way.

Keywords Security · Motorways · Invisible eye · Surveillance · Intelligent Transportation System (ITS) · Traffic

---

N. Patel (✉) · P. Panchal · Y. Shah · P. Sonawane · R. Mangrulkar
Computer Engineering Department, Dwarkadas J. Sanghvi College of Engineering, Mumbai, Maharashtra, India
e-mail: neelpatel3039@gmail.com

N. Dey et al. (eds.), *Internet of Things, Smart Computing and Technology: A Roadmap Ahead*, Studies in Systems, Decision and Control 266,
https://doi.org/10.1007/978-3-030-39047-1_17

1 Introduction

Transportation is a rudimentary aspect for every nation on the planet. It empowers trade between people, which in turn is crucial for the development of civilizations. Transport plays a critical part in globalization and financial germination. Though it is heavily subsidized by governments, careful planning of transport is quintessential to make traffic flow and smooth growth of any developing country.

Scaling up all 3 modes of transportation also brings in the need of reliable tools, technologies, and mechanisms for maintaining safety and security to the people and commodity traveling through it and even to the people who live by areas from which the vehicle carrying passengers it is going to pass through. Unsafe and insecure vents in any mode of transportation directly lead to either loss of life or drastic reduction in the quality of life [1].

Among all 3 modes motorways are the majorly used one in almost all the developed nations [2, 3]. Despite this large scale use of Motorways, assurance of safety on them is still a critical issue in developing countries like India.

Both technology and legislative wise there [4, 5] are considerable efforts being taken in the direction of rectifying the bad scenario. Also, authors have seen massive campaigns in India for the same which reflect on the importance of the issue [6]. Showing the importance of tech in road safety Dr. Nishi Mittal, Head, Traffic Engineering and Safety Central Road Research Institute, New Delhi said, "...Technology is an aid to traffic policemen, not a substitute. So, technology helps in curbing traffic violations because the people who deny that they did the violation they cannot deny because it's proof." [7]. Hence building reliable Intelligent Transportation Systems (ITS) is quite an evident need.

Technologies like Facial recognition and Internet of Things (IoT) [8] have immense potential in the police force's ammunition against fighting criminality. Additionally using image processing in criminal detection has been an efficient, effective and speedy way to catch malevolent individuals and activities [9, 10].

Processing of the images in real time using analytics algorithms is making it considerably more effective [11]. Many official bodies are suffering a future of lessened funding. Compelling new technologies render a way for forces to conduct and serve their populations more efficiently.

This chapter, A Cop Assisting ITS named TIE System is put forth. It attempts to identify and solve two loopholes of most current road security mechanisms (1) High knowability and (2) Chances of human error.

The rest of the chapter is organized as: Sect. 2 covers the related work, Sect. 3 delineates the model and design of TIE system, Sect. 4 poses a case study of the Pulwama Attack in the context of the proposed system, Sect. 5 contains conclusion of the chapter and finally the chapter ends with the future directions for further extension.

2 Related Work

Amongst all of the road safety and security problems like rash driving, breaking signals, etc. few common and current problems related to vehicle overloading and lawful use of license plates render inevitable.

Large democracies like India [12] and China [13] are facing a big challenge in rectifying truck overload problems in specific. Prior addressed problem of overloading vehicles causes serious harm to the roads plus reduce their life span to a noteworthy extent. Providentially the development of sensors that can sense the gross weight of the vehicles with 96–97% [14] accuracy favors prompt development of frameworks and systems that work towards keeping the weight of vehicles in legal limits.

License plate recognition is a big help to traffic forces since they unfold a lot of details about the vehicle, its owner and related commodities. Issues like non-uniformity of the license number plate models, the low resolution of the license plates for vehicles in video frames under typical surveillance systems, etc. are faced by many researchers [15].

Due to the physical placement constraints in many systems, the sole way to get more accurate results is by improving and innovating on the image processing aspect. On the other hand, if there is a possibility of changing the placement aspect of the cameras, its accuracy can be leveraged by appropriately placing the camera to face the region of interest as directly as possible.

Use of micro-controller based systems that are deployed on the motorways for various purposes and with various sensors and modules have yielded decent results in many past use cases. In supplement to producing decent results, they also make the system more ubiquitous with respect to the variety of motorways at a lower cost. And the cost of developing such systems is indeed a concern in some countries [16]. This low-cost option can sometimes render a security system useless because of their intermittent errors and high dependency. Hence to niche the system at its right position authors say such systems turn out, in case of a being a security system, to be useful "assisting systems" which leverage and fasten the security to a considerable extent on busy roads, doing which is quite important by bringing in technology [17].

3 Model and Design of TIE System

3.1 TIE System Positioning

In TIE System, positioning of the sensors and camera modules is an important issue since it has to accomplish the requirement of reducing the knowability about the presence and operation of the system.

The aim of the system is to position and deploy the sensors and camera modules such that they either seem (1) obvious to be present (e.g. presence of a CCTV Camera Module would be expected in a security checking scenario) or (2) Completely unnoticeable.

Figures 1, 2, 3 and 4 illustrate the placement of various modules in a Black Box manner.

VCP Sensor module is made unnoticeable by placing it in a Sewer man-hole like exterior. The sewer manhole will have a lattice-like cap making way for the smog to reach the VCP Sensor.

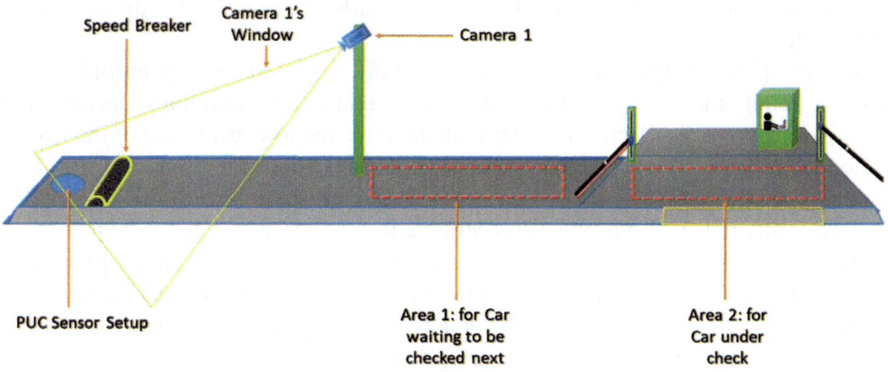

Fig. 1 Placements of VCP sensor setup, speed breaker and camera 1

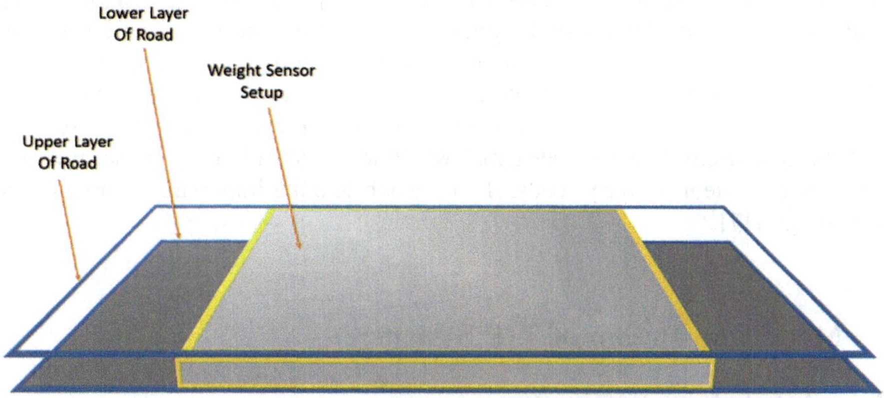

Fig. 2 Positioning of the weight sensor module beneath a slice of land surface

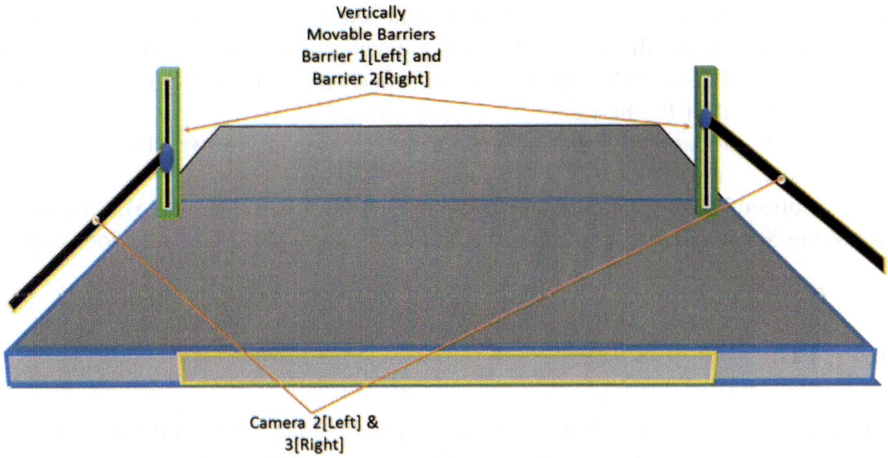

Fig. 3 Positioning of the barrier 1, barrier 2, camera 2 and camera 3

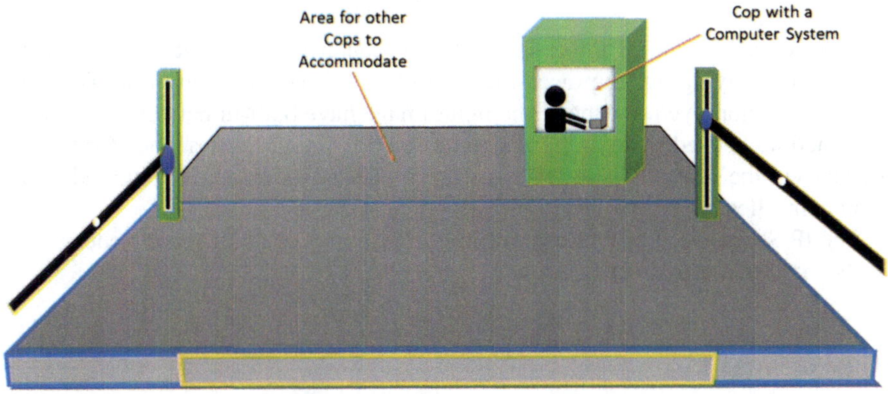

Fig. 4 Area where the cops would accommodate for manually checking

Speed breaker has been placed as shown in the Fig. 1 due to 3 major reasons or pros that accompany this placement decision:

1. A slight tilt of the vehicle upwards i.e. towards the Camera 1 will help in capturing a proper photo of the vehicle.
2. And that same tilt also encourages silencer of the vehicle to face more towards VCP Sensor Sink than otherwise.
3. To lower the speed of the car before the checking area. Area 2 and 1 are for the current and next vehicle under check respectively.

The weight sensor setup is sandwiched between two layers on the road i.e. the lower layer of road and the upper layer of the road. Both the layers get created when digging work is done for deploying the weight sensor setup. The lower layer is the earth surface whereas the upper layer acts alike sheath for the weight sensor setup.

Barrier 1 and 2 are both vertically movable and they hold the two cameras i.e. on the movable rod of the barrier.

In the Fig. 3 Camera 2 and 3 aim to capture backside license plate and driver image respectively.

The cop sitting with a computer system in the Fig. 4 will review the output of the Reporting Stage and then accordingly either let the car pass or take judicious action.

3.2 TIE System Stages

The TIE System is divided into 4 main stages based on the similar category of operations that it performs. Following subsections describe various stages.

3.2.1 Image Processing Stage or IP Stage

This stage carries out all the tasks related to capturing an image, sending to the server and applying image processing algorithms to extract specify objects. Hence the cameras employed to capture the input image have built-in networking module. This module is used to send the image to an HPC (High-Performance Computing Server) via the internet. The result of the requested operation is returned to the microcontroller.

The IP Stage executes its functions in two instances of a full vehicle check cycle (1) Before Area 1 and (2) In Area 2.

(1) Before Area 1

When the Vehicle Detector Module intersects with the front area of any vehicle it gets activated. This activation is as an indication that a vehicle has arrived and will very soon enter the Area 1 and 2. The activation signal from the Vehicle Detector will trigger the Camera 1. Additionally Vehicle Detector will also trigger the VCP Sensor.

Image captured by Camera 1 is sent to the HPC server via the internet for identifying the following things:

- Car Model Name
- Expected Driver Height (from ground level)
- Backside License Plate Positioned Height (from ground level)
- Expected Silencer Sound.

Fig. 5 Camera 1's position as a part of IP stage in top view

As the Vehicle Detector senses that a vehicle has left its detection area it is deactivated. The HPC Server returns that result to the microcontroller managing all the sensors.

Microcontroller writes the value of (1) Driver Height into the circular buffer residing inside Barrier 2's controller and (2) Backside License Plate Positional Height into that of Barrier 1's. Hence as a result of that when the current vehicle will get into checking area both the Barriers will automatically adjust their heights based on the dimensions of that particular vehicle. Once the current vehicle passes, the next values from the circular buffers are referred and the Barriers are position themselves accordingly (Fig. 5).

(2) In Area 2

In this part of the IP Stage, the barriers are positioned appropriately to capture the driver's face and backside license plate. In case, if the driver has learner's license the image of the person beside him/her is also captured and send for detection.

After the vehicle under check places itself appropriately in the checking area and the barriers take their vehicle-specific vertical positions, is when the cameras hidden amidst the barriers capture an image of their respective windows and send them to the HPC Server for the Image Processing Tasks.

Then after the HPC Server extracts the License Plate string from the image captured by Barrier 1 and the driver details from the image captured by Barrier 2. These extracted parameters go through a set of checks and eventually produce a Vulnerability Index (VI).

3.2.2 Sensing Stage

A vehicle's health is determined by many factors such as how frequently it is used, how it is driven i.e. how is the driver using the mechanisms e.g. in a harsh way, how regularly is it serviced etc. Its good health is not only important for the lives of people who travel inside the car but also for the roads and areas they are traveling through. At this stage, the sensors are used to inspect various safety and security-related aspects of vehicles.

The sensors are positioned in an almost unnoticeable way. It is also tried and made sure at a logical level that even if their functioning needs physical movement there is hiding from the knowledge of the existence of the sensing procedure and the sensors. Following subsections limn the sensor modules, their working and setup plan.

(1) Vehicle Detector Module

Using image processing to detect an arriving car has two major drawbacks in TIE Systems setup (1) a lot of processing power will be used upon in a long run and (2) the time delay introduced between capturing of images infinitely, then detecting presence of arriving car and then triggering appropriate modules leads to considerable lagging in many other modules. Hence authors choose a hardware-based approach instead.

In this module, Light Dependent Resistors (LDR) and Infrared Radiation (IR) Sensors are coupled. They work alternately in the day and night time to detect the presence of any arriving vehicle.

IR sensors work on the principle of infrared light. And hence IR sensors are ineffective mostly during day time i.e. under direct sunlight. On the other hand, LDRs react on the basis of the intensity of light falling on it. Hence they render ineffective at night time as there will be no light surrounding it.

Therefore at day time LDR will be active and used for detecting vehicles presence. So when the car will arrive and its shadow will fall on the LDR the intensity of light will change. On the other hand at night time IR will be active. As the vehicle arrives it will block the lights fired by the transmitter and receiver placed at opposite ends of the road will not receive any light.

Using the above mentioned method TIE system can detect vehicles during both day and night times (Fig. 6).

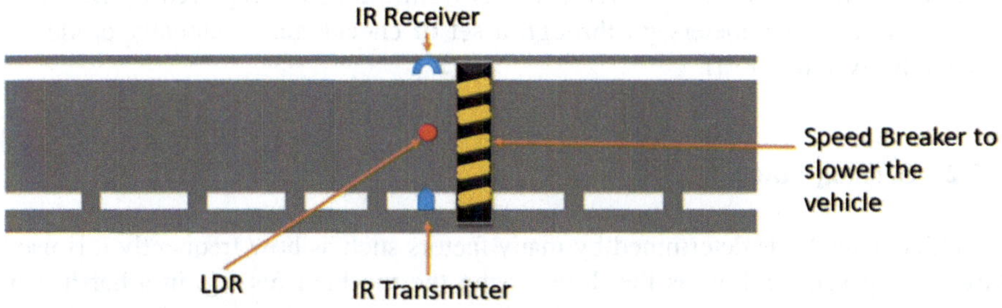

Fig. 6 Placement of the vehicle detector module from top view

(2) Weight Sensor Module

This sensor module is used to find the gross weight of a vehicle. Knowing the same also assists cops in sensing some weirdness in the weight of a vehicle. It is also used to detect the overweighting of vehicles.

Weight Sensor Module is composed of an array of weighing pads. To calculate the weight on each pad (i.e. each axle) the proposed system internally needs to find the sum of following 4 weight cells placed at 4 for corners of a weighing pad (Figs. 7 and 8).

Automobiles, especially trucks that are carrying excess weight can constitute a serious hazard for the driver and other drivers, which have led to the authorities cracking down on the issue and penalties in place.

Since all vehicles are not the same especially in case of trucks, the length and number of axles differ considerably; hence a straight-forward way of weighing multiple different types of vehicles has to be calculating the weight of all such vehicles on each axle. Therefore the authors came across a solution to create a set of weighing pads of a fixed length and breadth which can be placed one after other in an array-like fashion. The array can be extended linearly along the road lane to a point where it can accommodate the longest vehicle (horizontally) that is expected to pass by that road. This approach helps to find the weight of any type of vehicle

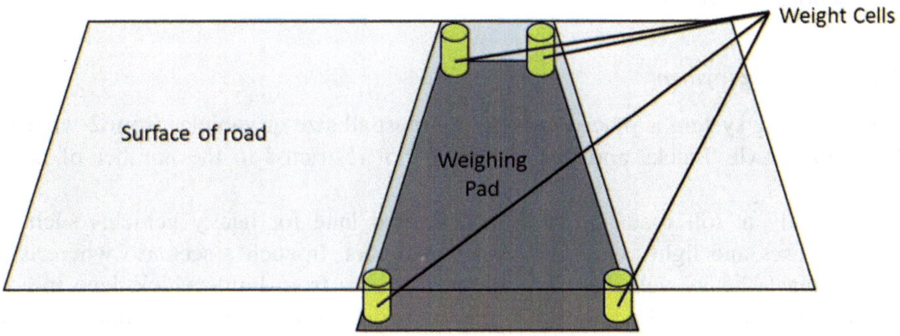

Fig. 7 Internal view of each weight pad

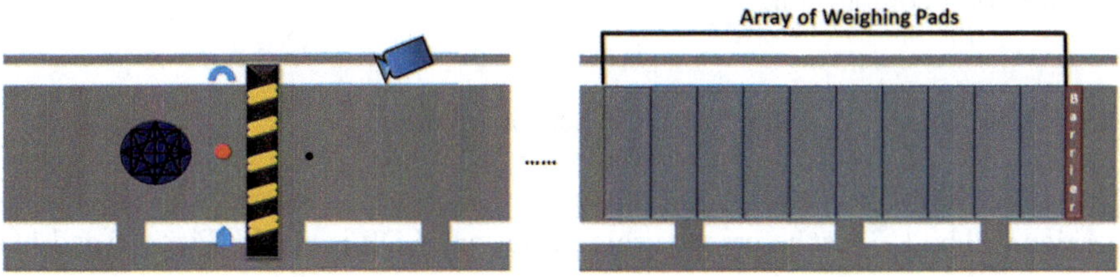

Fig. 8 Top view of placement of weighing pads

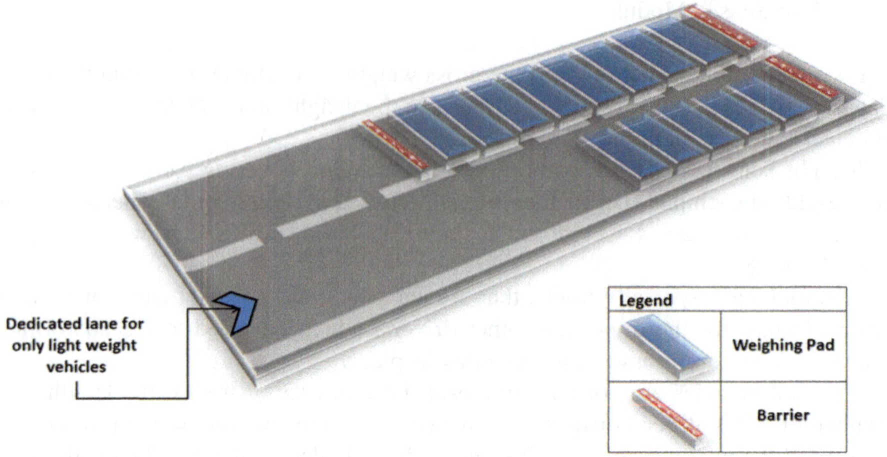

Fig. 9 Weight sensor setup using multi-lane deployment concept

from two-wheeler to multi axels truck. Once the weight of each axle is found then the total weight of the vehicle can be found out by adding all those weights.

Also, the benefit is every axle is designed to handle some maximum amount of weight. As a result, if the weight exceeds the system will come to know and alert the security personnel.

Mutli-lane Deployment

The weighing system is placed to weight almost all size of vehicles from 2-wheeler bikes to 5-Axle Trucks and this system is not restricted to the number of lanes (Fig. 9).

Typically at toll roads, there is a dedicated lane for heavy vehicles such as trucks, buses and light vehicles like personal cars. In such a scenario where it is certain that a heavy vehicle is not allowed to pass from light vehicle lane in any circumstances, the total number of weighing pads can be adjusted to reduce cost. In case where such segregation is not possible the array of weighing pads can be increased or decreased based on lengths of the expected vehicles.

This concept for Multi-lane deployment of the system to reduce cost is a niche use and not a compulsion.

(3) Vehicle Created Pollution or VCP Sensor Module

This sensor module will be placed under the road near the vehicle exhaust area; the position will be 1–3 feet before the vehicle detector's LDR is placed (Figs. 10 and 11).

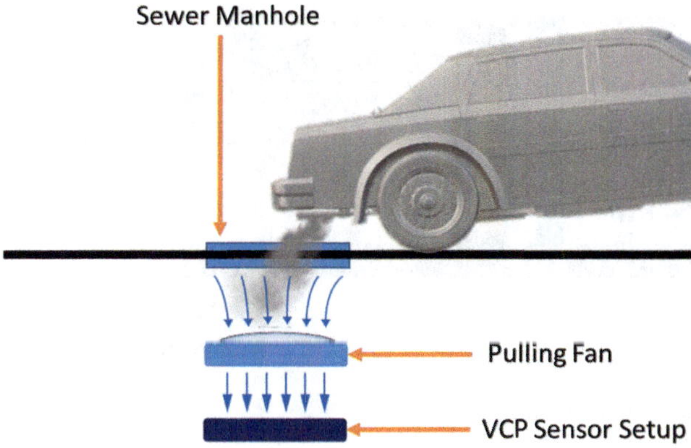

Fig. 10 Side view of VCP sensor module setup

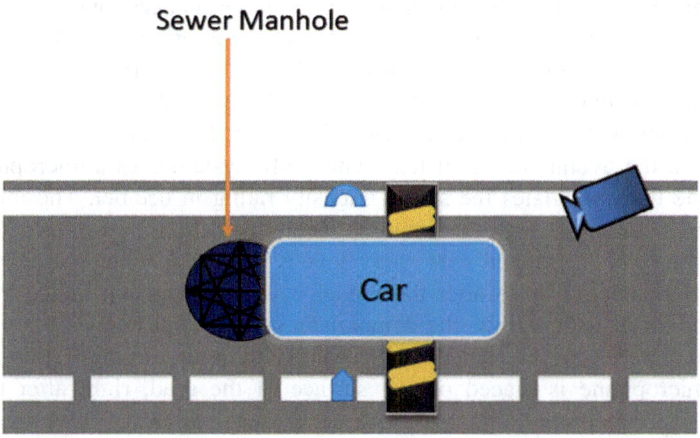

Fig. 11 Top view of VCP sensor module setup

To make the sensing process effective following arrangement are made:

- There will be a fan pulling the exhausted air of vehicle which is placed below the sewer manhole.
- This metal net will be parallel to the road and will be placed on the surface of the road.
- The air pulled by the fan will directly be thrown on the VCP Check Sensor which then finds the purity of air throw by any vehicle.

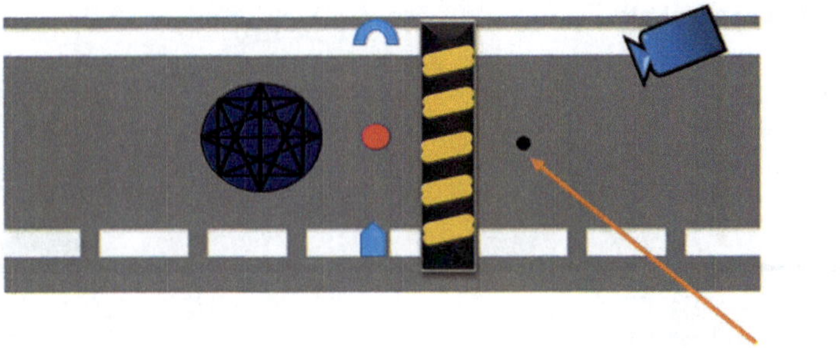

Fig. 12 Top view of placement of microphone

(4) Silencer Sound Module

Every vehicle is assigned some maximum limit on the silencer sound it can produce [18, 19]. If the sound is beyond that limit, it is illegal to drive that vehicle. To adapt to the standard method of me measuring the vehicle sound, some calibration mechanism will have to be used to calculate the equivalent value of the sound produced from 50 feet away of the vehicle's centre (Fig. 12).

To check the overall sound of the vehicle TIE system uses a microphone. The microphone then calculates the sound intensity rating in decibel. The microphone input is then compared with maximum allowed decibel.

Since the prior mentioned maximum limit is vehicle specific it is fetched from the vehicle details database after the model is extracted in the Image Processing Stage. And as a result of this, the comparison of the input intensity and the legal limit are compared and presented to the security personnel in the Reporting Stage.

This microphone is placed on the surface of the road, right after the speed breaker ends. Favourably, this is also the area where the vehicle is expected to get accelerated. Acceleration helps sense the intensity of sound more effectively.

3.2.3 VI Calculation Stage or Vulnerability Check Engine (VCE)

This is a software engine residing on the HPC Server. The goal of this engine is to consider all the sensor outputs and to inspect whether sensed parameters are within the legal boundaries or not.

Vulnerability Index or VI: It is a percentage value between 0 and 100%. The intended interpretation of the Vulnerability Index is as an indicator of how vulnerable the vehicle is in the context of the checks TIE System is performing in the VCE Stage.

The magnitude of VI is raised by a certain percentage when it satisfies the conditions mentioned in the VCE Subsystems. One peculiarity about the VI is that it doesn't indicate how safe a car is. That validation is completely left on the security personnel protocols and judgment which to a certain extent may be dependent on the TIE system's report.

The engine also uses the result of the IP Stage to check whether the vehicle and its driver are both legally sound.

Inputs to this stage are:

1. License Plate Number
2. Driver Name and Licence Number
3. Past Output for the Same Combination.

Appropriate weight is assigned to each inspected parameters based on their law based importance [20]. The weights have been calibrated based on the penalties and punishments associated with the performed check. In this way, if the intensity of the check being performed is legislatively weighed more than others, TIE System will assign it more weight. Assigning weights to VCE subsystem checks in this manner will help rudimentarily assess the vulnerability in a legislative manner and will render VI more impactful and legally sound.

Table 1 tabularizes the details about the various sections used from the Motor Vehicle Act (issued by the Indian government in 1988) that TIE System referred for probing the penalties imposed for particular unlawful activities.

Driving License Checks

This particular subsystem is responsible for conducting a series of checks related to the vehicle's license plate. Following cases are considered to be addressed:

1. Not Found: It means that no license plate was found after processing the image captured by camera 2.
2. Expired License: The driving license is renewed, yet the driver is driving the vehicle.
3. Past Criminal: The driver has past criminal cases associated with him/her.
4. Current Criminal: The driver has ongoing criminal cases associated with him/ her. This also means he/she is on bail or has to be caught (Fig. 13).
5. Wrong Use: This case is named as a consistency check in the Fig. 14. Consistency check here means to check whether the vehicle type detected from the captured image in the IP stage and category of vehicle(s) the license is assigned to, are same or not. If they differ it will be a case of "Wrong Use".
6. Unlawful Learner: This means that the driver has a learner's license yet he doesn't have a fully licensed driver seated beside him/her.

Table 1 Details of sections used from motor vehicle act for assigning weightage to VI calculation parameters

Violation scenario	Section	Penalty and/or punishment		
		1st time fine (Rupees)	Fine change after 1st time fined (Rupees)	Imprisonment (Years)
License invalid	177	100	300	–
Person with learner's license driving without any instructor besides	177	450	500	–
Improper license plate	177	100	300	–
PUC not done	177	100	300	–
Driving without license	180	1,000	–	Up to 3
Silencer sound is more than legal limit	190	500	–	–
Permit of the vehicle invalid	192	10,000	–	Up to 3
Overloaded vehicle	194	2,000 + 1,000 per extra ton of load		–

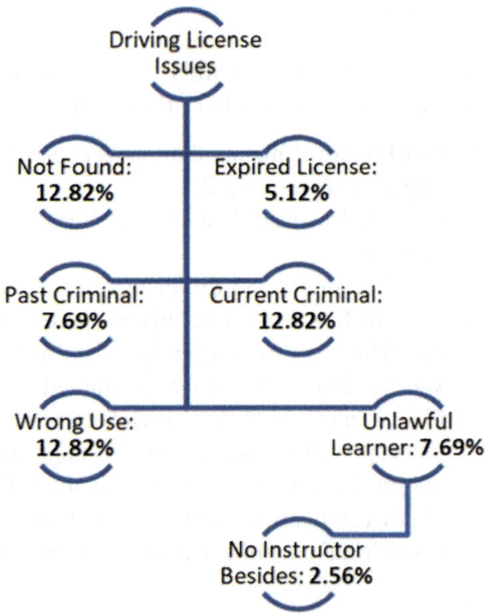

Fig. 13 Assigned weightage to driving license related checking

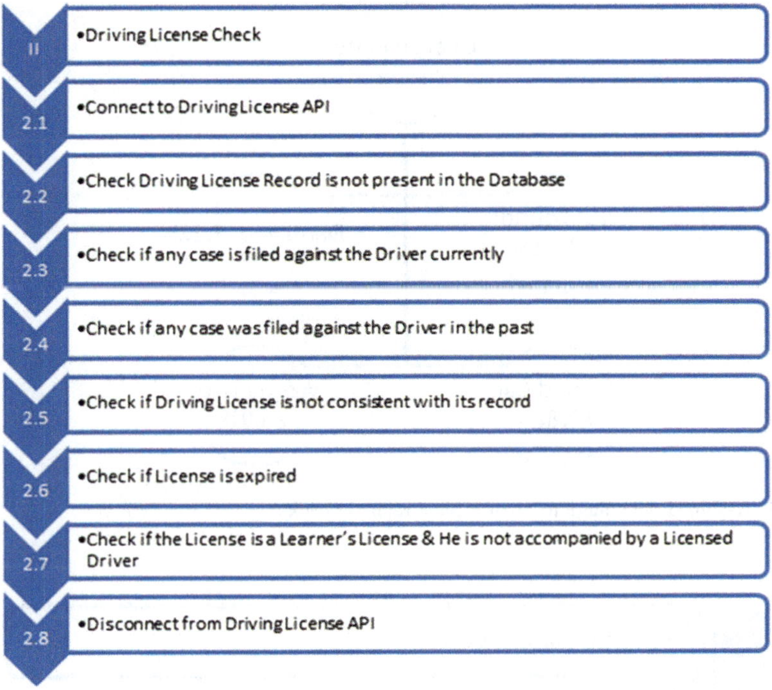

Fig. 14 Sequence of steps executed by driving license checks subsystem

License Plate Related Checks

This particular subsystem is responsible for conducting a series of checks related to the vehicle's license plate. This subsystem takes as input the string value of the license plate number which got extracted in the image processing stage after processing the image captured by Camera 2/Camera in the back barrier (Figs. 15 and 16).

Following cases are considered to be addressed:

1. Not Found: The license plate string was NULL i.e. No license plate was found.
2. Damaged: The string that was extracted from the image did not have proper length and/or had illegal characters on it.
3. Robbed Car: The license plate is of a car which has been reported robbed by the owner.
4. Wrong Use: The license plate is not used on the vehicle type it is assigned to.

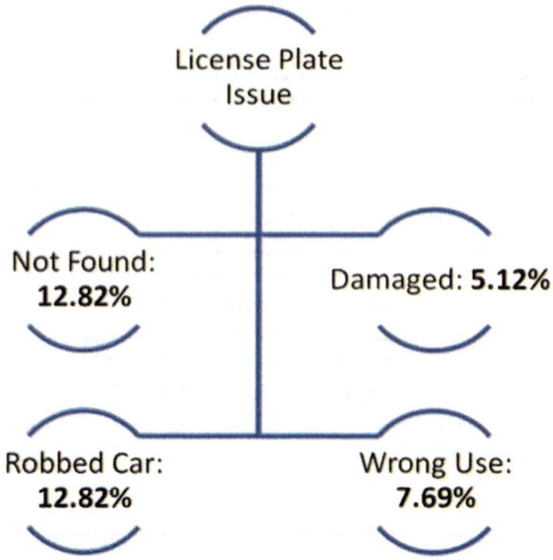

Fig. 15 Assigned weightage to license plate related checks

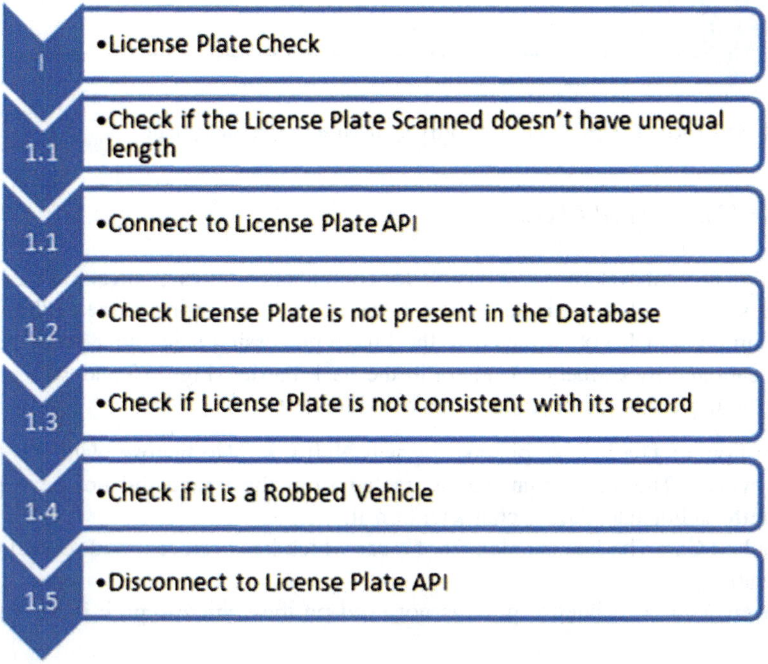

Fig. 16 Sequence of steps executed by license plate checks subsystem

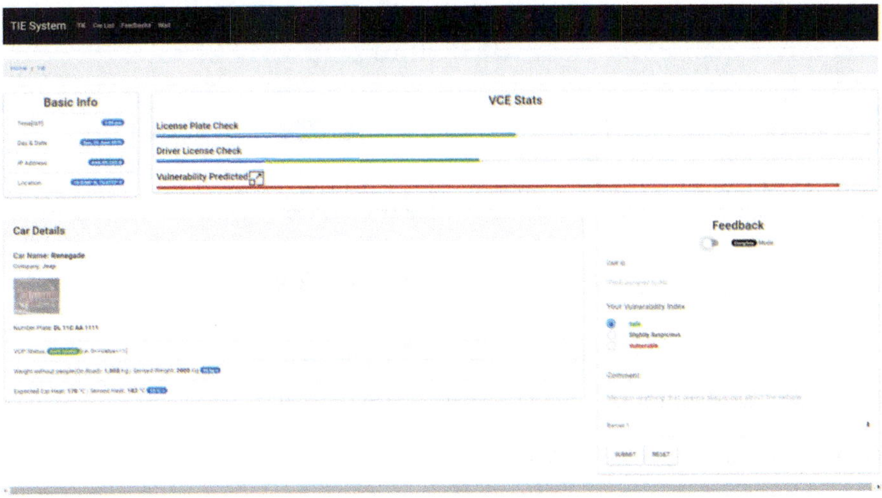

Fig. 17 Sample report produced by reporting stage

3.2.4 Reporting Stage

This stage is the last stage of the TIE System Life Cycle. It is meant to consolidate the results of all the sensed parameter and image processing based criminal association or history finder. It provides a comprehensive yet clear cut idea that out of the Parameter. A web-based report is prepared since that would allow the ability to share the report with the remote officials also who are doing surveillance tasks.

Above displayed Fig. 17 is a sample report that shows the basic look and feel of the output of the Reporting Stage. There are 4 sections in the generated report as shown in the Fig. 17:

1. Basic Info (Information) Section
2. VCE Stat (Statistics) Section
3. Vehicle Details Section
4. Feedback Section.

(1) Basic Info Section

The Fig. 18 shows the values that are automatically filled in the Basic Info Section. Time zone specific time, Current day and date, The Internet Protocol Address assigned to the Computer System being used, and Location of the deployment of the TIE System. All these parameters although rudimentary have a risk of being leaked and hence should not be stored. Due to this requirement, these values are all generated using only the browser-based libraries or frontend functions. These parameters, therefore, do not get communicated which lower the chances of leakage (Fig. 19).

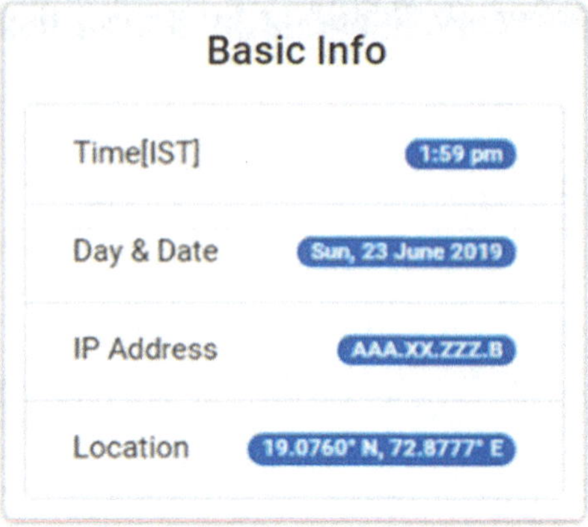

Fig. 18 Basic info section of sample report

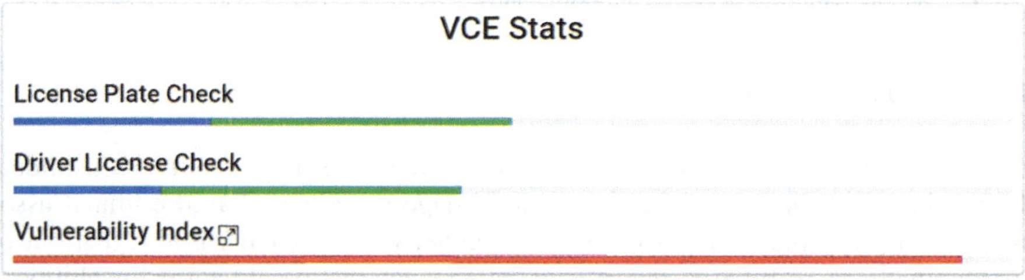

Fig. 19 VCE stats section of sample report

(2) VCE Stat Section

VCE Stats Section displays the outputs of the VCE Stage to the security personnel. Distinct colors in the progress bar below "License Plate Check" and "Driver License Check" indicate the weights of different checks that got satisfied in the respective subsystems. The Vulnerability Index's progress bar shows the sum total of both the subsystem's individual totals. The expansion icon beside the Vulnerability Index can be clicked to see the Vulnerability Index's breakdown. This breakdown is expected to help the cops find the exact vulnerable area of the vehicle under check. The Fig. 20 shows the pop-up window and the breakdown that appears when the expansion icon is clicked. The breakdown appeared should be interpreted using the following rules:

1. Red tick mark in the "Vulnerability Status" column should be interpreted as "the vulnerability check is satisfied" and so the intensity of suspicion was raised to a certain percentage (i.e. the weight assigned to that individual check).

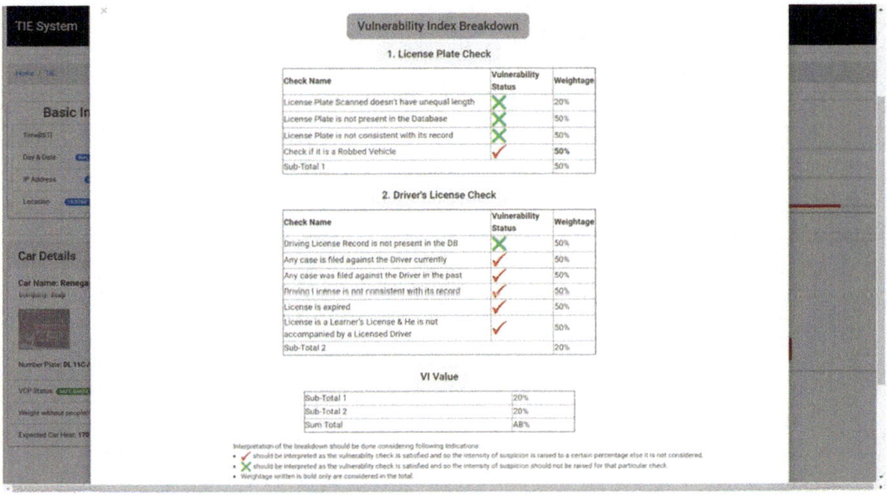

Fig. 20 Vulnerability breakdown pop-up

2. Green tick mark in the "Vulnerability Status" column should be interpreted as the vulnerability check is not satisfied and so the intensity of suspicion was not raised by the weight of that particular check.
3. Weights displayed is bold only are considered in the total.
4. The value written as the "Sum total" is the final value of the VI.

For e.g. breakdown like the one shown in Fig. 21 should be interpreted as there is "No vulnerability detected with respect to the checks performed." whereas the one alike Fig 22 should be interpreted as "The driver's license not renewed and hence he/she is unlawfully driving the vehicle."

(3) Vehicle Details Section

The car details section contains the subsections as shown in the above Fig. 23. Values of Vehicle name, company name, model name, license plate number, VCP Status, the weight of the vehicle are all encompassed in this section. Also, the difference between the corresponding values is displayed. A hover on the photo of the vehicle would enlarge it for viewing if needed.

(4) Feedback Section

Feedback Section is used to take asses the system's IP Stage and VCE Stage results. There are two modes that can be used to the give feedback mainly based on the level of traffic on that road but there can be other factors which lead to choosing one of them too (Figs. 24 and 25).

Vulnerability Index Breakdown

1. License Plate Check

Check Name	Vulnerability Status	Weightage
License Plate Scanned doesn't have unequal length	✗	20%
License Plate is not present in the Database	✗	50%
License Plate is not consistent with its record	✗	50%
Check if it is a Robbed Vehicle	✗	50%
Sub-Total 1		50%

2. Driver's License Check

Check Name	Vulnerability Status	Weightage
Driving License Record is not present in the DB	✗	50%
Any case is filed against the Driver currently	✗	50%
Any case was filed against the Driver in the past	✗	50%
Driving License is not consistent with its record	✗	50%
License is expired	✗	50%
License is a Learner's License & He is not accompanied by a Licensed Driver	✗	50%
Sub-Total 2		0%

VI Value

Sub-Total 1	0%
Sub-Total 2	0%
Sum Total	0%

Fig. 21 Example 1 of breakdown of VI

Complete Mode

This mode of the feedback section is used when there handleable traffic on the road. This section will take the following inputs from the security personnel viewing the report (Fig. 26):

1. Vehicle ID or Vehicle Identifier: This is the unique identifier assigned to the vehicle under check. The webpage itself will dynamically fill this value into the textbox from the Vehicle ID that it receives from the microcontroller.
2. Barrier Number: This parameter is used only in case of a security check arrangement with multiple barriers. It will indicate the barrier number where the TIE System is deployed.

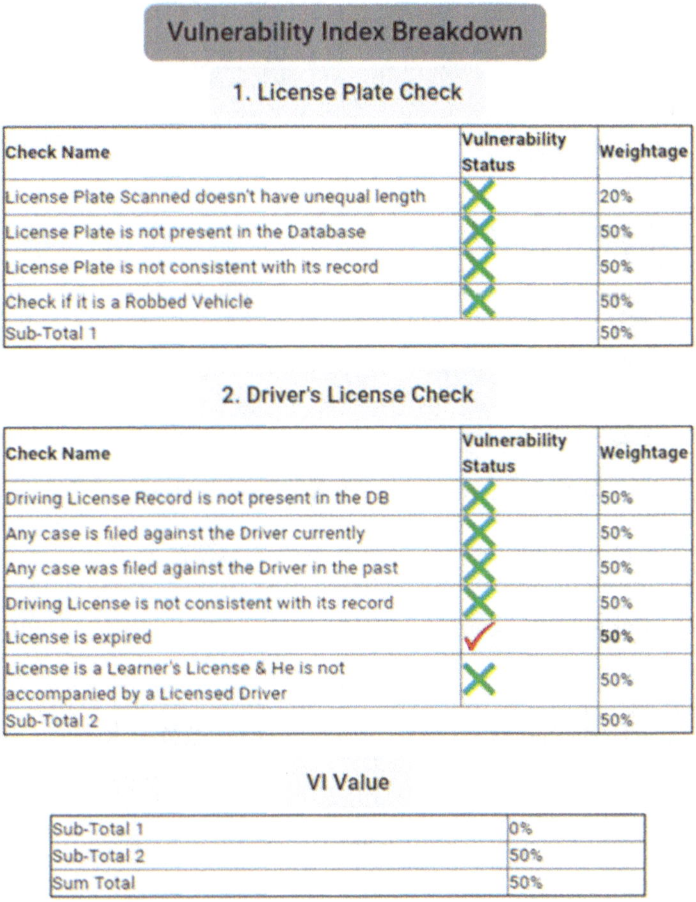

Vulnerability Index Breakdown

1. License Plate Check

Check Name	Vulnerability Status	Weightage
License Plate Scanned doesn't have unequal length	✗	20%
License Plate is not present in the Database	✗	50%
License Plate is not consistent with its record	✗	50%
Check if it is a Robbed Vehicle	✗	50%
Sub-Total 1		50%

2. Driver's License Check

Check Name	Vulnerability Status	Weightage
Driving License Record is not present in the DB	✗	50%
Any case is filed against the Driver currently	✗	50%
Any case was filed against the Driver in the past	✗	50%
Driving License is not consistent with its record	✗	50%
License is expired	✓	**50%**
License is a Learner's License & He is not accompanied by a Licensed Driver	✗	50%
Sub-Total 2		50%

VI Value

Sub-Total 1	0%
Sub-Total 2	50%
Sum Total	50%

Fig. 22 Example 1 of breakdown of VI

3. Comment: It is used to record what appeared as suspicious about the vehicle. It is expected to largely record those parameters or aspects related to vehicle and road security that is not examined by the TIE System. The comments can timely be then checked and analyzed to enhance TIE System and identify various patterns.
4. Actual VI: This is the most crucial part in the feedback section. This part has all the parameters shown in the Vulnerability Index Breakdown window excluding the (1) "totals" related rows in the "License Plate Check" and "Driver's License Check" and (2) the whole table showing "Sum total" in the last row. Actual VI has high importance because the value of this parameter will train the IP Stage and VCE Stage to improve their processing. And as a result of which the system will perform better in subsequent detections. This training cycling will continue infinite runs of the system.

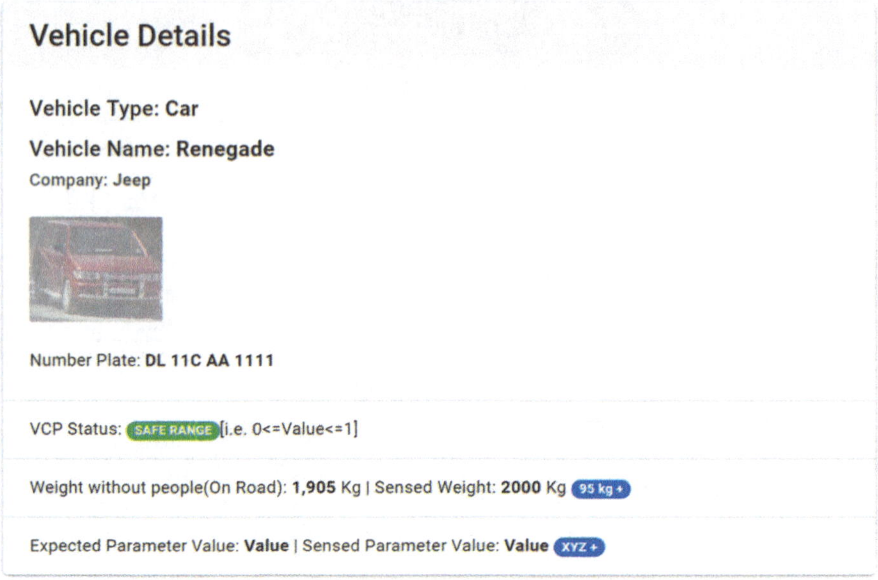

Fig. 23 Car details section of sample report

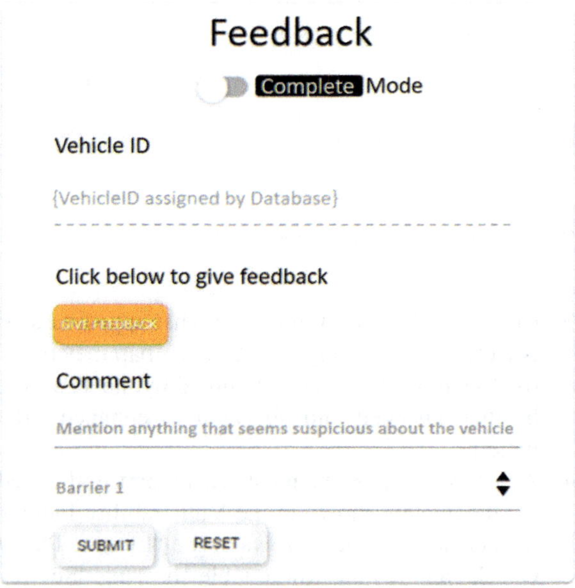

Fig. 24 Feedback section of the sample report configured in complete mode

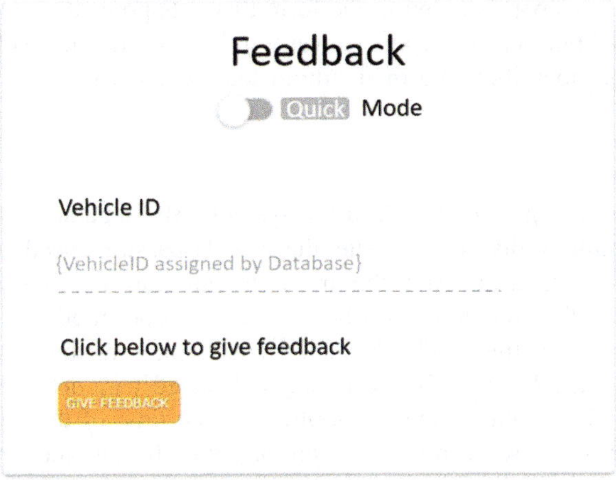

Fig. 25 Feedback section of the sample report configured in complete mode

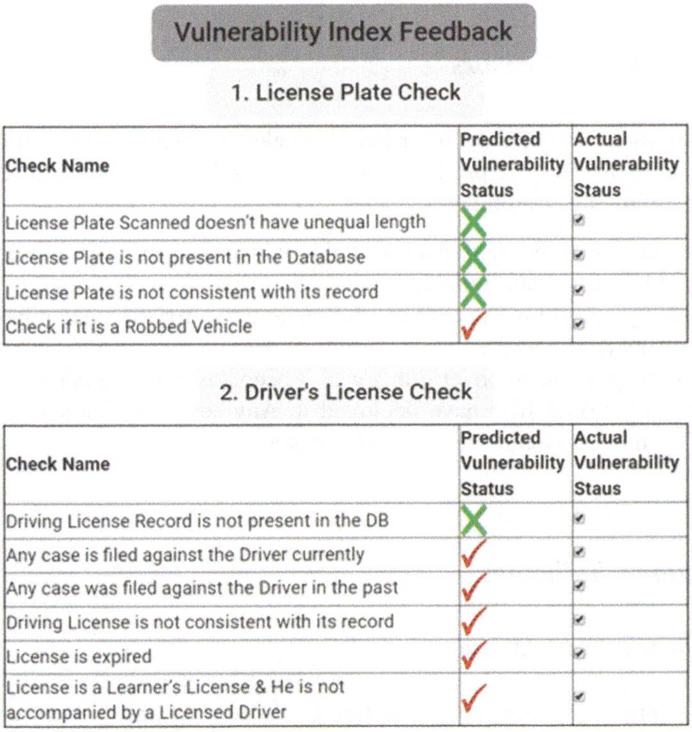

Vulnerability Index Feedback

1. License Plate Check

Check Name	Predicted Vulnerability Status	Actual Vulnerability Staus
License Plate Scanned doesn't have unequal length	✗	☑
License Plate is not present in the Database	✗	☑
License Plate is not consistent with its record	✗	☑
Check if it is a Robbed Vehicle	✓	☑

2. Driver's License Check

Check Name	Predicted Vulnerability Status	Actual Vulnerability Staus
Driving License Record is not present in the DB	✗	☑
Any case is filed against the Driver currently	✓	☑
Any case was filed against the Driver in the past	✓	☑
Driving License is not consistent with its record	✓	☑
License is expired	✓	☑
License is a Learner's License & He is not accompanied by a Licensed Driver	✓	☑

Fig. 26 Structure of feedback form

This input of this part is given in the form that gets popped up after clicking the "Give Feedback" button. The report viewer will have to just uncheck the parameter that was wrongly identified and then submit the feedback form.

Quick Mode

This mode is used to give feedback of the system's IP Stage and VCE Stage when there is more traffic or due to any issue, the vehicle passing speed via the checking zone has to be increased. And the most crucial part i.e. the Actual VI. All parameters inside the Actual VI will be the same in both modes.

For ease of giving the feedback all the checkboxes in the Actual VI for are already checked and hence only the wrongly detected parameter should be found and unchecked. This subtle strategy would save the time of the cops viewing the report since they are now freed from the job of going through the whole form every time they give their feedback.

Figure 27 presents a collective look at the all the TIE System Stages and their sequences.

3.3 TIE System Sensors

The authors make use of a set of criteria for selecting sensors that can be used to extend or scale up the system. These sensors are a subset of traditional sensors that satisfy the following conditions:

1. The sensor has proper datasheet available which also means it's officially produced and is tested and reliable.
2. It can stably operate in the temperature range tuned for the system based on the region of its deployment.
3. It should be possible to cover it in a way it becomes unnoticeable by any person other than the ones who have deployed it. Any sensor which satisfies the following condition is considered a TIE Sensor.

3.4 Major Challenges

3.4.1 Synchronizing All Stages

Timing the processes like Image Capturing, Processing and the sensors sensing appropriately so as to synchronize them in the time domain. In the proposed model the time delay of the IP Stage will take more time compared to the Sensing Stage and hence some tasks of this stage are executed first and a queue gets updated in the

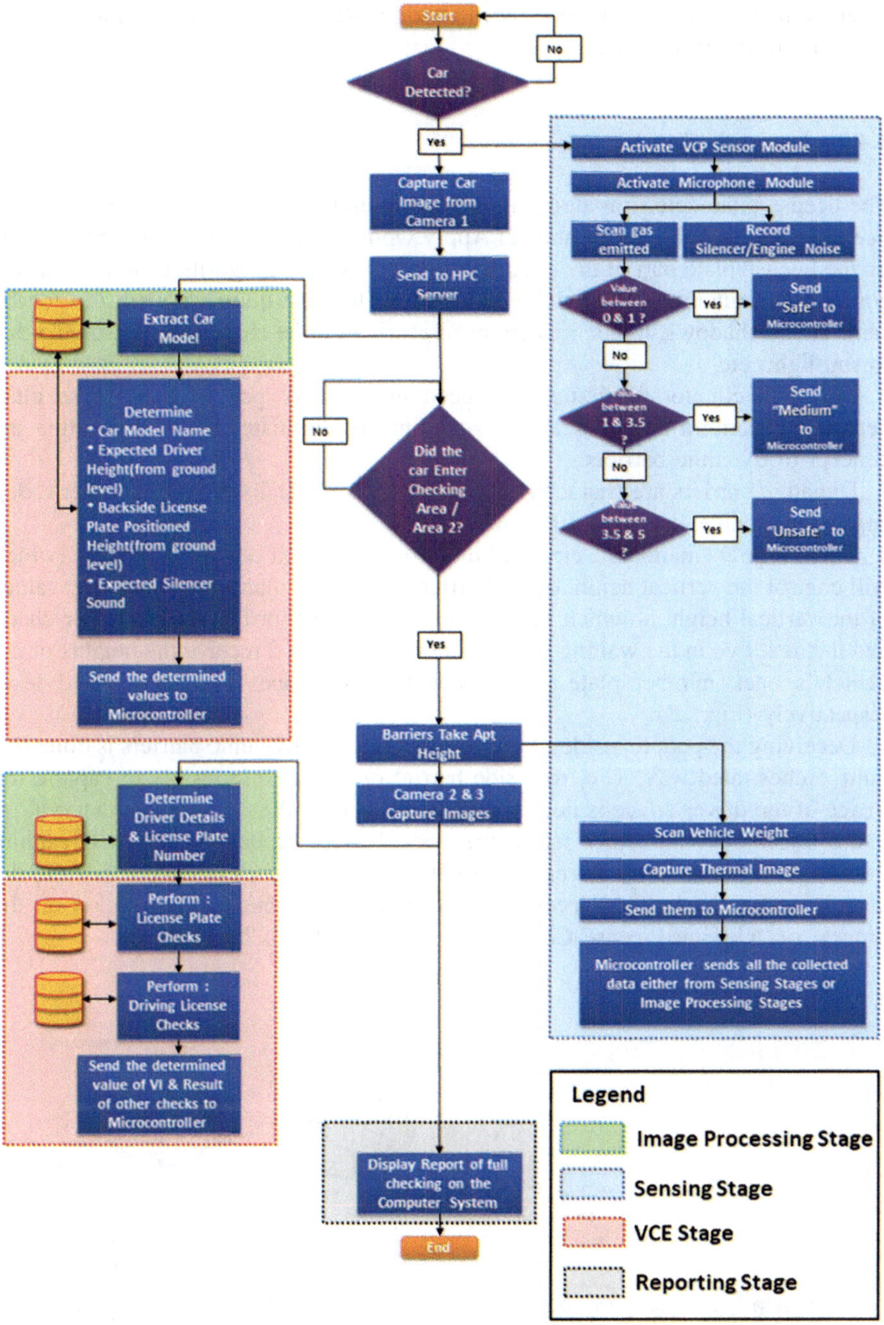

Fig. 27 Comprehensive diagram of operations performed by all 4 stages

local memory of the controllers placed inside barrier 1 and 2 corresponding to the order of checking the cars.

3.4.2 Dynamic Barrier

The license plate detection algorithm is mainly divided into 2 parts (1) Finding the license plate from the image and (2) Apply Optical Character Recognition or OCR on the license plate part of the image. But in reality, these algorithms usually face a considerable amount of issues in the prior step. There are quite a few reasons for the same as the shadows, lights, reflections appearing on the chassis of the vehicle due to sun light, etc.

Hence to counter the issue introduced in the prior part of the license plate detection algorithm that is creating uncertainty is eliminated by implementing the concept of dynamic barriers.

Dynamic barriers are characterized mainly by a vertically movable barrier rod, a controller module, and a circular buffer.

These barriers maintain a circular buffer locally in their controller module (which will control the vertical height of the barrier). These circular buffers store the values of the vertical height at which they have to rise or fall for the vehicle under check and the next two in the waiting queue. The barrier 1 and 2 receive the heights of the vehicle's back number plate and expected drivers face from the ground level respectively (Fig. 28).

Deceiving the people inside the vehicles from these dynamic barriers is done in a quite orchestrated way. The front side barrier or barrier responsible to capture the image of the driver's face is not very clearly visible to the next vehicle that will go under check. On the other hand, the back barrier or barrier that will capture the image of the rear side license plate will become completely unnoticeable to the vehicle under check. This deceiving will reduce the chances of suspicion on the process of "Driving License Check" and "License Plate Check".

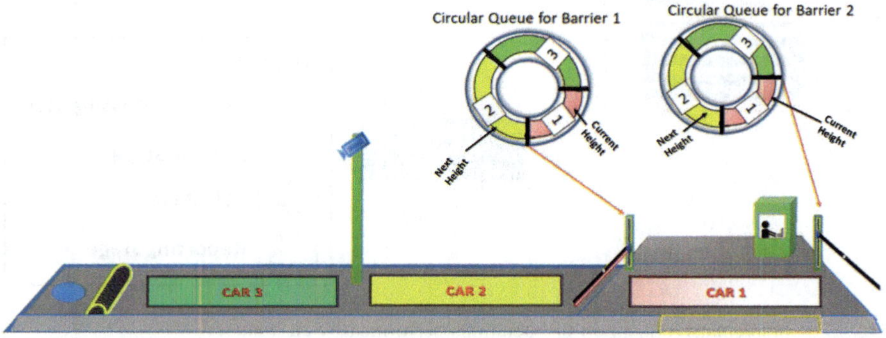

Fig. 28 Illustration of the circular queue working

4 Case Study

Authors saw a huge loss of human lives in the terrorist attack that took place on 14th Feb 2019 in Pulwama District of India [21]. Pointing out one of the reasons for the attack to be the removal of 3 barriers from the road where the incident happened Prof. Amita Singh (Professor of Law, Governance and Disaster Studies at Jawaharlal Nehru University) said, "The RDX filled vehicle could not be checked as the 3 check barriers were removed by Mehbooba Mufti…" [22, 23]. The potential correctness of the statement by Prof. Singh and other parameters e.g. inclinations of certain roads, traffic bottlenecks, etc. that directly or indirectly contribute to the vulnerability of the roads like the one where the discussed incident happened. Hence a system that assists the security personnel(s) in tracking the parameters unusual to the vehicles at the same time being unnoticeable by the malevolent and criminal minded individual/groups and is feasible to be implemented can be useful. Such a system would not only leverage the intensity with which certain parameters of the vehicle under check are examined but also aid in checking those parameters not directly known to the security personnel(s) or for that matter any individual at their first glance on the vehicle.

A vehicle-borne improvised explosive device (VBIED), also known as a truck bomb or car bomb, is an improvised explosive device installed inside a car or other automobile and then exploded. This very type of bomb was used in the attack mentioned above and many others [24].

Shielding against a car bomb requires keeping automobiles at a distance from exposed targets by using metal barriers, checkpoints, roadblocks and others alike them [25].

The TIE system would assist the security personnel positioned near the barriers to check either all or just the suspicious vehicles passing by the barriers. The system would capture the photo of the vehicle and sense the parameters simultaneously. TIE System will execute all its 4 stages and would send its report to the computer system of same or other security personnel. Particularly, the VCE Stage would find the unusualness in those parameters, assign a vulnerability index to the vehicle and show the results to the security personnel. This would also help when population of cars passing by the barriers is large enough the system will help in finding the ones that need to be thoroughly checked by the security personnel with respect VCE subsystem checks. It would help them to know the magnitude of parameters which one cannot find manually and feasibly in all circumstances.

5 Conclusion

Road security checks currently face 2 major problems (1) high knowability and (2) unautomated checking. Therefore it becomes crucial to make the checks in an unnoticeable way so as to deceive the malevolent individuals or groups to believe there are no clear obstacles that they have to overcome. The system is made almost unnoticeable by placing the sensors at unexpected positions on the site. Even if they are constrained to be operated in a moving manner, the placements are done in a way that ensures they are deployed and operated in an inconspicuous location and unexpected way respectively. The authors have consolidated a system that uses image processing; embedded system; IoT and web technology jointly to counter the prior mentioned loopholes in common security checks done currently.

6 Future Direction for Further Extension

A set of concepts and extensions can still be applied to make the system better in the aspects like reusability of sensors, turnaround time of checking(s), etc. Mentioned below are some of the specific ideas that can be worked upon to further extend the system:

1. The System can directly cut monetary penalties from the bank account of the person for the set of checks it satisfies.
2. TIE System Array can be developed which deploys TIE System clones that can reuse various sensors at the same time check multiple cars at a time.
3. To add more checks to the Learner's license checks like an "L" sticker present.
4. To use Piezoelectricity, solar energy and similar sciences and technologies to make TIE System as self-sufficient as possible with respect to power requirements.
5. To integrate mechanisms [26] that balance the tradeoff between security and efficiency involved in the communication between the various sensor modules employed.

References

1. The road safety crisis in India. NDTV-diageo road to safety. https://sites.ndtv.com/roadsafety/video-details-page/the-road-safety-crisis-in-india-479985/
2. Commodity flow survey 2017 | Bureau of transportation statistics. Bts.gov. https://www.bts.gov/newsroom/commodity-flow-survey-2017. Accessed 24 Jun 2019
3. Eurostat: Freight transport in the EU-28 modal split based on five transport modes (% of tonne-kilometres) (2018)
4. Modi highlights road safety, social issues in radio address. The Hindu. https://www.thehindu.com/news/national/mann-ki-baat-pm-speaks-on-road-safety-social-issues/article7466865.ece
5. Tamane, S., Solanki, V.K., Dey, N. (eds.): Privacy and security policies in big data. IGI Global (2017)
6. P S: Hard-hitting road safety campaigns that left a mark. Social Samosa (2018). http://www.socialsamosa.com/2018/01/hard-hitting-road-safety-campaigns-which-has-left-a-mark/
7. IIRIS consulting: India risk report EP49: road safety challenges (2018)

8. Schaller, A., Mueller, K.: Motorola's experiences in designing the internet of things. Int. J. Ambient. Comput. Intell. (IJACI) **1**(1), 75–85 (2009)
9. Overloaded trucks take heavy toll on highways. Tribune India News Service. https://www.tribuneindia.com/news/himachal/overloaded-trucks-take-heavy-toll-on-highways/778798.html
10. Hu, T.: A framework of truck overload intelligent monitoring system. In: 2011 Fourth International Symposium on Computational Intelligence and Design, p. 1. IEEE, Hangzhou (2014)
11. Dey, N., Wagh, S., Mahalle, P., Pathan, M.: Applied Machine Learning for Smart Data Analysis, 1st edn. CRC Press, Boca Raton (2019)
12. Toll weigh motion|Truck weighing systems|Vehicles weighing stations India. Essaedig.com. http://www.essaedig.com/truck-weigh-in-motion.php#overview
13. Challenges and overview of license plate character segmentation. Int. J. Comput. Sci. Eng. **3**, 1
14. Smart cameras to detect errant drivers|Noida news—Times of India. The Times of India. https://timesofindia.indiatimes.com/city/noida/smart-cameras-to-detect-errant-drivers/articleshow/69807281.cms
15. Evaluating the use of automated facial recognition technology in major policing operations. Phys.org. https://phys.org/news/2018-11-automated-facial-recognition-technology-major.html
16. Student seminar report & project report with presentation (PPT, PDF, DOC, ZIP)—Profile of jaseela123d. Studentbank.in. https://studentbank.in/author-jaseela123d
17. Manitha, P.V., Anandaraman, S.S., Manikumaran, K., Aswathaman, K.: Design and development of enhanced road safety mechanism using smart roads and energy optimized solar street lights. In: International Conference on Energy, Communication, Data Analytics and Soft Computing (ICECDS) (2017)
18. Motorcycle noise standards. Cga.ct.gov. https://www.cga.ct.gov/2003/olrdata/tra/rpt/2003-r-0676.htm. Accessed 24 June 2019
19. Vehicle-certification-agency.gov.uk. Cars and noise (2019). https://www.vehicle-certification-agency.gov.uk/fcb/cars-and-noise.asp
20. The motor vehicles act. Tn.gov.in. http://www.tn.gov.in/sta/Mvact1988.pdf
21. Das, S.: 44 CRPF jawans killed, 70 injured in Pulwama terror attack in J&K (2019). https://www.livemint.com, https://www.livemint.com/news/india/pulwama-terror-attack-deathtoll-rises-to-40-jem-claims-responsibility-1550143395449.html
22. JNU prof says Mehbooba Mufti ordered removal of barriers that led to Pulwama attack; PDP to take legal action. The New Indian Express. http://www.newindianexpress.com/nation/2019/feb/20/jnu-prof-saysmehbooba-mufti-ordered-removal-of-barriers-that-led-to-pulwamaattack-pdp-to-take-leg-1941055.html. Accessed 24 June 2019
23. Prof. Amita Singh (Chairperson). SCDR JNU, Special Centre of Disaster Research (2019). http://scdr.jnu.ac.in/team/prof-amita-singh-pi/
24. Jha, R.: Why car bombs are worrying our forces. The Economic Times (2019). https://economictimes.indiatimes.com/news/defence/why-car-bombs-are-worrying-our-forces/articleshow/68019980.cms. Accessed 25 June 2019
25. Mitigating the dangers of car bombs. Stratfor. https://worldview.stratfor.com/article/mitigating-dangers-car-bombs
26. Kimbahune, V.V., Deshpande, A.V., Mahalle, P.N.: Lightweight key management for adaptive addressing in next generation internet. Int. J. Ambient. Comput. Intell. (IJACI) **8**(1), 50–69 (2017)

MIX
Papier aus verantwortungsvollen Quellen
Paper from responsible sources
FSC® C105338

Printed by Books on Demand, Germany